From C–H to C–C Bonds
Cross-Dehydrogenative-Coupling

RSC Green Chemistry

Editor-in-Chief:
Professor James Clark, *Department of Chemistry, University of York, UK*

Series Editors:
Professor George A Kraus, *Department of Chemistry, Iowa State University, Ames, Iowa, USA*
Professor Dr Ir Andrzej Stankiewicz, *Delft University of Technology, The Netherlands*
Professor Peter Siedl, *Federal University of Rio de Janeiro, Brazil*
Professor Yuan Kou, *Peking University, China*

Titles in the Series:

How to obtain future titles on publication:
A standing order plan is available for this series. A standing order will bring
delivery of each new volume immediately on publication.

For further information please contact:
Book Sales Department, Royal Society of Chemistry, Thomas Graham House,
Science Park, Milton Road, Cambridge, CB4 0WF, UK
Telephone: +44 (0)1223 420066, Fax: +44 (0)1223 420247
Email: booksales@rsc.org
Visit our website at www.rsc.org/books

From C–H to C–C Bonds
Cross-Dehydrogenative-Coupling

Edited by

Chao-Jun Li
McGill University, Montreal, Canada
Email: cj.li@mcgill.ca

ROYAL SOCIETY OF CHEMISTRY

RSC Green Chemistry No. 26

Print ISBN: 978-1-84973-797-5
PDF eISBN: 978-1-78262-008-2
ISSN: 1757-7039

A catalogue record for this book is available from the British Library

Published by The Royal Society of Chemistry,
Thomas Graham House, Science Park, Milton Road,
Cambridge CB4 0WF, UK

Registered Charity Number 207890

For further information see our web site at www.rsc.org

Printed and bound by CPI Group (UK) Ltd, Croydon, CR0 4YY

Preface

The ability to achieve any goals in our society is largely dependent upon, and limited by, the state-of-the-art tools available. While the speed and ease of the construction of large objects, such as buildings, are rapidly enhanced by the creation of various heavy machines, established fundamental chemical reactions, in combination with our creative and intelligent designing skills, ultimately determine the synthesis of chemical products.

The development of novel chemical reactivities and reaction conditions that can improve resource efficiency, energy efficiency, product selectivity, operational simplicity, as well as environmental health and safety, represents both an ideology and an aspiration for generations of synthetic chemists, more so than ever at this time. Ever since the synthesis of urea by Friedrich Wöhler in 1828, organic chemistry has become increasingly important in modern society. The invention of organic reactions over the past two centuries has allowed us to create synthetic organic compounds and materials that have now touched essentially every corner of our life: from cosmetics to fashion, from pharmaceuticals to agrochemicals, from transportation to the interior of skyscrapers, and from electronics to genetic modification. However, the archetypical requirements of standard classical chemical transformations are the functional groups, which provide the platform for chemical conversions that have led to the synthesis of millions of both naturally existing and non-naturally existing molecules in just less than two centuries, only a blinking moment in human history. In spite of the great successes, there are still various shortcomings: the pre-functionalized starting materials need to be synthesized in separate steps, and the amount of waste associated with solvent usage, purification and isolation maneuvers, as well as the required manpower to perform a synthesis increases exponentially with the number of synthetic steps. With recent concerns regarding the adverse effects of chemical production processes, as well as the

RSC Green Chemistry No. 26
From C–H to C–C Bonds: Cross-Dehydrogenative-Coupling
Edited by Chao-Jun Li
© The Royal Society of Chemistry 2015
Published by the Royal Society of Chemistry, www.rsc.org

emphasis on green chemistry and chemical sustainability, various frontiers of synthetic chemistry have been explored. Among them, the concept of the cross-dehydrogenative-coupling (CDC) reaction was formulated in 2003. The direct generation of a C–C bond from two different C–H bonds constitutes an ideal that has game-changing potential in synthetic design. The subject has become a very rapidly expanding field. This edited book will cover some of the key developments in this area.

Chao-Jun Li
Montreal, Canada

Contents

RSC Green Chemistry No. 26
From C–H to C–C Bonds: Cross-Dehydrogenative-Coupling
Edited by Chao-Jun Li
© The Royal Society of Chemistry 2015
Published by the Royal Society of Chemistry, www.rsc.org

**Chapter 2   Dehydrogenative Heck-type Reactions:
              The Fujiwara–Moritani Reaction   33**
Tsugio Kitamura and Yuzo Fujiwara

**Chapter 3   Copper-Catalyzed Cross-Dehydrogenative-Coupling
              Reactions   55**
Xiaojian Zheng and Zhiping Li

CHAPTER 1

The Evolution of the Concept of Cross-Dehydrogenative-Coupling Reactions

SIMON A. GIRARD, THOMAS KNAUBER AND CHAO-JUN LI*

Department of Chemistry and FQRNT Centre for Green Chemistry and Catalysis (CCVC) McGill University, 801 Sherbrooke Street West, Montreal, Quebec, H3A 0B8, Canada
*Email: cj.li@mcgill.ca

1.1 Introduction

Among the countless reactions developed throughout the history of organic chemistry, carbon–carbon bond formation reactions are very special, as such reactions create the framework for organic molecules to build on and for functional groups to be attached to. Thus, the development of methods for forming C–C bonds plays a central role in the design and synthesis of organic matter: molecules and materials.[1] Historically, nucleophilic additions, substitutions, and Friedel–Crafts type reactions formed the pillars of methods to connect two simpler molecules *via* the formation of a C–C bond in acyclic structures.[2] The development of pericyclic reactions[3] laid the foundation for synthesizing cyclic structures. Over the past four decades, transition metal catalyses *via* cross-coupling and metathesis have overcome some limitations of the classical reactions, *e.g.*, nucleophilic substitutions involving sp^2 carbon centers, and have greatly increased the efficiency of C–C bond formations, especially those involving arenes and alkenes, in modern

RSC Green Chemistry No. 26
From C–H to C–C Bonds: Cross-Dehydrogenative-Coupling
Edited by Chao-Jun Li
© The Royal Society of Chemistry 2015
Published by the Royal Society of Chemistry, www.rsc.org

organic chemistry.[4] Their importance is attested by the awarding of Nobel Prizes in both 2005 and 2010.[5]

However, in spite of the great success of both classical C–C bond formation methods and the modern extraordinary achievements of transition metal catalysis, state-of-the-art C–C bond formation reactions must use pre-functionalized starting materials, which require extra steps (sometimes multiple steps) to synthesize. In many cases, during the core C–C bond formation processes, the pre-formed functional groups are simultaneously 'lost'. The necessity of these repetitive pre-functionalization and defunctionalization steps plus the associated isolations and purifications, ultimately diminishes the overall material efficiency in the synthesis of complex organic molecules and increases chemical waste. The reduction in efficiency is aggravated with an increase in the complexity of molecules, as exemplified by the E-factor of Sheldon.[6] To reduce the number of steps involved and increase the efficiency in synthetic chemistry, we must explore new frontiers of chemical reactions, in which various chemical bonds in widely available natural resources, petroleum, natural gas, biomass, N_2, CO_2, O_2, water, and others can be selectively transformed directly without affecting other bonds and without the need for excessive pre-activations. As part of this effort, the transition metal catalyzed C–H bond activation and subsequent C–C bond formations have, thus, attracted much interest in recent years.[7] Outstanding achievements have been made in this area and many complex compounds can be made much more rapidly. However, these reactions still require at least one functionalized partner in order to generate the desired C–C bond formation products.

Historically, the copper-mediated oxidative homodimerization of alkynes (the Eglinton reaction), an first reported over a century ago, represents the earliest success of directly generating a C–C bond from two C–H bonds.[8] The reaction requires a stoichiometric quantity of $Cu(OAc)_2$ as both mediator and oxidant. The Glaser–Hay coupling modified such oxidative homodimerization of alkynes by using a catalytic Cu(I) catalyst with oxygen as the terminal oxidant.[9] On the other hand, the oxidative homodimerization of electron-rich arenes has also become highly successful in generating arene dimers and polymers for a wide range of applications: from fine chemicals and pharmaceuticals to electronic materials.[10] Both types of reaction, however, are limited to homodimerizations and are beyond the present book.

In synthetic chemistry, what is very challenging and highly desirable is the selective formation of two different C–H bonds from two completely different compounds (or two chemically different sites within a molecule). As C–H bonds are generally relatively inert, compared to all other bonds in organic molecules, such cross-oxidative couplings involving only C–H bonds in the presence of, and without affecting other more reactive bonds, would be unthinkable within classical chemical knowledge.

Prior to the concept of cross-dehydrogenative-coupling (CDC), Moritani and Fujiwara developed the oxidative formation of Heck-type reaction products directly from arenes and alkenes, instead of aryl halides and alkenes, by using palladium as the catalyst.[11] This type of reaction is now referred to as the "Moritani–Fujiwara reaction". Although one can argue that an alkene is still a functional group, this is an early example of formal generation of a C–C bond from two different C–H bonds by removing two hydrogen atoms oxidatively. Since Chapter 2 is devoted entirely to this type of reaction, this chapter will only touch on them briefly.

Developing green chemistry methods[12] for chemical syntheses has been an objective of our laboratory over the past two decades. Over the years, we have explored various unconventional chemical reactions that could potentially simplify syntheses, decrease overall waste and maximize resource utilization. In our early studies, we focused on developing Grignard-type reactions in aqueous media in order to simplify protection–deprotection processes involved in organic synthesis, especially carbohydrates.[13] This led to the success of virtually all types of Barbier–Grignard reactions in water, as well as the synthesis of various natural products, both by us and many others. Since water is analogous to protonic functional groups such as hydroxyls, acids, and amines, these water-tolerant reactions allow a drastic reduction in the number of transformations in those syntheses by eliminating the protection and deprotection steps. Nevertheless, the pre-generation of organic halides and the requirement of stoichiometric quantities of metal will still lead to stoichiometric waste.

As an aspirational endeavor, we then shifted our attention to explore Barbier–Grignard type and other nucleophilic addition reactions by using C–H bonds as surrogates for organometallic reagents, to simplify the halogenation–dehalogenation process and to avoid the utilization of a stoichiometric amount of metal for such reactions. Furthermore, we would like to explore such reactions in water, combining the advantages of both simplifying the protection–deprotection processes as well as avoiding halogenation–dehalogenation processes. Our efforts have been highly fruitful. Among them, our laboratory has pioneered a wide range of direct catalytic additions of terminal alkynes to various electrophiles in water[14] and the most well-known one is the so-called "aldehyde–alkyne–amine coupling", often in water.[15]

The success of the above encouraged us to explore the ultimate question in 2003: can we generate C–C bonds selectively from two different C–H bonds of any type without having to convert either one into a pre-synthesized functional group in the first place, possibly even in water? The success of such reactions could potentially lead to chemical transformations beyond functional group-based transformations—a potential tool for the next generation of synthetic chemists. A general scheme for such a process would involve two different types of C–H bonds in any setting, and would form a C–C bond at the specific desirable sites by an overall formal loss of H_2 either in the form of an H_2 molecule or through the use of an oxidant (Scheme 1.1). To help our understanding, we termed these reactions CDCs. Despite the use of "dehydrogenative" in the name, the term is not limited to the generation of H_2 molecules. The CDC reaction has become one of the most active areas

$$C-H \ + \ H-C \ \xrightarrow[\text{[O]}]{\text{cat. M}} \ C-C$$

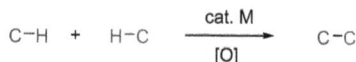

Scheme 1.1   Forming a C–C bond *via* cross-dehydrogenative-coupling (CDC).

of research and extensive progress has been made in all aspects of such reactions. This introductory chapter will only discuss briefly the early evolution of such reactions.

1.2   CDC Reactions Involving sp³ C–H Bonds and sp C–H Bonds

1.2.1   Reaction of sp³ C–H Bonds Adjacent to Nitrogen

As a starting point, we chose the formation of C–C bonds from α-sp³ C–H bonds of nitrogen in amines, and alkynyl sp C–H bonds to generate propargylic amines. This choice was based on three reasons: (1) propargylic amines are of great pharmaceutical interest and are synthetic intermediates for various nitrogen compounds;[16] (2) the sp³ C–H bond α to nitrogen in amines can be readily activated to generate iminium ions *via* single-electron-transfer (SET) processes, or by transition metals as described by Leonard[17] and Murahashi;[18] and (3) the aldehyde–alkyne–amine coupling (A³) reactions to afford propargyl amines described earlier, proceed *via* the formation of the same intermediate (Scheme 1.2). We reasoned that the CDC of the sp³ C–H α to nitrogen with a terminal alkyne should thus occur readily under oxidative conditions.

In the early 1990s, Miura observed the formation of a small amount of the alkynylation product in a complex product mixture when reacting amines with alkynes in the presence of CuCl₂ and oxygen; no further investigation was made.[19] As a prototype for the concept of the selective CDC reaction, we found that the desired CDC reaction product was obtained in good yield and high selectivity with the combination of a copper catalyst and *tert*-butyl hydroperoxide (TBHP) as the terminal oxidant. Various copper salts such as CuBr, CuBr₂, CuCl, and CuCl₂ were all effective for this transformation. Various alkynes were reacted with dimethylaniline derivatives to give the alkynylation products in 12–82% yields (Scheme 1.3)[20] and aromatic alkynes often provided better yields than aliphatic alkynes. The reactions tolerated various functional groups such as alcohols and esters.

As a potential synthetic application of CDCs, we found that various *p*-methoxyphenyl glycine amides could be directly alkynylated with phenylacetylene readily at room temperature (Scheme 1.4).[21] By hydrogenating the alkyne and removal of PMP, this methodology provides a versatile method for synthesizing homophenylalanine derivatives, an important synthon in many angiotensin-converting enzyme inhibitors.[22] A series of direct and site-selective peptide functionalizations was also realized by using CDC reactions (Scheme 1.5).[23]

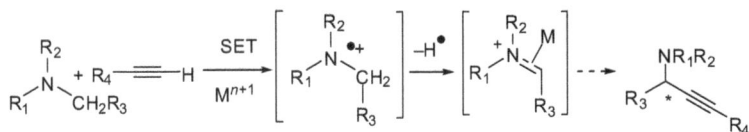

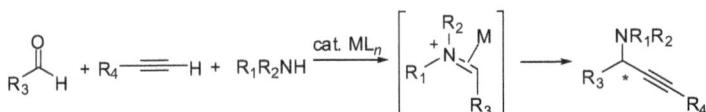

Scheme 1.2   Proposed alkynylation of an α-C–H bond of nitrogen in amines (top) and aldehdye–alkyne–amine (A³) coupling (bottom).

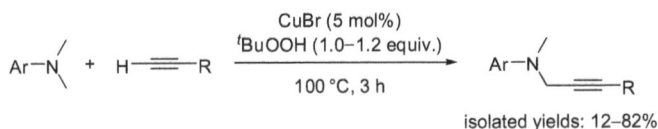

Ar = Ph, 4-MeC$_6$H$_4$, 2-MeC$_6$H$_4$, 4-BrC$_6$H$_4$

R = Ph, 4-MeOC$_6$H$_4$, 4-MeC$_6$H$_4$, 4-BrC$_6$H$_4$, 4-PhC$_6$H$_4$, 2-Py, HOCH$_2$, EtCO$_2$CH$_2$, CH$_3$OCO, Bu

Scheme 1.3   Copper-catalyzed alkynylation of *N,N*-dimethylanilines.

NR^{1}R^2 = NHMe, NHEt, NH(CH$_2$)$_3$CH$_3$, NH(CH$_2$)$_3$Ph, 1-pyrrolidinyl, OEt, Ph

R^3 = H, 4-Ph, 4-Br, 4-Me, 2-OMe

isolated yields: 50–78%

Scheme 1.4   Direct alkynylation of glycine amides *via* CDC.

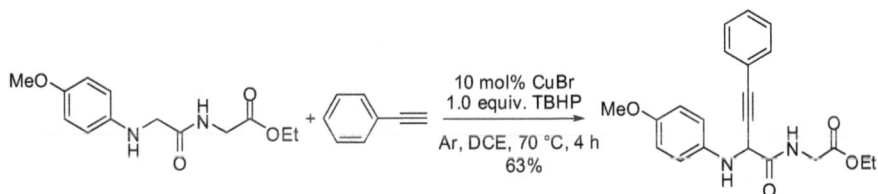

Scheme 1.5   Site-specific alkynylation of a dipeptide.

1.2.2 Asymmetric Reaction of sp³ C–H Bonds Adjacent to Nitrogen

The asymmetric synthesis of organic compounds is another major effort in modern organic chemistry. With our earlier high success it was intriguing to see if it is possible to achieve enantioselective C–C bond formations based on the direct reaction of prochiral CH$_2$ groups *via* CDC. Indeed, asymmetric alkynylation of tetrahydroisoquinolines (THIQs) to generate optically active C1-substituted derivatives was realized by using a copper salt together with pybox **1**, among others (Figure 1.1), as the chiral ligand. Both Cu(I) and Cu(II) were found to be effective catalysts, although slightly higher enantioselectivities were observed with Cu(I) catalysts (Scheme 1.6). The catalytic asymmetric CDC alkynylation also proceeded in water and without a solvent, however both the yields and the enantioselectivities were decreased.

1.2.3 Reaction of sp³ C–H Bonds Adjacent to Oxygen

Compared with the sp³ C–H bond adjacent to nitrogen, the sp³ C–H bond adjacent to oxygen is much less reactive. In our initial attempt to effect such CDC reactions, only the addition of an sp³ C–H bond across an alkyne was observed, and this was *via* a radical process.[24] Subsequently, we found that a

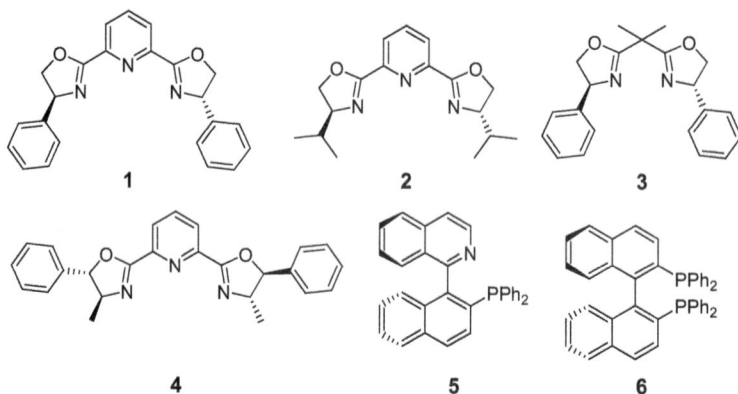

Figure 1.1   Examples of chiral ligands tested.

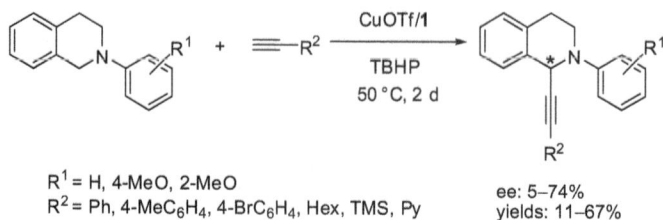

R^1 = H, 4-MeO, 2-MeO
R^2 = Ph, 4-MeC$_6$H$_4$, 4-BrC$_6$H$_4$, Hex, TMS, Py

ee: 5–74%
yields: 11–67%

Scheme 1.6   Asymmetric alkynylation of THIQs with terminal alkynes.

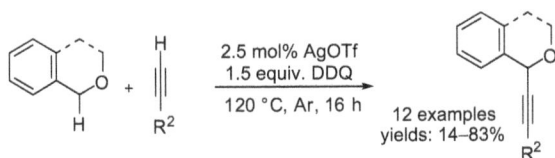

Scheme 1.7   CDC reaction between terminal alkynes and sp^3 C–H bonds adjacent to oxygen.

silver-catalyzed oxidative coupling of terminal alkynes and benzylic ethers using 2.5 mol% silver triflate, and 1.5 equiv. of 2,3-dichloro-5,6-dicyano-benzoquinone (DDQ) in a mixture of 4 : 1 toluene:chlorobenzene at 120 °C successfully provided the alkynylation of benzylic ether derivatives *via* the CDC process (Scheme 1.7).[25] However, poor yields were obtained with acyclic methyl benzyl ether.

1.2.4   Reaction of Benzylic sp^3 C–H Bonds

More recently, we demonstrated the first alkynylation of benzylic C–H bonds not adjacent to a heteroatom with 1 mol% of a CuOTf–toluene complex in the presence of 1.5 equiv. of DDQ. Various alkynes were successfully coupled with diphenylmethane derivatives (Scheme 1.8).[26] Aromatic alkynes were smoothly converted and the use of electron-rich derivatives resulted in a slightly improved yield, rationalized by the nucleophilicity of the substrates. However, aliphatic alkynes (*i.e.*, *n*-hexyne) did not give the corresponding CDC product under standard conditions. The mechanism was proposed to proceed *via* the generation of radical intermediates, which were converted into a benzylic cation in the presence of DDQ through two successive SET steps. The resulting hydroquinone subsequently then abstracted the acidic proton from the alkyne to form the copper acetylide, which added to the benzylic cation to afford the desired product.

1.3   CDC Reactions Involving sp^3 C–H Bonds and sp^2 C–H Bonds

1.3.1   Reaction of sp^3 C–H Bonds Adjacent to Nitrogen

The proposed iminium intermediate in the CDC-type alkynylation implies that other C–H based pronucleophiles, besides terminal alkynes, can also couple with an α-sp^3 C–H bond of a nitrogen in amines *via* the same process. Thus, we examined electron-rich arenes as one such nucleophile *via* a cross-dehydrogenative Friedel–Crafts type arylation. Indole derivatives were coupled with *N*-aryl-THIQs under the CuBr/TBHP system to produce the desired CDC reaction product in good-to-excellent yields (Scheme 1.9).[27] It is worth noting that the reaction was not sensitive to moisture or air, and that the desired product was obtained in reasonable yield even when the reaction was

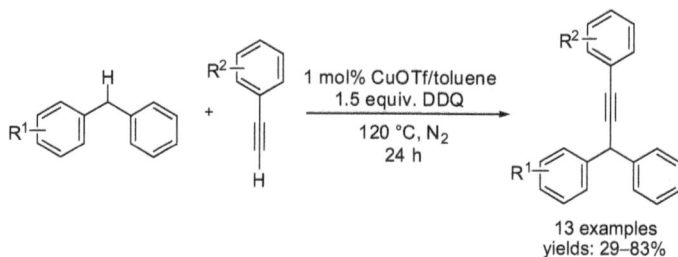

Scheme 1.8 CuOTf-Catalyzed CDC reaction of diphenylmethane derivatives and aromatic alkynes.

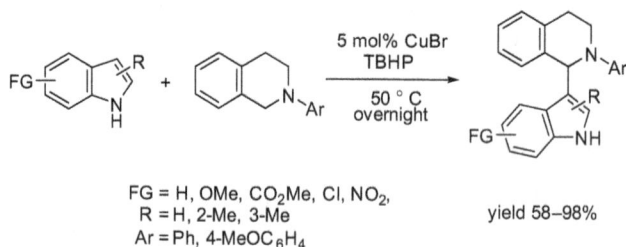

Scheme 1.9 CDC reactions of various indoles with *N*-aryl-THIQs.

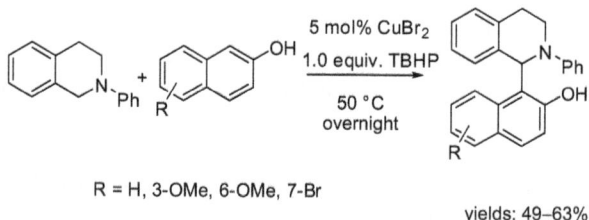

Scheme 1.10 CDC reaction of *N*-phenyl-THIQ with 2-naphthol derivatives.

carried out in water under an atmosphere of air. The reactions selectively occurred at the C3-position of the indoles, if both the C2- and C3-positions of the indoles were unoccupied, and the C2-substituted products were obtained when the C3-position of the indoles was substituted.

2-Naphthol is another electron-rich aromatic compound which can also lead to sp³–sp² CDC-type products. Thus, a new type of Betti base was formed *via* the CDC reaction of *N*-phenyl-THIQ with 2-naphthol derivatives under our CuBr/TBHP system with a small amount of homocoupled 2,2′-binaphthol (BINOL) (Scheme 1.10).[28] Subsequently, the scope of cross-dehydrogenative Friedel–Crafts type arylations was significantly improved by the development of highly efficient catalyst systems, and an intramolecular Cu-catalyzed aerobic synthesis of functionalized cinnolines *via* a Friedel–Crafts-type CDC arylation was reported by Zhang *et al.*[29]

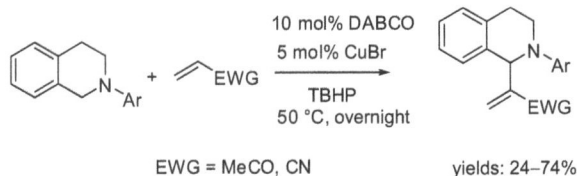

EWG = MeCO, CN   yields: 24–74%

Scheme 1.11   Aza-Baylis–Hillman CDC reaction.

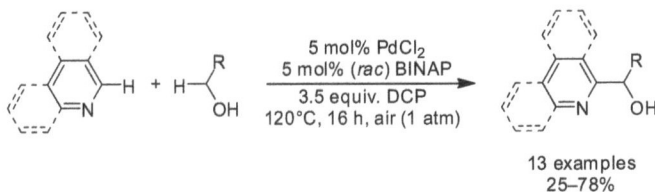

13 examples
25–78%

Scheme 1.12   CDC reaction between an arene and a C–H bond adjacent to a hydroxyl group. BINAP = 2,2′-bis(diphenylphosphino)-1,1′-binaphthalene; DCP = dicumyl peroxide.

We then investigated the CDC reaction between α-sp^3 C–H bonds of nitrogen in THIQs and sp^2 C–H bonds of electron-deficient alkenes. The reaction generated the Morita–Baylis–Hillman (MBH) reaction product by using 1,4-diazabicyclo[2.2.2]octane (DABCO) as a catalyst (Scheme 1.11). Another common MBH catalyst, triphenylphosphine, was found to be nearly ineffective due to the generation of triphenylphosphine oxide during the reaction.

1.3.2   Reaction of sp^3 C–H Bonds Adjacent to Oxygen

In addition to the sp^3 C–H bond adjacent to nitrogen, sp^3 C–H/sp^2 C–H CDC reactions have also been successful adjacent to oxygen. We discovered a palladium-catalyzed coupling of N-heterocycles with simple alcohols initiated by dicumyl peroxide (Scheme 1.12).[30]

Subsequently, the radical coupling of benzothiazoles, benzoxazoles and benzimidazoles with alcohols or ethers in the presence of excess TBHP was reported by He *et al.*[31]

1.3.3   Reaction of Benzylic and Allylic sp^3 C–H Bonds

Allylic compounds as well as diphenylmethanes have also been disclosed as substrate classes for CDC-type arylations.[32] A PdCl$_2$-catalyzed cross-dehydrogenative allylation reaction of indole derivatives was introduced by the group of Bao in 2009 using DDQ as the stoichiometric oxidant (Scheme 1.13). Shi also reported a FeCl$_2$-catalyzed benzylation reaction of electron-rich arenes with diphenylmethanes (Scheme 1.14).[32]

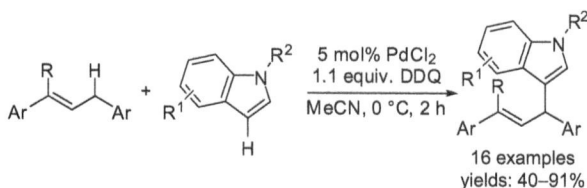

Scheme 1.13   CDC reaction between an indole derivative and an allylic C–H bond.

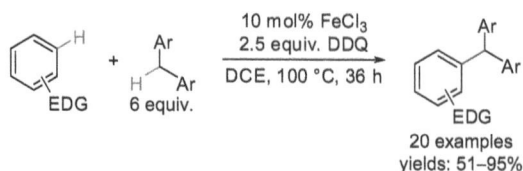

Scheme 1.14   CDC reaction between an arene and a benzylic C–H bond.

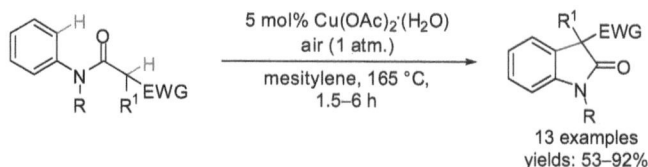

Scheme 1.15   Intramolecular CDC reaction between an arene and a C–H bond adjacent to a carbonyl group.

1.3.4   Reaction of α-sp³ C–H Bonds of Carbonyl Groups

The cross-dehydrogenative synthesis of oxindole derivatives *via* intramolecular cyclization of acetanilide radicals was independently explored by the groups of Kündig[33] and Taylor[34] in 2009. Phenylacetic acid anilides were converted into the corresponding heterocycles in the presence of stoichiometric quantities of copper salts. In 2010, the group of Taylor also succeeded in developing a Cu(OAc)$_2$-catalyzed aerobic variant.[35] The absence of basic additives and performing the reaction in a non-polar, high-boiling-point solvent such as mesitylene was the key to achieve a catalytic turnover (Scheme 1.15). Recently, Li extended this strategy to the synthesis of various 3-alkylated oxindole derivatives.[36]

Two procedures have independently been introduced to allow a formal CDC of sp² C–H bonds with α-sp³ C–H bonds of carbonyl groups, *via* the generation of an unsaturated intermediate that is subsequently coupled with an arene or an olefin.[37] The group of Hong developed two complementary Pd(OTFA)$_2$-catalyzed protocols for the arylation and olefination of a series of chromanones and dihydroquinolinones (Scheme 1.16).[37]

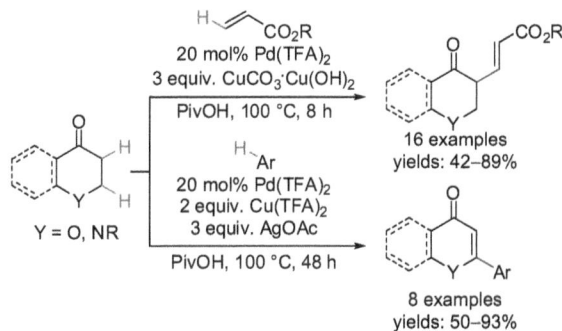

Scheme 1.16 CDC reaction of chromanones and dihydroquinolinones.
TFA = trifluoroacetate; PivOH = pivalic acid.

1.3.5   Reaction of Alkane sp³ C–H Bonds

Among all the potential CDC reactions, the reaction with simple alkanes (without any functional groups) is the most challenging. Among the many methods for the direct transformation of alkane C–H bonds, Fenton chemistry[38] and the Gif process[39] are the classical methods and allow the conversion of aliphatic C–H bonds into C–O bonds under mild conditions by using peroxides catalyzed by various iron catalysts. We hypothesized that these classical processes might be intercepted by carbon-based reactive intermediates and thus diverted to form C–C bonds. Our first success came with 1,3-dicarbonyl compounds (to be discussed Section 1.4). Another possibility is the alkyl–aryl CDC coupling. As a first step to model the coupling of an arene with a radical intermediate during the reaction of alkane C–H bonds, we reacted 2-phenylpyridine with dicumyl peroxide (with the objective of producing a methyl radical *in situ*) with 10 mol% Pd(OAc)₂ as the catalyst at 130 °C under an atmosphere of nitrogen; the reaction generated mono-methylation and *bis*-methylation products efficiently.[40] Other peroxides and palladium catalysts could also be used, but generated lower yields of the methylation products. When benzo[*h*]quinoline was used for the reaction, 76% of the mono-methylation product was obtained. Other substituted aromatic compounds such as acetanilides were also effective in this transformation, generating the methylation product in moderate yields. The mechanism of the reaction was proposed to proceed *via* a methylpalladium species (generated from the fragmentation of the peroxide), which underwent a nitrogen-assisted aryl C–H activation followed by reductive elimination to give the methylation product (Scheme 1.17, top). The success of this methylation gave us hope for CDC reactions involving simple alkanes. Thus, we reacted 2-phenylpyridine with cyclooctane in the presence of *tert*-butyl peroxide (TBP) under palladium-catalyzed reaction conditions, however the reaction only gave a trace (<1%) amount of the desired product. We then examined other transition metal catalysts and found that a 42–75% yield of the corresponding CDC products was obtained between various

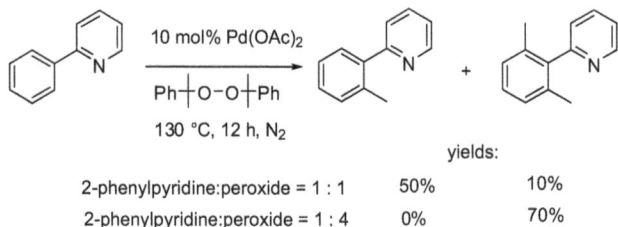

yields:

2-phenylpyridine:peroxide = 1 : 1	50%	10%
2-phenylpyridine:peroxide = 1 : 4	0%	70%

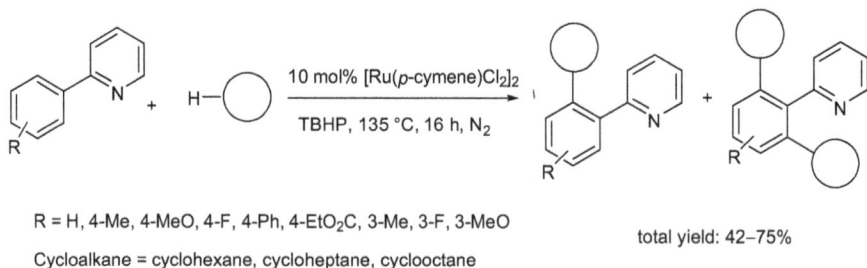

R = H, 4-Me, 4-MeO, 4-F, 4-Ph, 4-EtO$_2$C, 3-Me, 3-F, 3-MeO

Cycloalkane = cyclohexane, cycloheptane, cyclooctane

total yield: 42–75%

Scheme 1.17   Methylation of 2-phenylpyridine with dicumyl peroxide (top) and CDC reaction between 2-arylpyridines and cycloalkanes (bottom).

2-arylpyridines and cycloalkanes by using 10 mol% [Ru(p-cymene)Cl$_2$]$_2$ as the catalyst, and TBHP as the hydrogen acceptor, at 135 °C for 16 h under an atmosphere of air (Scheme 1.17, bottom).[41]

The mechanism of this CDC was proposed to involve a ruthenium-catalyzed aryl C–H activation[42] followed by an H–alkyl exchange mediated by the peroxide most likely *via* an alkyl radical intermediate as in the palladium-catalyzed methylation reaction. Then, reductive elimination of this intermediate generated the arene–cycloalkane coupling product and re-generated the active ruthenium catalyst (Scheme 1.18). A large negative kinetic isotope effect was observed using deuterated starting materials. The results suggested that the ruthenium-catalyzed aryl C–H activation is a fast equilibrium and the H–alkyl exchange is the rate-limiting step.

Interestingly, we also found that using Ru$_3$(CO)$_{12}$ in combination with 1,4-bis(diphenylphosphino)butane (DPPB) in the presence of di-TBP provided the *para*-selective CDC reaction of arenes and cycloalkanes.[43] A wide range of arenes were functionalized with simple cycloalkanes (Scheme 1.19). Both electron-withdrawing groups (EWG) and electron-donating groups (EDG) on the arene partner were suitable for this reaction to give the coupling product in good yields and high *para*-selectivity, even with chelating *ortho*-directing substituents. The ring size has a dramatic influence on the reaction yield, with the lowest yield obtained with cyclopentane. No kinetic isotope effect ($k_H/k_D = 1.00$) was observed when using chlorobenzene-d$_5$ as the substrate, which suggests the possibility of a radical mechanism. The regioselectivity was rationalized by the stabilization of the radical intermediate by both

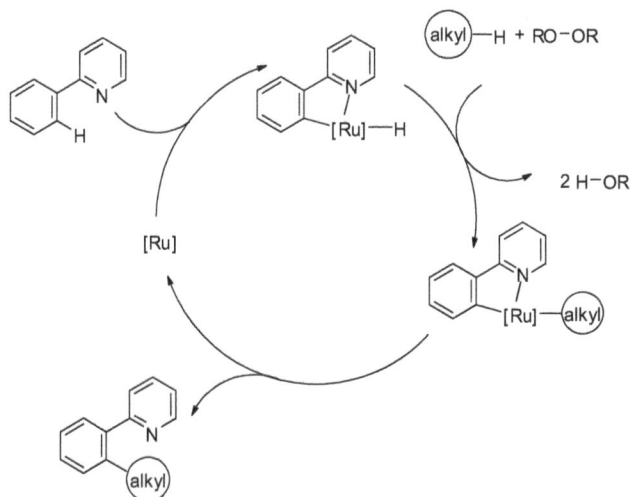

Scheme 1.18 Proposed mechanism for the ruthenium-catalyzed cycloalkylation of arenes *via* CDC.

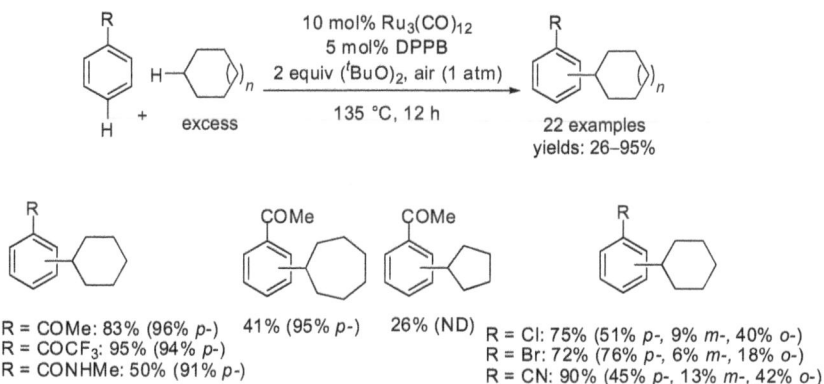

22 examples
yields: 26–95%

R = COMe: 83% (96% *p*-)   41% (95% *p*-)   26% (ND)
R = COCF$_3$: 95% (94% *p*-)
R = CONHMe: 50% (91% *p*-)

R = Cl: 75% (51% *p*-, 9% *m*-, 40% *o*-)
R = Br: 72% (76% *p*-, 6% *m*-, 18% *o*-)
R = CN: 90% (45% *p*-, 13% *m*-, 42% *o*-)

Scheme 1.19 *para*-Selective ruthenium-catalyzed cycloalkylation of arenes *via* CDC.

electron-donating and electron-withdrawing groups through frontier molecular orbital (FMO) interactions.[44] Recently, the ruthenium-catalyzed CDC was modified to functionalize nucleotides.[45]

In subsequent efforts, we succeeded in introducing significantly improved variants of the Minisci reaction. Pyridine-*N*-oxide proved to be reactive enough to undergo radical alkylation with cyclic hydrocarbons even in the absence of an activator (Scheme 1.20).[46]

Several Pd-catalyzed intramolecular CDC cyclizations have been developed that involve the connection of alkyl sp^3 C–H bonds with the sp^2 C–H bonds of heterocycles (Scheme 1.21):[47] the group of Fagnou introduced an aerobic

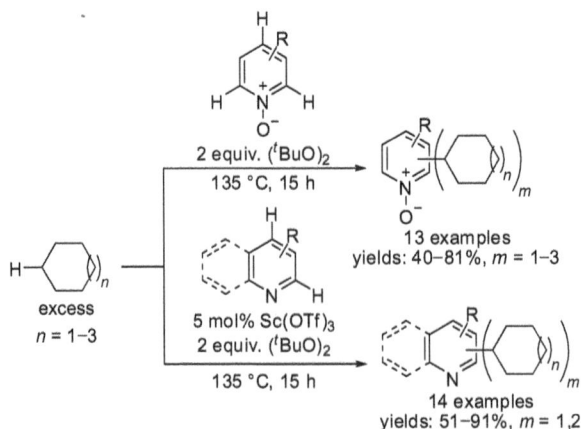

Scheme 1.20   CDC reaction of pyridine-*N*-oxide with cycloalkanes.

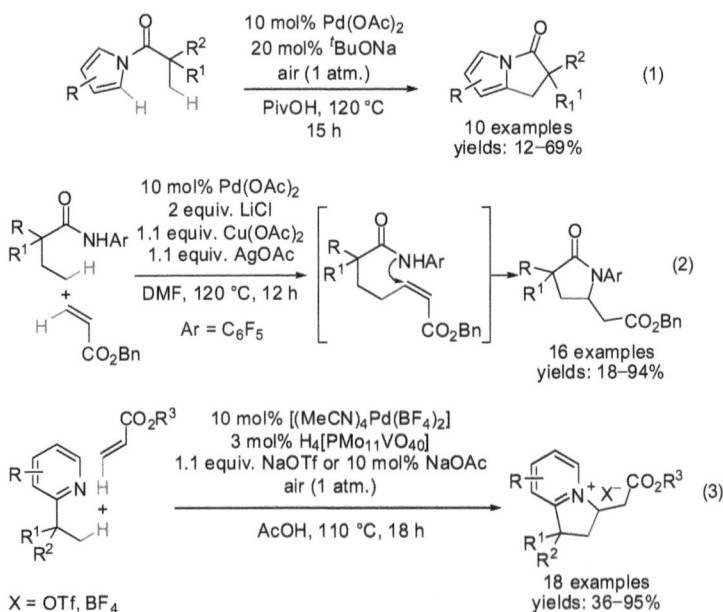

Scheme 1.21   Intramolecular CDC reaction of arenes/alkenes with alkyl sp³ C–H
bonds.

cyclization of *N*-pivaloyl pyrroles; the group of Yu disclosed a one-pot
sequence for the synthesis of functionalized 2-pyrrolidinones that starts
with an intramolecular olefination of the pivaloyl amides, followed by a 1,4-
conjugate addition step; and the group of Sanford provided an intramolecular aerobic Pd-catalyzed synthesis of various 2,3-dihydroindolizinium
salts.

1.4 CDC Reactions Involving sp³ C–H Bonds and sp³ C–H Bonds

1.4.1 Reaction of sp³ C–H Bonds Adjacent to Nitrogen

The success of the sp³ C–H/sp C–H and sp³ C–H/sp² C–H couplings led us to explore the possibility of the even more challenging CDC reaction between sp³ C–H and sp³ C–H bonds. We started with the reaction of an α-sp³ C–H bond of nitrogen in amines with nitroalkanes, which would provide aza-Henry-type reaction products. Using CuBr as the catalyst and TBHP as the terminal oxidant, various β-nitroamine derivatives were generated by this new methodology (Scheme 1.22).[48a]

Dialkyl malonates are another type of important synthon with relatively reactive sp³ C–H bonds, and can thus react with THIQs similarly in the presence of 5 mol% CuBr and TBHP at room temperature to give the CDC products, β-diester amine derivatives, in high yields (Scheme 1.23).[48b] Using malononitrile as the pronucleophile generated β-dicyano-THIQs under standard reaction conditions. Meldrum's acid can also be used as the pronucleophile, with which the CDC reaction of free 1,2,3,4-THIQ is possible. A subsequent major achievement, by Sodeoka and co-workers, was the CDC reaction of malonate with N-Boc-protected THIQ asymmetrically using a chiral palladium catalyst together with DDQ as the dehydrogenating reagent to generate the desired product in 86% enantiomeric excess (ee) (Scheme 1.24).[48c]

Peroxide is potentially hazardous in large-scale reactions and replacing peroxides with molecular oxygen would offer a safer and more

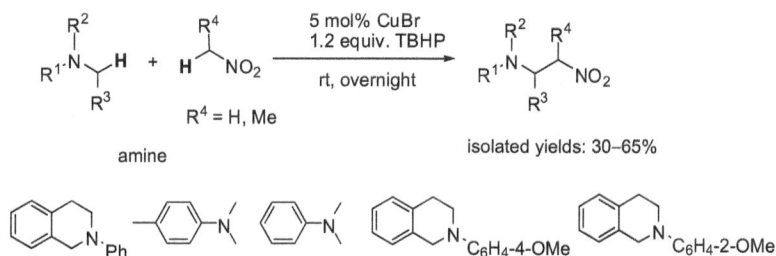

Scheme 1.22 CDC reaction of tertiary amines with nitroalkanes.

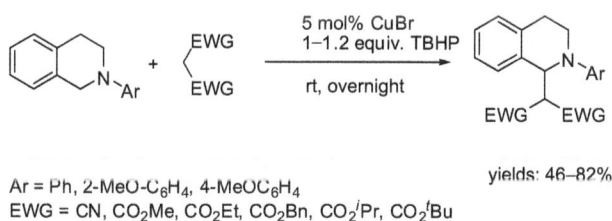

Scheme 1.23 CDC reaction of N-aryl-THIQs with malonates.

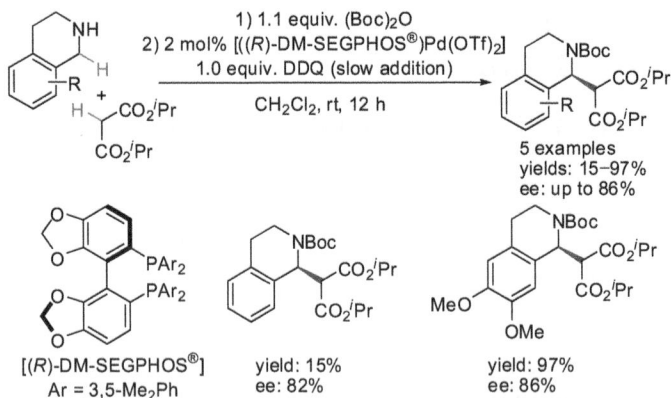

Scheme 1.24   Asymmetric CDC reaction of tertiary amines with malonates. DM-SEGPHOS = 5,5′-Bis(diphenylphosphino)-4,4′-bi-1,3-benzodioxole.

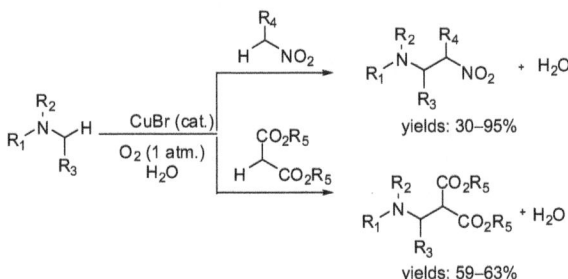

Scheme 1.25   CDC reaction of tertiary amines with oxygen in water.

atom-economical process. We found that, in water, molecular dioxygen (or even simply air atmosphere) can efficiently serve as the hydrogen acceptor for the CDC reaction. Both the nitroalkane reaction and the malonate reaction gave the corresponding CDC products in excellent yields catalyzed by CuBr under an oxygen atmosphere in water (Scheme 1.25).[49] As a green chemistry effort, we also investigated recoverable heterogeneous nanoparticles as catalysts in aza-Henry-type CDC reactions. By using magnetic Fe$_2$O$_3$ nanoparticles, highly efficient aerobic coupling of N-arylated THIQs with nitroalkanes was obtained under neat conditions with oxygen as the terminal oxidant.[50] The nanoparticle catalysts easily separated from the reaction mixture with a magnet and could be re-used without significant loss of reactivity.

Interestingly, we found that the CuBr-catalyzed aza-Henry-type CDC reaction also proceeds well in ionic liquids under an oxygen atmosphere. The catalyst-containing ionic liquid can be re-used without loss of reactivity after extraction of the product with diethyl ether. Since ionic liquids are highly polar and excellent media for conducting electricity,[51] we subsequently demonstrated that the CDC of N-phenyltetrahydroisoquinoline with nitromethane is also feasible under electrochemical conditions.[52] The reaction was performed with a large-area Pt electrode in an electrochemical H-cell using the ionic liquid [BMIm][BF$_4$] as the solvent and triethylamine to trap the protons generated. The reaction furnished the product in 80% chemical yield and 93% Faradaic yield.

The subsequent phenomenal work by Stephenson and others on using photo-oxidation catalysts in cross-dehydrogenative aza-Henry reactions which permit the use of visible light as a primary oxidant are covered in detail in a separate chapter in this book. Many other variations of this reaction have also been reported.

1.4.2 Reaction of sp^3 C–H Bonds Adjacent to Oxygen

With the initial success of amines, we began to explore the functionalization of the α-C–H bond of oxygen in ethers, which is more challenging due to the former's higher oxidation potential. Thus, a stronger oxidant than TBHP or molecular oxygen would be required. A good choice is DDQ, which is known to react with benzyl ether to generate oxonium ions. Thus, the CDC reaction of an sp^3 C–H bond adjacent to an oxygen atom with an sp^3 C–H bond in pronucleophiles proceeds efficiently to give β-diester ethers in the presence of DDQ by using a combination of indium and copper as catalysts. In this reaction, InCl$_3$ is proposed to further activate DDQ by increasing its oxidative potential while the copper catalyst activates the malonates (Scheme 1.26).[53]

However, the In/Cu/DDQ reaction system is limited to CDC reactions between benzyl ethers and relatively reactive malonate; and simple ketones do not react. To our surprise then, we found that in the absence of any metal catalyst, a CDC reaction between benzyl ethers and simple ketones mediated by DDQ proceeds efficiently (Scheme 1.27).[54]

The mechanism for the coupling is proposed in Scheme 1.28 in which an SET from the benzyl ether to DDQ generates a radical cation and a DDQ radical anion. The radical oxygen of the DDQ radical anion then abstracts an

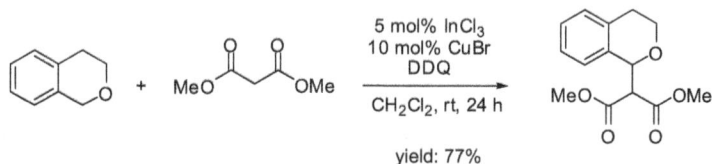

yield: 77%

Scheme 1.26 CDC reaction of isochroman with dimethyl malonate.

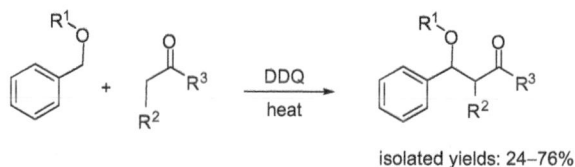

isolated yields: 24–76%

Scheme 1.27 CDC reaction between benzyl ethers and simple ketones.

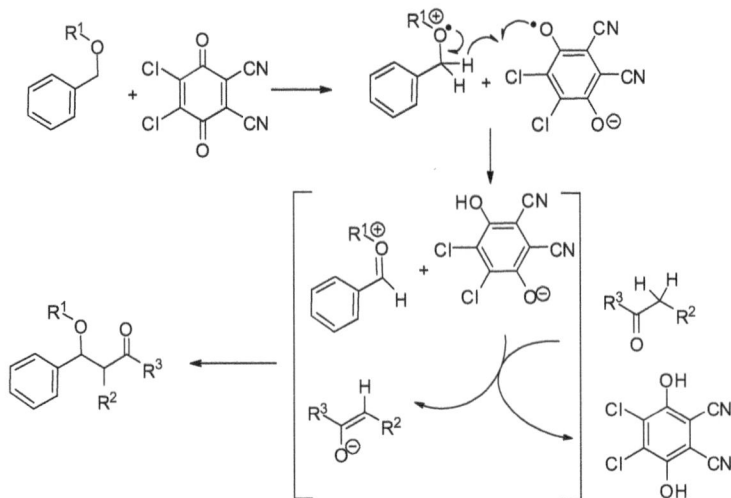

Scheme 1.28 Proposed mechanism for the CDC reaction of benzyl ethers with ketones mediated by DDQ.

H atom from the radical cation and generates a benzoxy cation, and the anionic oxygen of the DDQ radical anion abstracts an α-hydrogen from the ketone to generate an enolate. Finally, the enolate attacks the benzoxy cation to generate the CDC product.

1.4.3   Reactions with Allylic and Benzylic C–H Bonds

1.4.3.1   *Allylic Alkylation (sp³–sp³)*

The Tsuji–Trost palladium-catalyzed allylic alkylation (Scheme 1.29) is an important reaction in modern organic synthesis.[55] However, in general, a carboxylate (or another leaving group) is required at the allylic position. The direct utilization of an allylic C–H bond rather than an allylic functional group would be more desirable. Trost and co-workers reported the formation of an allylic alkylation from an allylic sp³ C–H in two steps in the late 1970s, directly using allylic C–H bonds to form π-allyl palladium complexes.[56] However, this reaction was stoichiometric with respect to Pd(II), as it served as both the catalyst and the oxidant because the *in situ* re-oxidation of the reduced Pd(0) to Pd(II) is difficult.

To explore the direct Tsuji–Trost CDC reaction, we found that by using a combination of CuBr (2.5 mol%) and CoCl$_2$ (10 mol%) as a catalyst, various 1,3-dicarbonyl compounds reacted smoothly with cyclohexene by directly using allylic sp³ C–H and methylenic sp³ C–H bonds (Scheme 1.30).[57]

When cycloheptatriene was reacted with 2,4-pentadione, the corresponding tropylacetylacetone was obtained in 41% isolated yield. If cyclopentadiene was used, the major product obtained was a dihydrofuran

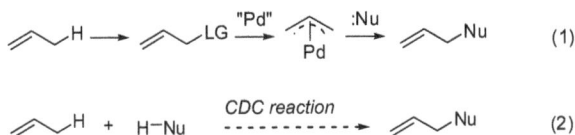

(1)

(2)

Scheme 1.29   Tsuji–Trost reaction (1) and allylic CDC reaction (2).

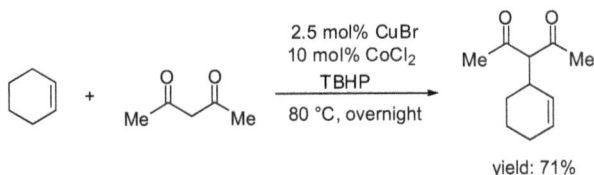

yield: 71%

Scheme 1.30   Allylic alkylation *via* CDC.

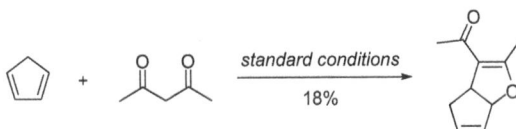

Scheme 1.31   Tandem allylic alkylation–cyclization *via* CDC.

derivative due to the further transformation of the alkylation product *in situ* (Scheme 1.31).

1.4.3.2   Benzylic Alkylation (sp³–sp³)

To explore the CDC reaction of benzylic C–H bonds, diphenylmethane was reacted with benzoylacetone. We found that FeCl$_2$ in combination with TBP (instead of TBHP) is an effective catalyst in this case, giving the corresponding CDC products cleanly in good yields (Scheme 1.32).[58]

1.4.4   Reactions with Alkanes

Various activated methylene substrates were reacted with cyclohexane, cyclopentane, cycloheptane, cyclooctane, norbornane, and adamantane to give the corresponding CDC products in good yields in most cases by using 10 mol% FeCl$_2$ · 4H$_2$O as the catalyst and TBP the oxidant at 100 °C for 12 h under an atmosphere of nitrogen (Scheme 1.33).[59] A mechanism analogous to the Gif process was proposed. This mechanism involves an Fe(II)-catalyzed decomposition of the peroxide to give an Fe-enolate and an RO radical. The RO radical then reacts with the cyclohexane to give a cyclohexyl radical. The cyclohexyl radical then reacts with the enolate to form the alkylated β-ketoester and the re-generated Fe(II) for further reactions (Scheme 1.34).

yields: 25–87%

R^2, R^3 = Ph, OMe, OEt, Me, ^{t}Bu, 4-ClC$_6$H$_4$, 4-MeOC$_6$H$_4$, 2-MeOC$_6$H$_4$

Benzyl =

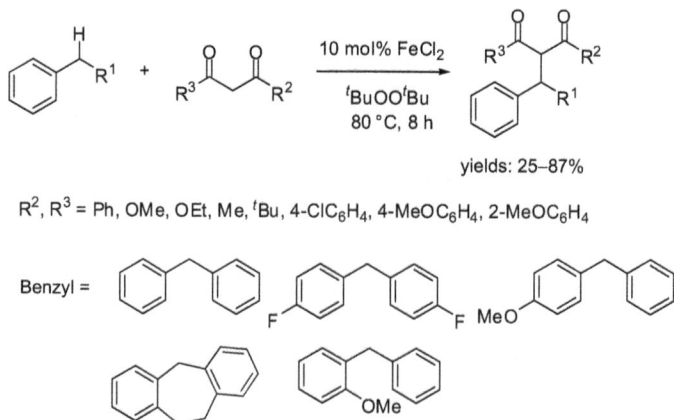

Scheme 1.32 Benzylic alkylation *via* CDC.

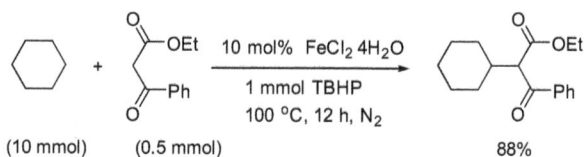

(10 mmol)   (0.5 mmol)   88%

Scheme 1.33 Alkane alkylation *via* CDC.

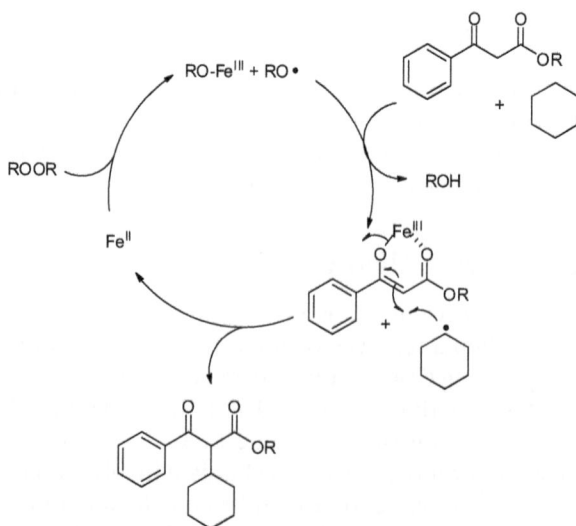

Scheme 1.34 Tentative mechanism for Fe-catalyzed alkylation with simple alkanes.

1.5   CDC Reactions Involving sp² C–H Bonds

Over the past few decades, substantial research has led to the introduction of efficient procedures that allow for the selective connection of two sp² C–H bonds. Direct arylations and oxidative olefinations date back to the pioneering investigations of van Helden and Verberg,[60] and Fujiwara and Moritani[61] as well as other contributors[62] who demonstrated that aromatic C–H bonds are efficiently activated for coupling with a second sp² C–H bond in the presence of stoichiometric quantities of Pd salts. However the products were usually obtained as isomeric mixtures in moderate yields (Scheme 1.35).

The groundbreaking work of Shue,[63] de Vries and van Leeuwen[64] and others[65] on oxidative olefinations and the contributions of Lu,[66] Fagnou,[67] DeBoef[68] and others[69] on direct arylations laid the foundation for the development of an impressive number of transition-metal-catalyzed selective cross-dehydrogenative sp² C–H couplings. Notably, the Fujiwara–Moritani reaction is also commonly termed the "oxidative Mizoroki–Heck reaction".[70–72] Progress in this area has been extensively reviewed and will thus only be briefly mentioned here, notwithstanding their great importance.

The addition of free sp² C radicals to aromatic compounds followed by a formal single electron oxidation step represents a third strategy for cross-dehydrogenative sp² C–C bond formations. This research area dates back to the pioneering studies of Fenton, Gif and others.[72] The first cross-dehydrogenative acylation of protonated nitrogen-containing heterocycles with aromatic and aliphatic aldehydes was introduced by the group of Minisci in the late 1960s (Scheme 1.36).[73]

The reaction is mediated by a combination of TBHP and stoichiometric quantities of $FeSO_3$ in strongly acidic media. Electron-deficient substrates proved to be far more reactive than their corresponding electron-rich derivatives and consequently, over-acylation was frequently observed. The

van Helden and Verberg in 1965:

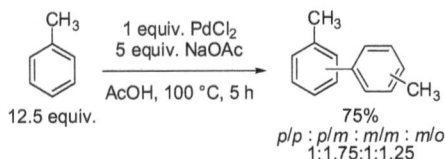

Fujiwara and Moritani in 1968:

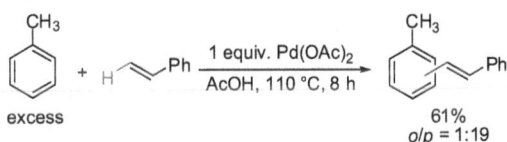

Scheme 1.35   Pioneering aromatic C–H functionalizations.

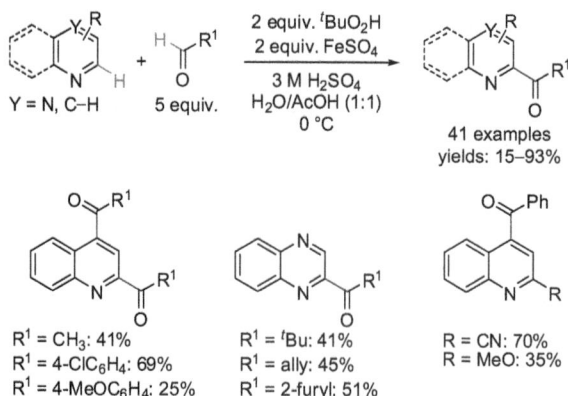

Scheme 1.36   Radical acylation of heterocycles.

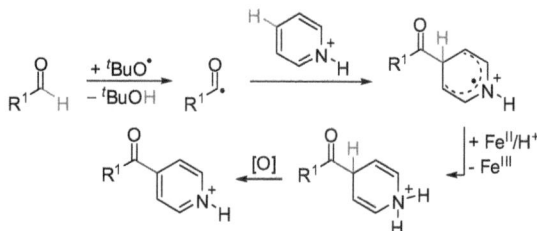

Scheme 1.37   Proposed mechanism of the radical acylation reaction of heterocycles.

acylation occurs preferentially in the *ortho-* and *para-*positions for pyridine derivatives and thus complements the scope of Friedel–Crafts-type acylations.[74] The authors proposed a reaction mechanism that starts with the generation of a *tert*-butoxy radical which subsequently abstracts the hydrogen atom from the carbonyl group of the aldehyde (Scheme 1.37).

The resulting nucleophilic acyl radical adds onto the protonated heterocycle and the resulting heterocyclic radical is reduced by the Fe(II) salt into the corresponding dihydropyridine derivative that is subsequently oxidized into the corresponding products. The use of a TBHP/Ti(III) redox system allowed for the isolation and characterization of the dihydropyridine intermediate.[34j]

Efficient procedures for the cross-dehydrogenative synthesis of aryl ketones have been introduced by complementing the radical acylation with transition-metal-catalyzed activation of aromatic C–H bonds. The first cross-dehydrogenative Pd-catalyzed acylation of 2-phenylpyridines was developed independently by the groups of Cheng[75] and Li.[76] The latter demonstrated that various aliphatic, alicyclic and some aromatic aldehydes are smoothly converted in the presence of 5 mol% of $Pd(OAc)_2$ in combination with 1.5 equiv. of TBHP at 120 °C under an air atmosphere (Scheme 1.38).

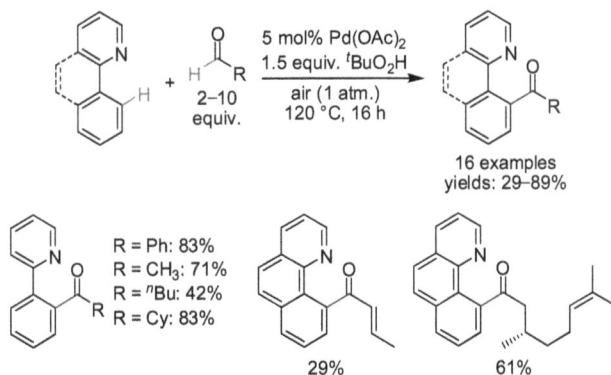

Scheme 1.38 Pd-Catalyzed acylation with aliphatic aldehydes.

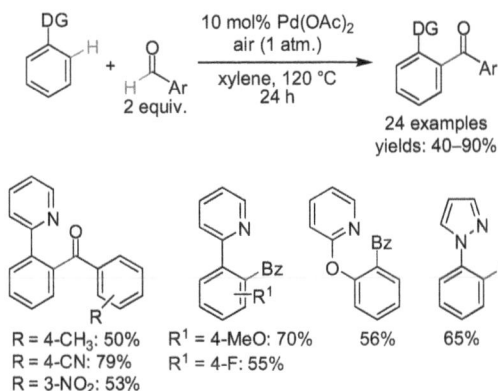

Scheme 1.39 Pd-Catalyzed acylation with aromatic aldehydes. DG = directing group.

The reaction conditions proved to be remarkably mild and even highly sensitive substrates such as (*E*)-croton aldehyde could be converted into the corresponding aryl ketone. The procedure is also applicable for the arylation of naturally occurring aldehydes, and this has been demonstrated by the coupling of enantiomerically pure (*S*)-citronellal. The group of Cheng developed a complementary Pd(OAc)$_2$-catalyzed aerobic protocol that is highly efficient for the acylation of 2-phenylpyridine derivatives with electron-rich, electron-deficient and heterocyclic aromatic aldehydes (Scheme 1.39).[75]

In ensuing contributions, phenyl oximes,[77] anilides[78] and 2-phenylbenzothiazoles[79] have been successfully opened up as substrates. The group of Yu proposed a mechanism[78a] for the acylation of pivaloyl anilides (Scheme 1.40) on the basis of deuterium labeling experiments ($k_H/k_D = 3.6$ for the C–H palladation), a Hammett correlation study with a series of *meta*-substituted pivaloyl anilides, and the studies of Li *et al.* with pre-formed Pd complexes.[76]

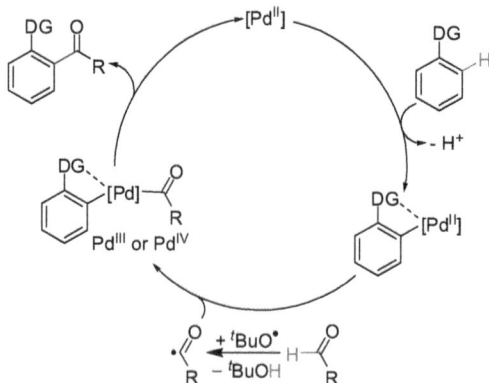

Scheme 1.40   Proposed mechanism of the Pd-catalyzed acylation.

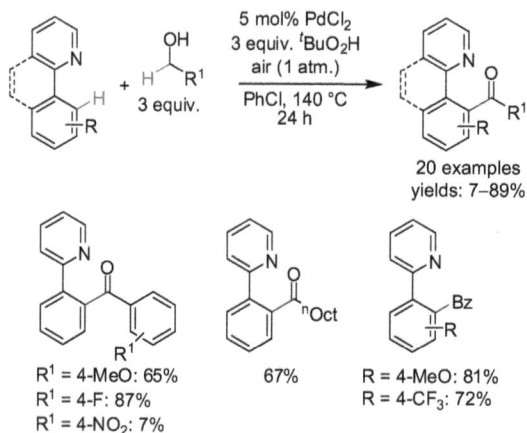

Scheme 1.41   Pd-Catalyzed acylation with alcohols.

The reaction starts with the thermal generation of *tert*-butoxy radicals that subsequently abstract a hydrogen atom from the carbonyl group of the aldehyde. The generated acyl radicals oxidize the palladacycles into either a Pd(IV) complex[80] or a dimeric Pd(III) species[81] which are generated by a rate-determining C–H palladation step. Reductive elimination liberates the product and closes the catalytic cycle.

In 2011, the groups of Deng and Li demonstrated that various aliphatic and benzyl alcohols could serve as aldehyde surrogates in this reaction (Scheme 1.41). The alcohols are oxidized *in situ* into the corresponding aldehydes, which are subsequently converted into aryl ketones.[82]

The utilization of *ortho*-phenoxy benzaldehydes allowed the group of Li to establish an intramolecular xanthone synthesis (Scheme 1.42).[83] The oxidative cyclization was efficiently mediated by a RhCl$_3$/PPh$_3$ catalyst system in chlorobenzene at 160 °C.

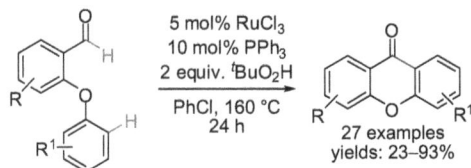

Scheme 1.42   Rh-Catalyzed dehydrogenative xanthone synthesis.

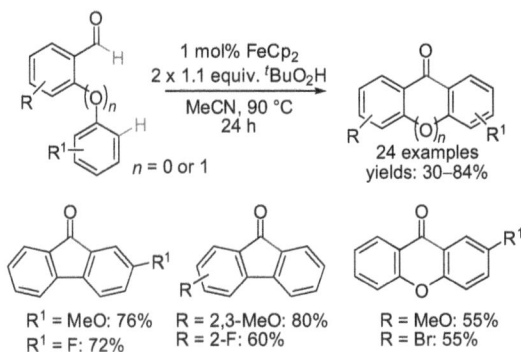

Scheme 1.43   Fe-Catalyzed radical cyclization.

The scope of this procedure was recently improved by the group of Studer with the introduction of ferrocene as a radical chain initiator (Scheme 1.43).[84]

Under optimized reaction conditions (Scheme 1.43) various 2-phenoxy- as well as 2-phenylbenzaldehydes were smoothly converted into the corresponding cyclic ketones in the presence of only 1 mol% ferrocene. Other Fe salts such as $FeCl_2$, $Fe(OAc)_2$ and $FeSO_4$ have also been shown to initiate the reaction but unreliable variations in yield were observed. The TBHP is best added in two batches to ensure reproducible results. The authors proposed a mechanism in which the Fe catalyst initiates a radical chain reaction by generating a *tert*-butoxy radical that subsequently abstracts the hydrogen atom from the aldehyde (Scheme 1.44).

The resulting acetyl radical adds onto the adjacent arene ring and the resulting aryl radical is oxidized into the corresponding product by a second molecule of peroxide.

Based on the studies of Gottschalk and Neckers in 1985,[85] the group of Lei developed a radical Cu-catalyzed cross-dehydrogenative olefination reaction that gives direct access to α,β-unsaturated ketones (Scheme 1.45).[86]

Various aromatic and heteroaromatic aldehydes were coupled with styrenes in the presence of catalytic quantities of $CuCl_2$ and TBHP at 80 °C. The procedure is remarkably selective, and the products detected arise from the acylation of the α,β-unsaturated ketone or radical polymerization of double bonds. The mechanism proceeds *via* the Cu-assisted generation of an acyl radical that adds onto the double bond of the olefin. Oxidation by the Cu

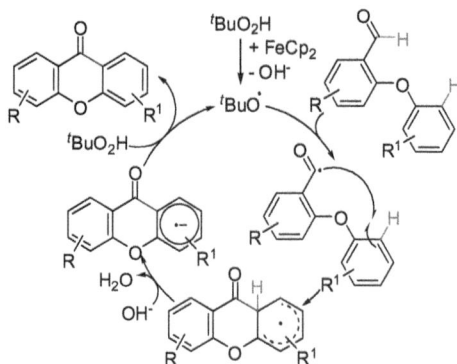

Scheme 1.44 Mechanism of the Fe-catalyzed radical cyclization.

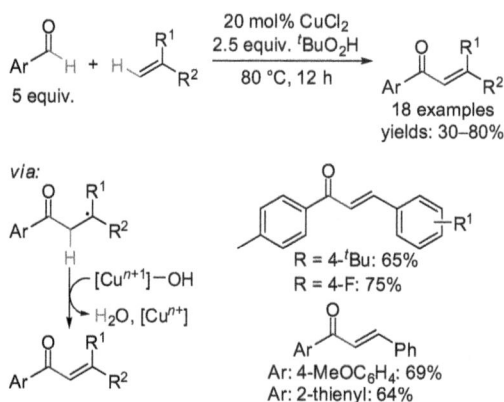

Scheme 1.45 Cu-Catalyzed radical synthesis of α,β-unsaturated ketones.

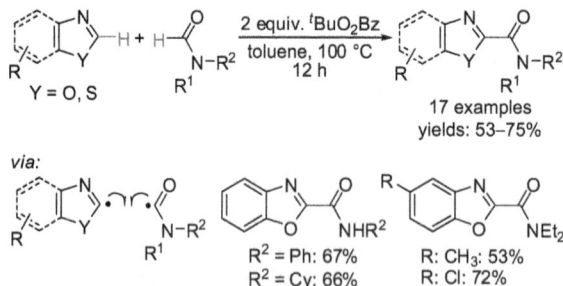

Scheme 1.46 Radical amidation.

catalyst and proton elimination yields the product and re-generates the Cu-catalyst.

In parallel contributions, two metal-free acylation reactions have been reported. In 2011, the group of Wang introduced a *tert*-butyl perbenzoate mediated amidation of thiazoles and oxazoles with a series of formamides

(Scheme 1.46).[87] The groups of Qu and Guo recently applied our method of the di-TBP mediated alkylation to various purines and purine glycosides with cycloalkanes.[88]

1.6 Conclusion and Outlook

As an endeavor to explore novel chemical reactions and to search new tools for more efficient chemical synthesis and green chemistry, a new concept in forming C–C bonds, cross-dehydrogenative-coupling, evolved. Representative examples illustrated in this chapter have shown that various C–C bonds can be generated directly from C–H and C–H bonds under oxidative conditions. This concept is continuing to evolve and many fascinating examples are being reported. These reactions will lay the foundation for the next generation of synthetic chemists with an eye on green chemistry.

Acknowledgements

I am indebted to my colleagues, whose names are cited in the references, and who made this research possible. I also thank the Canada Research Chair (Tier I) Foundation, the CFI, NSERC, and the (US) NSF-EPA Joint Program for a Sustainable Environment for their partial support of this research.

References

1. E. J. Corey and X. M. Cheng, *The Logic of Chemical Synthesis*, John Wiley & Sons, New York, 1989, pp. 1–91.
2. P. Y. Bruce, *Organic Chemistry*, Pearson Education, New Jersey, 4th edn, 2004.
3. I. Fleming, *Pericyclic Reactions*, Oxford University Press, New York, 1999, pp. 1–89.
4. (a) R. H. Crabtree, *The Organometallic Chemistry of Transition Metals*, Wiley Interscience, New York, 4th edn, 2005, pp. 1–560; (b) F. J. McQuillin, D. G. Parker and G. R. Stephenson, *Transition Metal Organometallics for Organic Synthesis*, Cambridge University Press, Cambridge, 1991, pp. 1–614; (c) J. Tsuji, *Transition Metal Reagents and Catalysts: Innovations in Organic Synthesis*, Wiley, Chichester, 2002, pp. 1–496.
5. http://www.nobelprize.org/nobel_prizes/chemistry/laureates/.
6. R. A. Sheldon, *ChemTech*, 1994, 38.
7. (a) V. Ritleng, C. Sirlin and M. Pfeffer, *Chem. Rev.*, 2002, **102**, 1731; (b) G. Dyker, *Angew. Chem., Int. Ed.*, 1999, **38**, 1698; (c) H. Chen, S. Schlecht, T. C. Semple and J. F. Hartwig, *Science*, 2000, **287**, 1995; (d) L. Ackermann, *Top. Organomet. Chem.*, 2007, **24**, 35.
8. G. Eglinton and A. R. Galbraith, *Chem. Ind.*, 1956, 737.
9. (a) C. Glaser, *Ber. Dtsch. Chem. Ges.*, 1869, **2**, 422; (b) A. S. Hay, *J. Org. Chem.*, 1962, **27**, 3320.

10. A. A. O. Sarhan and C. Bolm, *Chem. Soc. Rev.*, 2009, **38**, 2730.
11. I. Moritani and Y. Fujiwara, *Tetrahedron Lett.*, 1967, **12**, 1119.
12. C.-J. Li and B. M. Trost, *Proc. Natl. Acad. Sci. U. S. A.*, 2008, **105**, 13197.
13. (a) C.-J. Li, *Chem. Rev.*, 2005, **105**, 3095; (b) C.-J. Li, *Tetrahedron*, 1996, **52**, 5643; (c) C.-J. Li and T. H. Chan, *Comprehensive Organic Reactions in Aqueous Media*, John Wiley & Sons, New York, 2007.
14. C.-J. Li, *Acc. Chem. Res.*, 2010, **43**, 581.
15. (a) C. Wei, J. T. Mague and C.-J. Li, *Proc. Natl. Acad. Sci. U. S. A.*, 2004, **101**, 5749; (b) C. Wei, Z. Li and C.-J. Li, *Synlett*, 2004, 1472; (c) W.-J. Yoo, L. Zhao and C.-J. Li, *Aldrichimica Acta*, 2011, **44**, 43.
16. H. Nakamura, T. Kamakura, M. Ishikura and J.-F. Biellmann, *J. Am. Chem. Soc.*, 2004, **126**, 5958.
17. N. J. Leonard and G. W. Leubner, *J. Am. Chem. Soc.*, 1949, **71**, 3408.
18. S.-I. Murahashi, N. Komiya and H. Terai, *Angew. Chem., Int. Ed.*, 2005, **44**, 6931.
19. S. Murata, K. Termoto, M. Muira and M. Nomura, *J. Chem. Res., Miniprint*, 1993, 2827.
20. (a) Z. Li and C.-J. Li, *J. Am. Chem. Soc.*, 2004, **126**, 11810; (b) C.-J. Li and Z. Li, *Pure Appl. Chem.*, 2006, **78**, 935.
21. L. Zhao and C.-J. Li, *Angew. Chem., Int. Ed.*, 2008, **47**, 7075.
22. M. J. Wyvratt, *Clin. Physiol. Biochem*, 1988, **6**, 217.
23. (a) A. I. Meyers, *Aldrichimica Acta*, 1985, **18**, 59; (b) P. Beak, W. J. Zajdel and D. B. Reitz, *Chem. Rev.*, 1984, **84**, 471.
24. Y. Zhang and C.-J. Li, *Tetrahedron Lett.*, 2004, **45**, 7581.
25. C. A. Correia and C.-J. Li, *Heterocycles*, 2010, **82**, 555.
26. C. A. Correia and C.-J. Li, *Adv. Synth. Catal.*, 2010, **352**, 1446.
27. Z. Li and C.-J. Li, *J. Am. Chem. Soc.*, 2005, **127**, 6968.
28. Z. Li, D. S. Bohle and C.-J. Li, *Proc. Natl. Acad. Sci. U. S. A.*, 2006, **103**, 8928.
29. G. Zhang, J. Miao, Y. Zhao and H. Ge, *Angew. Chem., Int. Ed.*, 2012, **51**, 8318.
30. C. A. Correia, L. Yang and C.-J. Li, *Org. Lett.*, 2011, **13**, 4581.
31. T. He, L. Yu, L. Zhang, L. Wang and M. Wang, *Org. Lett.*, 2011, **13**, 5016.
32. (a) H. Mo and W. Bao, *Adv. Synth. Catal.*, 2009, **351**, 2845; (b) Y.-Z. Li, B.-J. Li, X.-Y. Lu, S. Lin and Z.-J. Shi, *Angew. Chem., Int. Ed.*, 2009, **48**, 3817.
33. Y.-X. Jia and E. P. Kündig, *Angew. Chem., Int. Ed.*, 2009, **48**, 1636.
34. A. Perry and R. J. K. Taylor, *Chem. Commun.*, 2009, 3249.
35. J. E. M. N. Klein, A. Perry, D. S. Pugh and R. J. K. Taylor, *Org. Lett.*, 2010, **12**, 3446.
36. W.-T. Wei, M.-B. Zhou, J.-H. Fan, W. Liu, R.-J. Song, Y. Liu, M. Hu, P. Xie and J.-H. Li, *Angew. Chem., Int. Ed.*, 2013, **52**, 3638.
37. (a) Y. Moon, D. Kwon and S. Hong, *Angew. Chem., Int. Ed.*, 2012, **51**, 11333; (b) K.-T. Yip, R. Y. Nimje, M. V. Leskinen and P. M. Pihko, *Chem.–Eur. J*, 2012, **18**, 12590.

38. (a) D. T. Sawyer, A. Sobkowiak and T. Matsushita, *Acc. Chem. Res.*, 1996, **29**, 409; (b) C. Walling, *Acc. Chem. Res.*, 1998, **31**, 155.
39. D. H. R. Barton and D. Doller, *Pure Appl. Chem.*, 1991, **63**, 1567.
40. Y. Zhang, J. Feng and C.-J. Li, *J. Am. Chem. Soc.*, 2008, **130**, 2900.
41. G. Deng, L. Zhao and C.-J. Li, *Angew. Chem., Int. Ed.*, 2008, **47**, 6278.
42. N. Chatani, Y. Ie, F. Kakiuchi and S. Murai, *J. Org. Chem.*, 1997, **62**, 2604.
43. X. Guo and C.-J. Li, *Org. Lett.*, 2011, **13**, 4977.
44. I. Fleming, *Frontier Orbitals and Organic Chemical Reactions*, Wiley, New York, 1976.
45. R. Xia, H.-Y. Niu, G.-R. Qu and H.-M. Guo, *Org. Lett.*, 2012, **14**, 5546.
46. G. Deng, K. Ueda, S. Yanagisawa, K. Itami and C.-J. Li, *Chem.–Eur. J*, 2009, **15**, 333.
47. (a) B. Liégault and K. Fagnou, *Organometallics*, 2008, **27**, 4841; (b) M. Wasa, K. M. Engle and J.-Q. Yu, *J. Am. Chem. Soc.*, 2010, **132**, 3680; (c) K. J. Stowers, K. C. Fortner and M. S. Sanford, *J. Am. Chem. Soc.*, 2011, **133**, 6541.
48. (a) Z. Li and C.-J. Li, *J. Am. Chem. Soc.*, 2005, **127**, 3672; (b) Z. Li, D. S. Bohle and C.-J. Li, *Proc. Natl. Acad. Sci., U. S. A.*, 2006, **103**, 8928; (c) N. Sasamoto, C. Dubs, Y. Hamashima and M. Sodeoka, *J. Am. Chem. Soc.*, 2006, **128**, 14010.
49. O. Basle and C.-J. Li, *Green Chem.*, 2007, **9**, 1047.
50. T. Zeng, G. Song, A. Moores and C.-J. Li, *Synlett*, 2010, 2002.
51. (a) S. S. Libendi, Y. Demizu and O. Onomura, *Org. Biomol. Chem.*, 2009, **7**, 351; (b) N. Shankaraiah, R. A. Pilli and L. S. Santos, *Tetrahedron Lett.*, 2008, **49**, 5098; (c) T. Tajima and A. Nakajima, *J. Am. Chem. Soc.*, 2008, **130**, 10496; (d) N. Girard and J.-P. Hurvois, *Tetrahedron Lett.*, 2007, **48**, 4097; (e) E. Le Gall, J.-P. Hurvois, T. Renaud, C. Moinet, A. Tallec, P. Uriac, S. Sinbandhit and L. Toupet, *Liebigs Ann.*, 1997, 2089; (f) S. Michel, E. Le Gall, J.-P. Hurvois, C. Moinet, A. Tallec, P. Uriac and L. Toupet, *Liebigs Ann.*, 1997, 259; (g) F. Effenberger and H. Kottmann, *Tetrahedron*, 1985, **41**, 4171; (h) G. Bidan, M. Geniès and R. Renaud, *Electrochim. Acta*, 1981, **26**, 275; (i) G. Bidan and M. Genies, *Tetrahedron*, 1981, **37**, 29.
52. O. Baslé, N. Borduas, P. Dubois, J. M. Chapuzet, T.-H. Chan, J. Lessard and C.-J. Li, *Chem.–Eur. J*, 2010, **16**, 8162.
53. Y. Zhang and C.-J. Li, *Angew. Chem.*, 2006, **45**, 1949.
54. Y. Zhang and C.-J. Li, *J. Am. Chem. Soc.*, 2006, **128**, 4242.
55. B. M. Trost and M. L. Crawley, *Chem. Rev.*, 2003, **103**, 2921.
56. B. M. Trost, P. E. Strege, L. Weber, T. J. Fullerton and T. J. Dietsche, *J. Am. Chem. Soc.*, 1978, **100**, 3407.
57. Z. Li and C.-J. Li, *J. Am. Chem. Soc.*, 2006, **128**, 56.
58. Z. Li, L. Cao and C.-J. Li, *Angew. Chem., Int. Ed.*, 2007, **46**, 6505.
59. Y. Zhang and C.-J. Li, *Eur. J. Org. Chem.*, 2007, 4654.
60. R. van Helden and G. Verberg, *Recl. Trav. Chim. Pays-Bas*, 1965, **84**, 1263.
61. (a) I. Moritani and Y. Fujiwara, *Tetrahedron Lett.*, 1967, **12**, 1119; (b) Y. Fujiwara, I. Moritani, M. Matsuda and S. Teranishi, *Tetrahedron*

Lett., 1968, **5**, 633; (c) Y. Fujiwara, I. Moritani, M. Matsuda and S. Teranishi, *Tetrahedron Lett.*, 1968, **35**, 3863; (d) Y. Fujiwara, I. Moritani and M. Matsuda, *Tetrahedron Lett.*, 1968, **24**, 4819; (e) Y. Fujiwara, I. Moritani, R. Asano and S. Teranishi, *Tetrahedron Lett.*, 1968, **57**, 6015; (f) Y. Fujiwara, I. Moritani, S. Danno, R. Asano and S. Teranishi, *J. Am. Chem. Soc.*, 1969, **91**, 7166; (g) S. Danno, I. Moritani and Y. Fujiwara, *Tetrahedron*, 1969, **25**, 4809; (h) Y. Fujiwara, I. Moritani, R. Asano, H. Tanaka and S. Teranishi, *Tetrahedron*, 1969, **25**, 4815; (i) S. Danno, I. Moritani and Y. Fujiwara, *Tetrahedron*, 1969, **25**, 4819.

62. (a) J. M. Davidson and C. Triggs, *J. Chem. Soc. A*, 1968, 1324; (b) J. M. Davidson and C. Triggs, *J. Chem. Soc. A*, 1968, 1331; (c) M. O. Unger and R. A. Fouty, *J. Org. Chem.*, 1969, **34**, 18; (d) Y. Fujiwara, I. Moritani, K. Ikegami, R. Tanaka and S. Teranishi, *Bull. Chem. Soc. Jpn.*, 1970, **43**, 863; (e) H. Yoshimoto and H. Itatani, *Bull. Chem. Soc. Jpn.*, 1973, **46**, 2490; (f) L. Eberson, L. Gomez-Gonzalez, U. Strand, H. Jalonen, B. Lüning and C.-G. Swahn, *Acta Chem. Scand.*, 1973, **27**, 1249; (g) F. R. S. Clark, R. O. C. Norman, C. B. Thomas and J. S. Willson, *J. Chem. Soc., Perkin Trans. 1*, 1974, 1289; (h) A. Shiotani and H. Itatani, *Angew. Chem., Int. Ed.*, 1974, **13**, 471; (i) T. Itahara, M. Hashimoto and H. Yumisashi, *Synthesis*, 1984, 255; (j) T. Itahara, *J. Org. Chem.*, 1985, **50**, 5272; (k) T. Jintoku, H. Taniguchi and Y. Fujiwara, *Chem. Lett.*, 1987, 1865; (l) T. Matsumoto and H. Yoshida, *Chem. Lett.*, 2000, 1064; (m) H. Weissman, X. Song and D. Milstein, *J. Am. Chem. Soc.*, 2001, **123**, 337.

63. (a) R. S. Shue, *J. Chem. Soc. Chem. Commun.*, 1971, 1510; (b) R. S. Shue, *J. Am. Chem. Soc.*, 1971, **93**, 7116; (c) R. Shue, *J. Catal*, 1972, **26**, 112.

64. M. D. K. Boele, G. P. F. van Strijdonck, A. H. M. de Vries, P. C. J. Kamer, J. G. de Vries and P. W. N. M. van Leeuwen, *J. Am. Chem. Soc.*, 2002, **124**, 1586.

65. (a) M. Rönn, P. G. Andersson and J.-E. Bäckvall, *Tetrahedron Lett.*, 1997, **38**, 3603; (b) J.-J. Li, T.-S. Mei and J.-Q. Yu, *Angew. Chem., Int. Ed.*, 2008, **47**, 6452; (c) Y.-H. Zhang, B.-F. Shi and J.-Q. Yu, *J. Am. Chem. Soc.*, 2009, **131**, 5072; (d) D.-H. Wang, K. M. Engle, B.-F. Shi and J.-Q. Yu, *Science*, 2009, **327**, 315.

66. R. Li, L. Jiang and W. Lu, *Organometallics*, 2006, **25**, 5973.

67. (a) D. R. Stuart and K. Fagnou, *Science*, 2007, **316**, 1172; (b) D. R. Stuart, E. Villemure and K. Fagnou, *J. Am. Chem. Soc.*, 2007, **129**, 12072; (c) B. Liégault, D. Lee, M. P. Huestis, D. R. Stuart and K. Fagnou, *J. Org. Chem.*, 2008, **73**, 5022; (d) B. Liégault and K. Fagnou, *Organometallics*, 2008, **27**, 4841.

68. (a) T. A. Dwight, N. R. Rue, D. Charyk, R. Josselyn and B. DeBoef, *Org. Lett.*, 2007, **9**, 3137; (b) S. Potavathri, A. S. Dumas, T. A. Dwight, G. R. Naumiec, J. M. Hammann and B. DeBoef, *Tetrahedron*, 2008, **49**, 4050.

69. (a) K. L. Hull, E. L. Lanni and M. S. Sanford, *J. Am. Chem. Soc.*, 2006, **128**, 14047; (b) K. L. Hull and M. S. Sanford, *J. Am. Chem. Soc.*, 2007, **129**, 11904; (c) B.-J. Li, S.-L. Tian, Z. Fang and Z.-J. Shi, *Angew. Chem., Int.*

Ed., 2008, **47**, 1115; (d) G. Brasche, J. García-Fortanet and S. L. Buchwald, *Org. Lett.*, 2008, **10**, 2207; (e) K. L. Hull and M. S. Sanford, *J. Am. Chem. Soc.*, 2009, **131**, 9651; (f) H. B. Zhang, L. Liu, Y. J. Chen, D. Wang and C.-J. Li, *Adv. Cat. Syn*, 2006, **348**, 229; (g) H. B. Zhang, L. Liu, Y. J. Chen, D. Wang and C.-J. Li, *Eur. J. Org. Chem.*, 2006, 869.

70. (a) R. F. Heck, *J. Am. Chem. Soc.*, 1968, **90**, 5518; (b) T. Mizoroki, K. Mori and A. Ozaki, *Bull. Chem. Soc. Jpn.*, 1971, **44**, 581; (c) R. F. Heck and J. P. Nolley, *J. Org. Chem.*, 1972, **37**, 2320; (d) H. A. Dieck and R. F. Heck, *J. Am. Chem. Soc.*, 1974, **96**, 1133; (e) A. De Meijere and F. E. Meyer, *Angew. Chem.*, 1994, **106**, 2473; (f) A. De Meijere and F. E. Meyer, *Angew. Chem., Int. Ed.*, 1995, **33**, 2379; (g) W. Cabri and I. Candiani, *Acc. Chem. Res.*, 1995, **28**, 2; (h) C. Amatore and A. Jutand, *Acc. Chem. Res.*, 2000, **33**, 314; (i) I. P. Beletskaya and A. V. Cheprakov, *Chem. Rev.*, 2000, **100**, 3009; (j) E. J. Farrington, J. M. Brown, C. F. J. Barnard and E. Rowsell, *Angew. Chem., Int. Ed.*, 2002, **41**, 169; (k) A. B. Dounay and L. E. Overman, *Chem. Rev.*, 2003, **103**, 2945; (l) J. Le Bras and J. Muzart, *Chem. Rev.*, 2011, **111**, 1170.

71. (a) L. Ackermann, *Modern Arylation Methods*, Wiley-VCH, Weinheim, 1st edn, 2009; (b) F. Bellina and R. Rossi, *Chem. Rev.*, 2010, **110**, 1082; (c) F. Bellina and R. Rossi, *Chem. Rev.*, 2010, **110**, 3850; (d) D. A. Colby, R. G. Bergman and J. A. Ellman, *Chem. Rev.*, 2010, **110**, 624; (e) J. Roger, A. Gottumukkala and H. Doucet, *ChemCatChem*, 2010, **2**, 20; (f) M. Wasa, K. M. Engle and J.-Q. Yu, *Isr. J. Chem.*, 2010, **50**, 605; (g) T. W. Lyons and M. S. Sanford, *Chem. Rev.*, 2010, **110**, 1147; (h) E. Clot, O. Eisenstein, N. Jasim, S. A. Macgregor, J. E. McGrady and R. N. Perutz, *Acc. Chem. Res.*, 2011, **44**, 333; (i) C. Verrier, P. Lassalas, L. Théveau, G. Quéguiner, F. Trécourt, F. Marsais and C. Hoarau, *Beilstein J. Org. Chem.*, 2011, **7**, 1584; (j) S. R. Neufeldt and M. S. Sanford, *Acc. Chem. Res.*, 2012, **45**, 936; (k) J. Yamaguchi, A. D. Yamaguchi and K. Itami, *Angew. Chem., Int. Ed.*, 2012, **51**, 8960; (l) L. G. Mercier and M. Leclerc, *Acc. Chem. Res.*, 2013, **46**, 1597; (m) J. J. Mousseau and A. B. Charette, *Acc. Chem. Res.*, 2013, **46**, 412.

72. (a) H. J. H. Fenton, *J. Chem. Soc.*, 1894, **65**, 899; (b) J. H. Baxendale and J. Magee, *Discuss. Faraday Soc*, 1953, **14**, 160; (c) R. G. R. Bacon, R. Grime and D. J. Munro, *J. Chem. Soc.*, 1954, 2275; (d) DeLos F. DeTar and R. A. J. Long, *J. Am. Chem. Soc.*, 1958, **80**, 4742; (e) D. H. Hey, M. J. Perkins and G. H. Williams, *J. Chem. Soc.*, 1963, 5604; (f) W. L. Carrick, G. L. Karapinka and G. T. Kwiatkowski, *J. Org. Chem.*, 1969, **34**, 2388; (g) R. A. Abramovitch and T. Takaya, *J. Chem. Soc. Chem. Commun.*, 1969, 1369; (h) T. Tezuka and N. Narita, *J. Am. Chem. Soc.*, 1979, **101**, 7413; (i) I. V. Kozhevnikov, V. I. Kim, E. P. Talzi and V. N. Sidelnikov, *J. Chem. Soc. Chem. Commun.*, 1985, 1392; (j) D. H. R. Barton and D. Doller, *Pure Appl. Chem.*, 1991, **63**, 1567.

73. (a) F. Minisci, R. Galli, M. Cecere, V. Malatesta and T. Caronna, *Tetrahedron Lett.*, 1968, **9**, 5609; (b) G. P. Gardini, G. Palla, A. Arnone and R. Galli, *Tetrahedron Lett.*, 1971, **12**, 59.

74. (a) E. Ador and J. Crafts, *Ber*, 1877, **10**, 2173; (b) N. O. Calloway, *Chem. Rev.*, 1935, **17**, 327; (c) P. H. Gore, *Chem. Rev.*, 1955, **55**, 229; (d) J. K. Groves, Chem. *Soc. Rev.*, 1972, **1**, 73.
75. X. Jia, S. Zhang, W. Wang, F. Luo and J. Cheng, *Org. Lett.*, 2009, **11**, 3120.
76. O. Baslé, J. Bidange, Q. Shuai and C.-J. Li, *Adv. Synth. Catal.*, 2010, **352**, 1145.
77. C.-W. Chan, Z. Zhou, A. S. C. Chan and W.-Y. Yu, *Org. Lett.*, 2010, **12**, 3926.
78. (a) C.-W. Chan, Z. Zhou and W.-Y. Yu, *Adv. Synth. Catal.*, 2011, **353**, 2999; (b) Y. Wu, B. Li, F. Mao, X. Li and F. Y. Kwong, *Org. Lett.*, 2011, **13**, 3258; (c) J. Weng, Z. Yu, X. Liu and G. Zhang, *Tetrahedron Lett.*, 2013, **54**, 1205.
79. A. Banerjee, S. K. Santra, S. Guin, S. K. Rout and B. K. Patel, *Eur. J. Org. Chem.*, 2013, **2013**, 1367.
80. (a) A. R. Dick, J. W. Kampf and M. S. Sanford, *J. Am. Chem. Soc.*, 2005, **127**, 12790; (b) S. R. Whitfield and M. S. Sanford, *J. Am. Chem. Soc.*, 2007, **129**, 15142.
81. (a) D. C. Powers and T. Ritter, *Nat. Chem*, 2009, **1**, 302; (b) D. C. Powers, M. A. L. Geibel, J. E. M. N. Klein and T. Ritter, *J. Am. Chem. Soc.*, 2009, **131**, 17050.
82. F. Xiao, Q. Shuai, F. Zhao, O. Baslé, G. Deng and C.-J. Li, *Org. Lett.*, 2011, **13**, 1614.
83. P. Wang, H. Rao, R. Hua and C.-J. Li, *Org. Lett.*, 2012, **14**, 902.
84. S. Wertz, D. Leifert and A. Studer, *Org. Lett.*, 2013, **15**, 928.
85. P. Gottschalk and D. C. Neckers, *Org. Chem*, 1985, **50**, 3498.
86. J. Wang, C. Liu, J. Yuan and A. Lei, *Angew. Chem., Int. Ed.*, 2013, **52**, 2256.
87. T. He, H. Li, P. Li and L. Wang, *Chem. Commun.*, 2011, **47**, 8946.
88. R. Xia, H.-Y. Niu, G.-R. Qu and H.-M. Guo, *Org. Lett.*, 2012, **14**, 5546.

Dehydrogenative Heck-type Reactions: The Fujiwara–Moritani Reaction

TSUGIO KITAMURA*[a] AND YUZO FUJIWARA[b]

[a] Department of Chemistry and Applied Chemistry, Graduate School of Science and Engineering, Saga University, Honjo-machi, Saga 840-8502, Japan; [b] Kyushu University, 6-10-1 Hakozaki, Fukuoka 812-8581, Japan
*Email: kitamura@cc.saga-u.ac.jp

2.1 Introduction

It is widely recognized that there is an increasing demand for environmentally benign reactions in organic synthesis. The use of catalysis and non-toxic reagents is desirable for green and sustainable chemistry. Hydrocarbons are abundant in nature but not used frequently in organic synthesis as substrates because their C–H bonds lack reactivity. If less reactive C–H bonds could couple directly with each other, it would be an ideal C–C bond formation reaction. Such a coupling reaction between C–H bonds is termed a "cross-dehydrogenative-coupling" (CDC) reaction.[1] A formal CDC reaction is shown in Scheme 2.1. Since the C–H bonds are inert, they need an activator for the coupling reaction. Among the CDC reactions, the coupling reaction between aromatic C–H bonds and vinylic C–H bonds is termed a "Fujiwara–Moritani reaction" or a "dehydrogenative Heck-type cross-coupling reaction" (Scheme 2.2).[2] This coupling reaction is promoted or catalyzed by Pd compounds.

RSC Green Chemistry No. 26
From C–H to C–C Bonds: Cross-Dehydrogenative-Coupling
Edited by Chao-Jun Li
© The Royal Society of Chemistry 2015
Published by the Royal Society of Chemistry, www.rsc.org

$$R^1\text{-}H + H\text{-}R^2 \longrightarrow R^1\text{-}R^2$$

Scheme 2.1   The CDC reaction.

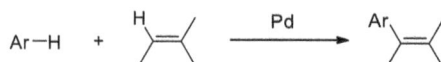

Scheme 2.2   The Fujiwara–Moritani reaction or dehydrogenative Heck-type cross-coupling reaction.

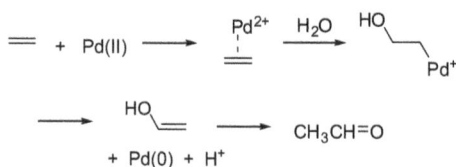

Scheme 2.3   The Wacker reaction.

2.2   Stoichiometric CDC Reactions

Historically, the study of the Fujiwara–Moritani reaction goes back to 1967.[3] It was known that substitution of olefins with nucleophiles takes place *via* the catalysis of Pd salts, as represented by the Wacker process (Scheme 2.3).[4] It was believed that a nucleophile attacks an electron-deficient olefinic carbon co-ordinated with the Pd catalyst and the resulting Pd complex undergoes elimination of palladium hydride to form a vinylic compound. The study by Fujiwara and Moritani was initiated to learn about the stereochemistry of the substitution reaction. To examine the stereochemistry, the reaction of a palladium chloride/styrene complex with acetic acid was conducted in benzene solution. Since the substitution of a hydrogen atom on the olefinic double bond by an aromatic compound had not been observed in those days, benzene was thought to be inert in the reaction. However, surprisingly, the initially expected β-acetoxystyrene was not observed, but *trans*-stilbene **2a** and 1-acetoxy-1-phenylethane **3** were obtained in 26 and 13% yields, respectively (Scheme 2.4).[3,5] Furthermore, the reaction of **1** in toluene afforded *para*-methylstilbene **2b** and 1-acetoxy-1-phenylethane **3** in 25 and 14% yields, respectively. These results indicate that aromatic solvents such as benzene and toluene react with styrene. The formation of the stilbenes means that a CDC reaction takes place between an aromatic C–H bond and an olefinic C–H bond. Carboxylic acids such as acetic acid and chloroacetic acid were essential, suggesting that acetic acid, especially the acetate ion, played an important role in this reaction.

Addition of sodium acetate produced a dramatic increase in the yield of stilbene, as shown in Table 2.1.[6] Using 20 equiv. of NaOAc to PdCl$_2$ resulted in almost quantitative formation of stilbene **2a**. The reaction of styrene with benzene proceeded when PdCl$_2$ and NaOAc were used instead of the palladium/styrene complex **1**.

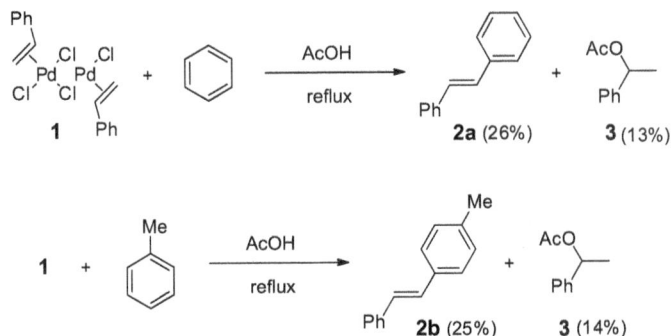

Scheme 2.4   Stilbene formation from the reaction of a PdCl$_2$/styrene complex with aromatics.

Table 2.1   Effect of NaOAc on stilbene formation from **1** and benzene.

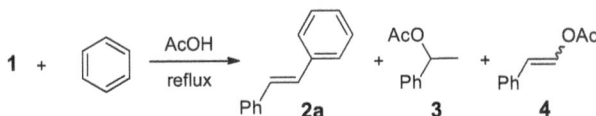

Entry	1/g	NaOAc/equiv.[a]	Benzene/mL	AcOH/mL	Product and yield/%		
					2a	3	4
1	9.1	0.36	340	80	25	8.0	1.5
2	4.0	2.0	150	36	86	1.8	4.7
3	9.1	7.2	340	340	90	0	1
4	3.0	20	113	27	97	0	1

[a]Molar ratio of NaOAc to palladium chloride.

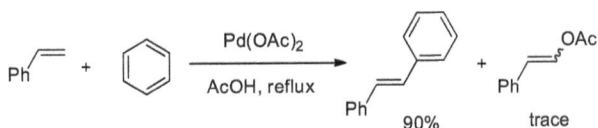

Scheme 2.5   Pd(OAc)$_2$-Promoted reaction of styrene with benzene.

The presence of the acetate ion accelerated the reaction of styrene with benzene. From this result, it was speculated that ligand exchange between chloride and acetate ions took place to form Pd(OAc)$_2$ as an active catalyst. In fact, as shown in Scheme 2.5, use of Pd(OAc)$_2$ promoted the reaction without complex **1** and stilbene was obtained in high yield.[6]

The Pd(OAc)$_2$-promoted reaction of styrene with benzene was applied to mono-substituted benzenes such as toluene, ethylbenzene, anisole, chloro-benzene and nitrobenzene.[7] The results are given in Table 2.2. Methyl, ethyl, and methoxy groups are *ortho–para*-directing, while the nitro group is *meta*-directing.

The reactivity of various mono-substituted benzenes is confirmed by competitive reactions.[2] For example, when ethylbenzene and nitrobenzene

Table 2.2 Reaction of styrene with mono-substituted benzenes.

$$\text{Ph} \diagup\!\!= + \;\; \bigcirc\!\!-R \;\xrightarrow[\text{AcOH, reflux}]{\text{Pd(OAc)}_2}\; \text{Ph}\diagdown\!\!=\!\!\diagup\bigcirc\!\!-R$$

Entry	Aromatic	Stilbenes	Yield/%
1	Toluene	o-Methylstilbene	17
		m-Methylstilbene	24
		p-Methylstilbene	33
2	Ethylbenzene	o-Ethylstilbene	11
		m-Ethylstilbene	23
		p-Ethylstilbene	31
3	Anisole	o-Methoxystilbene	30
		m-Methoxystilbene	5
		p-Methoxystilbene	48
4	Nitrobenzene	o-Nitrostilbene	4
		m-Nitrostilbene	29
		p-Nitrostilbene	4
5	Chlorobenzene	o-Chlorostilbene	12
		m-Chlorostilbene	20
		p-Chlorostilbene	30

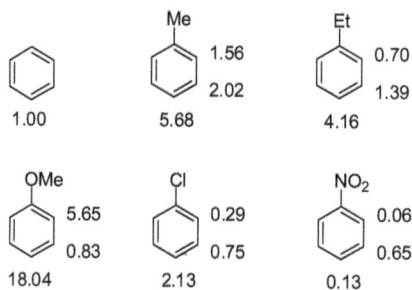

Benzene: 1.00

Me: 1.56 (ortho), 2.02 (meta), 5.68 (para)

Et: 0.70 (ortho), 1.39 (meta), 4.16 (para)

OMe: 5.65 (ortho), 0.83 (meta), 18.04 (para)

Cl: 0.29 (ortho), 0.75 (meta), 2.13 (para)

NO$_2$: 0.06 (ortho), 0.65 (meta), 0.13 (para)

Figure 2.1 Partial factors in the substitution of mono-substituted benzenes with styrene.

are allowed to react with styrene, nitro- and ethylstilbenes are formed in 4.3 and 56% yields, respectively. The partial rate factors are obtained from several competitive reactions where benzene is employed as a standard (Figure 2.1).

In the course of styrene–arene coupling reactions, Pd(OAc)$_2$ was found to be most effective. The coupling reaction was applied to ethene,[8] 1,1-diphenylethene,[8,9] 1,1,2-triphenylethene,[8,9] 1,2-dichloroethene,[10] allyl esters,[11] chalcone,[12] acrylonitrile,[13] and related derivatives. As shown in Scheme 2.6, phenylation of ethane and phenyl-substituted ethenes takes place in the presence of Pd(OAc)$_2$. Even in the case of electron-deficient alkenes such as acrylonitrile and chalcone, phenylation occurs. Allyl esters undergo

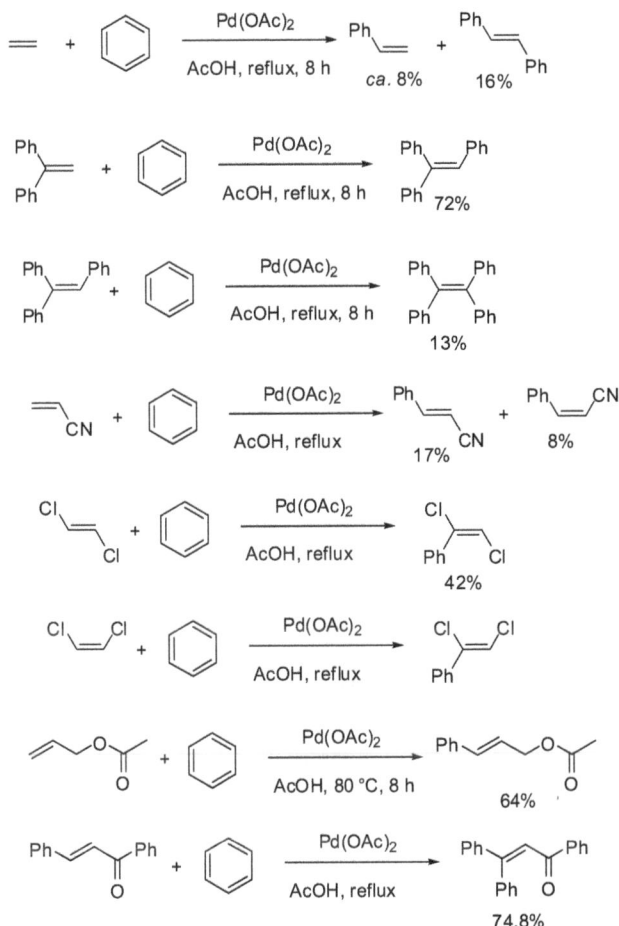

Scheme 2.6 Pd(OAc)₂-Mediated coupling reactions of benzene with alkenes.

phenylation without isomerization. Interestingly, phenylation of *cis-* and *trans-*1,2-dichloroethenes takes place with the retention of stereochemistry.

Pd(OAc)₂-Mediated coupling reactions of arenes with 1,4-benzoquinones and 1,4-naphthoquinones proceed to afford aryl-substituted 1,4-benzoquinones and 1,4-naphthoquinones (Scheme 2.7).[14]

Arylation accompanying double bond isomerization has been observed in the case of glycals[15] and 1-phenyl-1-propenes[16] (Scheme 2.8). This process was also applied to the asymmetric arylation of cyclohexene derivatives.[17]

Coupling reactions of alkenes occur even in the case of non-benzenoid aromatics.[18–20] Tricarbonyl(η⁴-cyclobutadiene)iron[18] and ferrocene[19] undergo coupling reactions with styrene and various alkenes to give the corresponding styryl derivatives, although the yields are not high (Scheme 2.9).

The CDC reaction of arenes and alkenes can be applied to heteroaromatic compounds. Coupling with furan, thiophene, pyrrole, benzofuran,

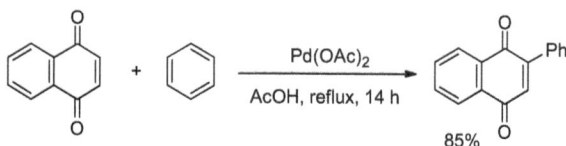

Scheme 2.7   Pd(OAc)$_2$-Mediated   coupling   reaction   of   benzene   with   1,4-naphthoquinone.

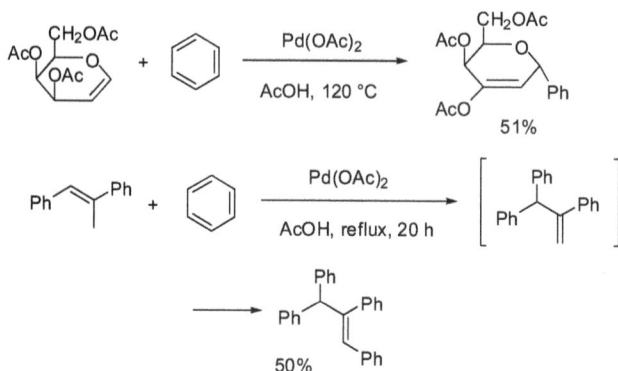

Scheme 2.8   Phenylation with double bond isomerization.

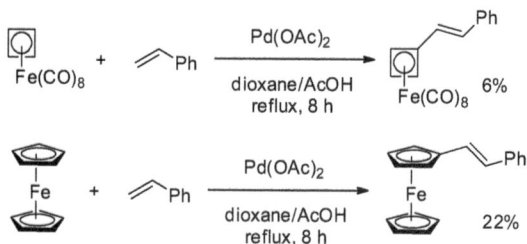

Scheme 2.9   Coupling reactions of non-benzenoid aromatics.

benzothiophene, indole and their derivatives was examined.[14,21–26] Pd(OAc)$_2$-Mediated coupling reaction of benzofurans with various alkenes including 1-phenylalkenes, acrylates, and acrylonitriles proceeds smoothly to give 2-alkenylated benzofurans. In the case of benzofuran, ethyl-2-benzofuranylacrylate is obtained in 70% yield (Scheme 2.10).[27] In a similar reaction using a mixed solvent of dioxane and AcOH, single- and double-alkenylated benzofurans, benzothiophenes, and *N*-acetylindole are formed. Simple heteroarenes such as furan, thiophene, selenophene, and *N*-methyl-pyrrole undergo a coupling reaction with styrene, acrylonitrile, and methyl acrylate in the presence of Pd(OAc)$_2$ in a mixed solvent of dioxane and AcOH, to afford 2-alkenylated and 2,5-bis-alkenylated products. The example of furan and styrene is depicted in Scheme 2.11.[28]

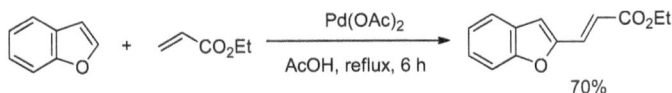

Scheme 2.10   Coupling reaction of benzofuran with ethyl acrylate.

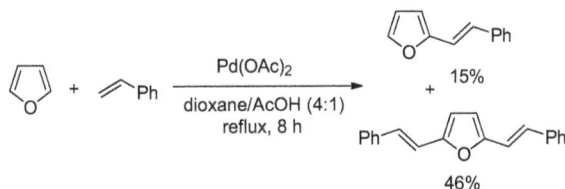

Scheme 2.11   Coupling reaction of furan and styrene.

C–H bond palladation

(1) Oxidative addition

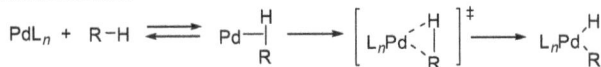

(2) σ-Bond metathesis

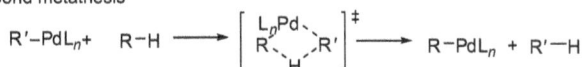

(3) Electrophilic substitution

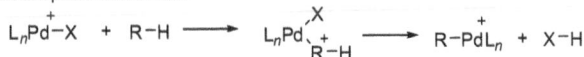

(4) Carboxylate-assisted palladation

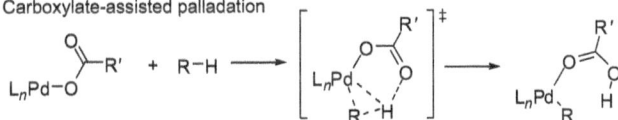

Alkene insertion and elimination of HPdOAc

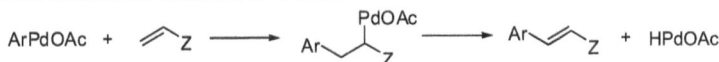

Scheme 2.12   C–H Bond palladation and subsequent reactions.

As discussed in a recent review,[29] there are four modes of C–H bond metalation: (1) oxidative addition with electron-rich late transition metals; (2) σ-bond metathesis with early transition metals; (3) electrophilic activation with electron-deficient late transition metals; and (4) carboxylate-directed metalation. The last mode has been only recently reported and proceeds *via* a continuum of electrophilic, ambiphilic, and nucleophilic interactions with the assistance of a bifunctional ligand bearing an additional Lewis-basic heteroatom, such as carboxylate. Possible mechanisms for the C–H bond palladation are shown in Scheme 2.12.

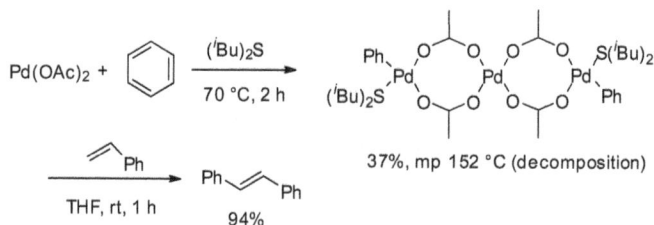

Scheme 2.13 Formation and reaction of a Ph–Pd complex.

As shown by the above, the details of the mechanism of C–H bond palladation may change with the nature of the metal or the reaction conditions. The important role of carboxylates has been proven by recent theoretical studies[30] together with experimental results (the electrophilic nature of Pd species and a kinetic isotope effect of $k_H/k_D = 4.5$–5.1).[31] Formation of a phenyl–Pd(II) complex by the C–H bond metalation of benzene is followed by a Heck-type reaction *via* alkene insertion (Scheme 2.12). The Ph–Pd species is isolated as a diphenyltripalladium complex by reaction of benzene with $Pd(OAc)_2$ in the presence of dialkyl sulfides. The reaction of the Ph–Pd complex with styrene in THF affords *trans*-stilbene in 94% yield (Scheme 2.13).[32] Using electrospray ionization mass spectrometry (ESI-MS) studies, mono-nuclear and dinuclear Pd(II) species were indicated to be involved in the CDC of furans with acryrates.[33]

2.3 Catalytic CDC Reactions

The coupling reaction with $Pd(OAc)_2$ proceeds with the C–H bond activation of aromatic compounds, followed by the formation of ArPdOAc species, then alkene insertion into the Ar–Pd bond, and finally elimination of HPdOAc leading to arylalkenes. The resulting HPdOAc decomposes to Pd(0) and AcOH, and terminates the reaction. If the Pd(0) species can be re-oxidized into Pd(II) species, $Pd(OAc)_2$, by an oxidant, it will have the chance to participate in the coupling reaction again. Such a catalytic cycle is depicted in Scheme 2.14. The original catalytic reaction was performed using $Pd(OAc)_2$ as the catalyst and AgOAc as the re-oxidant.[34]

In the coupling reaction of benzene with ethylene or styrene, several oxidants are used as re-oxidizing agents, including AgOAc, $Cu(OAc)_2$, heteropolyacids, and oxygen. Although $Cu(OAc)_2$ gives better results than AgOAc, it is observed that carrying out the reaction in the presence of O_2 improves the catalytic efficiency. Heteropolyacids such as $HPMo_{11}V$ also show good catalytic activity. Representative examples are given in Scheme 2.15.[8,35]

In the coupling reactions of unsubstituted and β-phenyl-substituted α,β-unsaturated esters, ketones, and aldehydes, many catalytic systems of $Pd(OAc)_2$ are used with the following effective re-oxidizing combinations: O_2,[36] acetylacetone/$HPMo_{11}V$/NaOAc/O_2,[37] $PhCO_3{}^tBu$,[40] BQ/^tBuOOH,[40] acetylacetone/$H_4PMo_{11}VO_{40}$/NaOAc/O_2,[40] 2,6-bis(2-ethylhexyl)pyridine/Ac_2O/O_2,[41] $Mn(OAc)_3$/O_2,[36] AgOAc or AgOBz,[39] BQ/MnO_2,[39] BQ/H_2O_2,[39] and

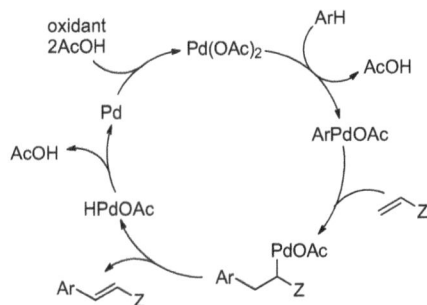

Scheme 2.14 Catalytic CDC reaction of arenes and alkenes.

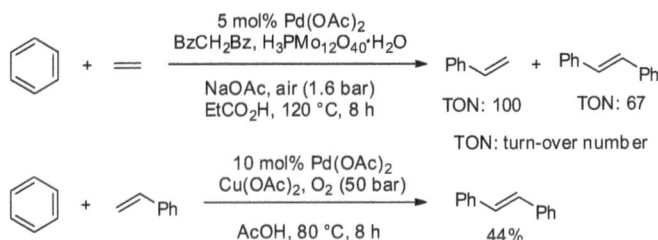

Scheme 2.15 Catalytic CDC reactions of benzene with ethylene or styrene.

$Cu(OAc)_2/^tBuOOH$, where BQ stands for benzoquinone.[39] Representative examples are given in Scheme 2.16.

As discussed in the stoichiometric reactions before, the catalytic coupling reaction of mono-substituted benzenes also affords a mixture of *o*-, *m*-, and *p*-isomers, the ratios of which depend on the nature of the substituents. CDC reactions with a high *meta*-selectivity have been observed in the coupling reaction of electron-deficient arenes bearing CO_2Et, CO_2Me, CF_3, COMe, or NO_2 groups in the presence of $Pd(OAc)_2$ and a sterically bulky pyridine ligand, 2,6-bis(2-ethylhexyl)pyridine.[41] The coupling reaction of ethyl benzoate and ethyl acrylate under these conditions gives no *ortho*-coupling products but a mixture of 4 : 1 of *meta*-to-*para* coupling products. Similarly, 1,3-bis-(trifluoromethyl)benzene undergoes selective coupling with ethyl acrylate at the *meta*-position to give ethyl-3,5-bis(trifluoromethyl)cinnamate in 62% yield. These reactions are given in Scheme 2.17. In addition, pyridine ligands such as 3,5-dichloropyridine can enhance the rate, yield, substrate scope, and site selectivity of arene C–H olefination reactions.[42]

Interestingly, electron-deficient fluorobenzenes undergo Pd-catalyzed coupling reactions with mono-substituted alkenes, as shown in Scheme 2.18.[43]

Heteroaromatic compounds show similar behavior to benzene derivatives. Furan, pyrrole, thiophene, benzofuran, indole, benzothiophene, and their related compounds undergo CDC reactions with alkenes. In the case of furan, the coupling reactions with acrylates, acrylonitrile, styrenes, and

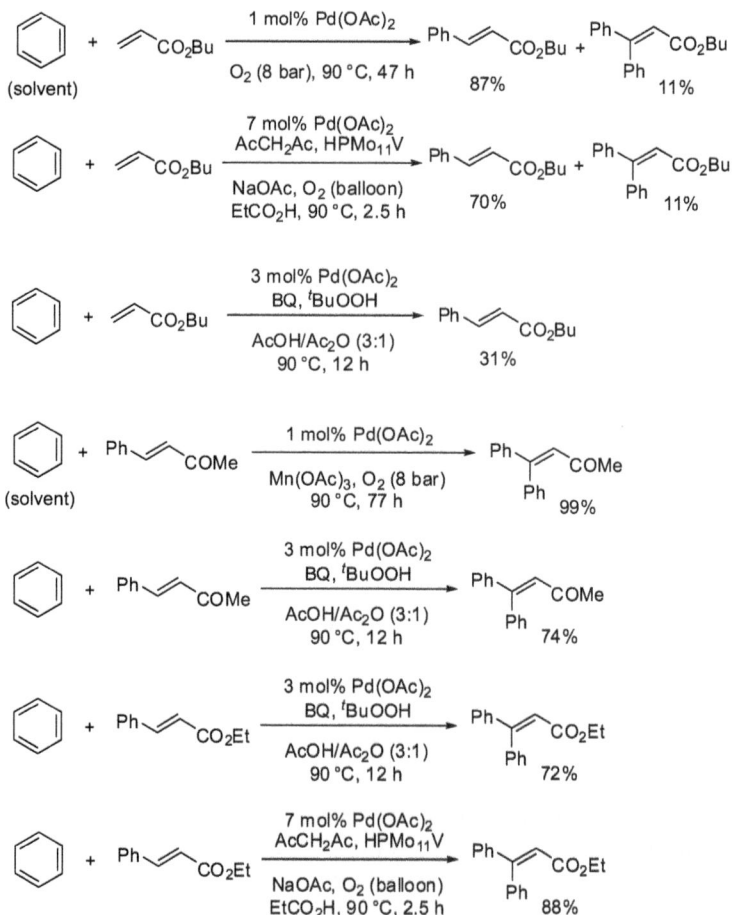

Scheme 2.16 Catalytic CDC reactions of benzene.

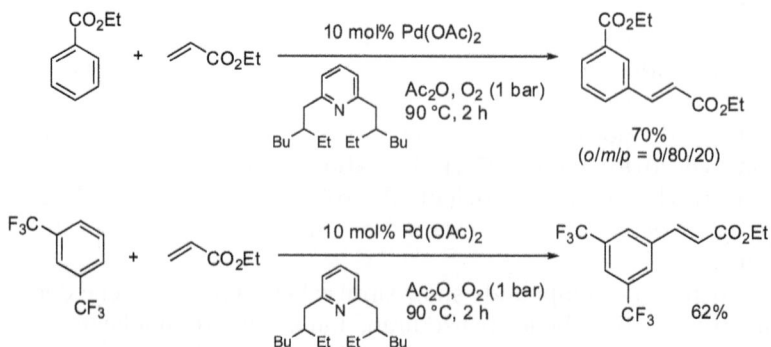

Scheme 2.17 CDC reactions with high *meta*-selectivity.

DMF: *N*,*N*-dimethylformamide, DMSO: dimethyl sulfoxide

Scheme 2.18   CDC of pentafluorobenzene with ethyl acrylate.

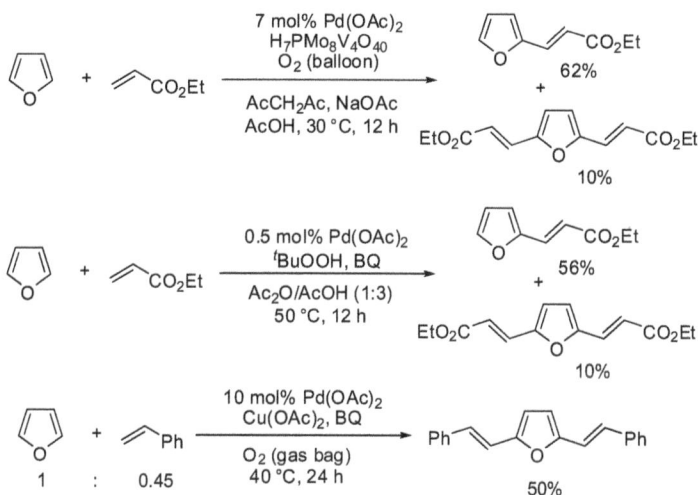

Scheme 2.19   CDC of furan and ethyl acrylate or styrene.

methylvinylketone are reported to give good results. Successful catalytic systems include $Pd(OAc)_2/H_7PMo_8V_4O_{40}/O_2$/acetylacetone/NaOAc,[40] $Pd(OAc)_2/$ $^tBuOOH/BQ$,[39] $Pd(OAc)_2/PhCO_3{}^tBu$,[38] $Pd(OAc)_2/AgOAc$/pyridine,[44]   and $Pd(OAc)_2/Cu(OAc)_2/BQ/O_2$.[45] Generally, the coupling reactions of furan provide double-alkenylated products as a minor product in addition to mono-alkenylated products. By using an excess of alkene substrate, the double-alkenylated furans are obtained in good yield. Representative examples are given in Scheme 2.19.

Furans substituted at the 2-position efficiently undergo coupling to give 2-alkenylated furans in good yields (Scheme 2.20).[39,44,45] Selected alkenes as the coupling partner are acrylates, acrylamides, methylvinylketone, acrylaldehyde, acrylonitrile, and styrenes.

The catalytic system of $Pd(OAc)_2/BQ/^tBuOOH$ has been applied to the coupling of benzofuran and acrylates (Scheme 2.21).[39]

The coupling of pyrroles has also been studied with various alkenes, and the nitrogen of pyrroles is protected in most cases. The coupling reaction of 1-(2,6-dichlorobenzoyl)pyrrole and methyl acrylate has been conducted in the presence of $Pd(OAc)_2$ as catalyst with AgOAc, $Cu(OAc)_2$, $Na_2S_2O_8$, or $NaNO_2$ as re-oxidant under air atmosphere, giving a mixture of 2-alkenylated and 2,5-bis -alkenylated pyrroles in good yields (Scheme 2.22).[26]

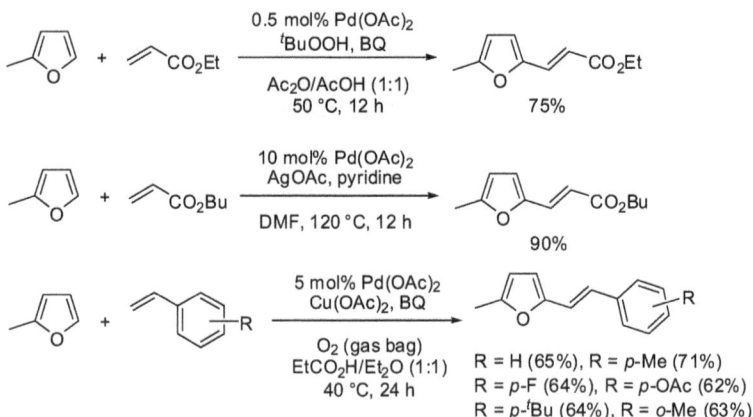

Scheme 2.20   CDC of 2-methylfuran and acrylates or styrenes.

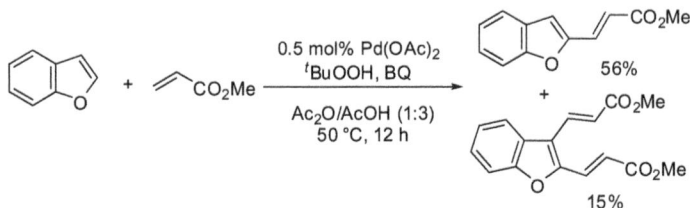

Scheme 2.21   CDC of benzofuran and methyl acrylate.

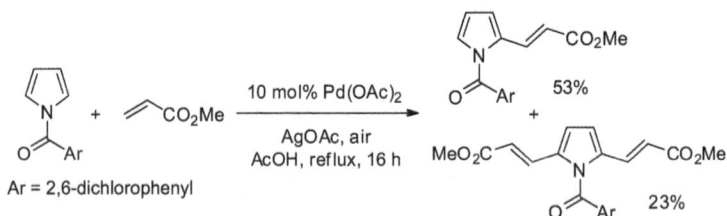

Scheme 2.22   CDC of a pyrrole and methyl acryate.

It is noteworthy that the coupling reaction is controlled by the nature of the substituent on the nitrogen of the pyrrole. In the coupling reaction with benzyl acrylate using the catalytic system of Pd(OAc)$_2$/PhCO$_3$^{t}Bu, pyrroles with *N*-Boc, *N*-Ac, or *N*-Ts groups react at the 2-position to give 2-alkenylated pyrroles, while 1-(tri-isopropylsilyl)pyrrole reacts at the 3-position to afford 3-alkenylated pyrroles selectively (Scheme 2.23).[46,47] It is understood that the sterically bulky tri-isopropylsilyl group blocks the reaction at the neighboring 2-position of the pyrrole and causes the reaction at the 3-position. Similarly, a high regioselectivity is observed for various alkenes such as acrylates and α,β-unsaturated carbonyl compounds.

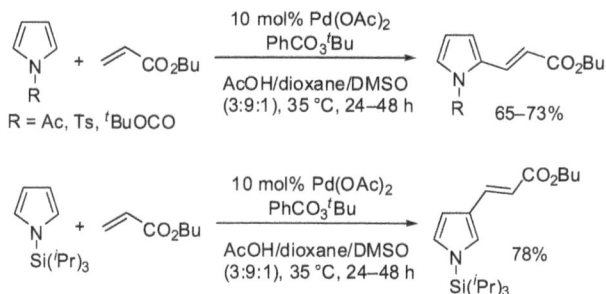

Scheme 2.23 Regioselectivity on dehydrogenative coupling of pyrroles and acrylates.

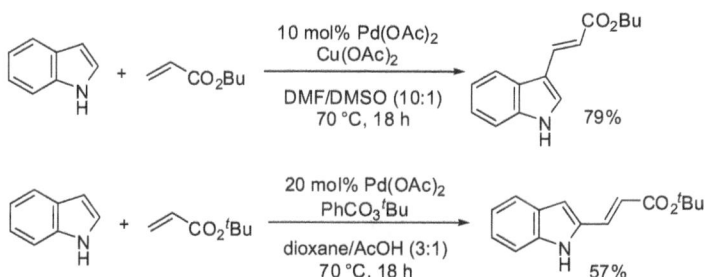

Scheme 2.24 CDC of indole and acrylates: regioselectivity.

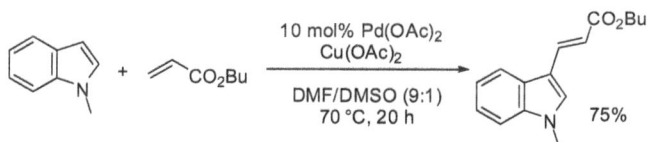

Scheme 2.25 CDC of *N*-methylindole and butyl acrylate.

The coupling reactions of indoles and alkenes also take place both at the 2- and 3-positions and the selectivity depends on the solvent used in the reaction, as shown in Scheme 2.24.[48] The coupling reaction of indole and butyl acrylate using the catalytic system of Pd(OAc)$_2$/Cu(OAc)$_2$ selectively affords 3-alkenylated indoles in DMSO, DMF, and a mixture of DMF and DMSO (10:1).

N-Substituted indoles bearing *N*-2,6-dichlorobenzoyl,[23] *N*-phenylsulfonyl,[24] or *N*-methyl groups[48] undergo coupling reactions with acrylates at the 3-position of indole (Scheme 2.25). The coupling reaction in AcOH also occurs at the 3-position of indole to give 3-alkenylated indoles.

In the cases of thiophene and 2-substituted thiophenes, the coupling reaction proceeds with alkenes in the presence of a Pd(OAc)$_2$ catalyst. Good yields of the coupling products are obtained using a Pd(OAc)$_2$/pyridine catalytic system in DMF (Scheme 2.26).[44]

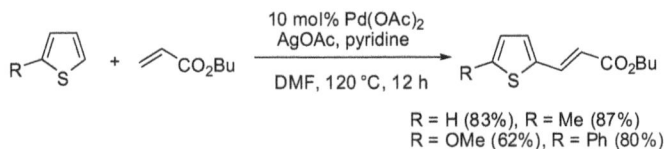

Scheme 2.26   CDC of thiophenes and butyl acrylate.

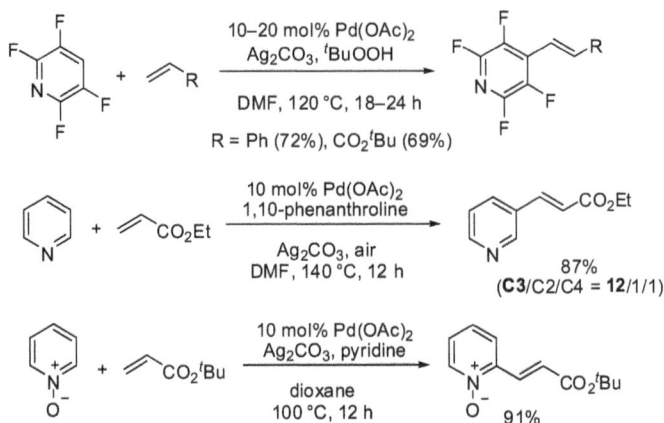

Scheme 2.27   Regioselective coupling of pyridine and ethyl acrylate.

Pd-Catalyzed CDC reactions have also been applied to pyridine. The coupling has been conducted by the reaction of 2,3,5,6-tetrafluoropyridine with styrene or *tert*-butyl acrylate in the presence of Pd(OAc)$_2$/Ag$_2$CO$_3$/^{t}BuOOH in DMF.[43a] The C3-selective akenylation of pyridines with alkenes takes place using 1,10-phenanthroline as the ligand and the catalytic system of Pd(OAc)$_2$/Ag$_2$CO$_3$/air.[49] The effect of the bidentate 1,10-phenenthroline ligand is explained by a strong *trans* effect from the bis-pyridine ligand that forces the N co-ordination of the Pd(II) complex to transform a reactive π-coordination. Pyridine-*N*-oxides undergo CDC with alkenes in the presence of Pd(OAc)$_2$.[50] A representative example is given in Scheme 2.27.

2.4   Ligand-Directing CDC Reactions

Intramolecularly ligand-directing palladation in stoichiometric reactions of *N,N*-dimethylbenzylamine, *N,N*-dimethylaminomethylnaphthalene, benzanilide, and their derivatives has been studied (Scheme 2.28). Reaction of *N,N*-dimethylbenzylamine with Li$_2$PdCl$_4$ in MeOH gives a 95% yield of di-μ-chloro-bis(*N,N*-dimethylbenzylamine-2-*C,N*)dipalladium(II).[51] Similarly, 1-(*N,N*-dimethylaminomethyl)naphthalene undergoes the palladation at the 2-position and the resulting palladium complex reacts with ethyl acrylate to give the corresponding 2-alkenylated naphthalene derivative.[52] The reaction of acetanilide with Pd(OAc)$_2$ in refluxing toluene affords an *ortho*-palladated

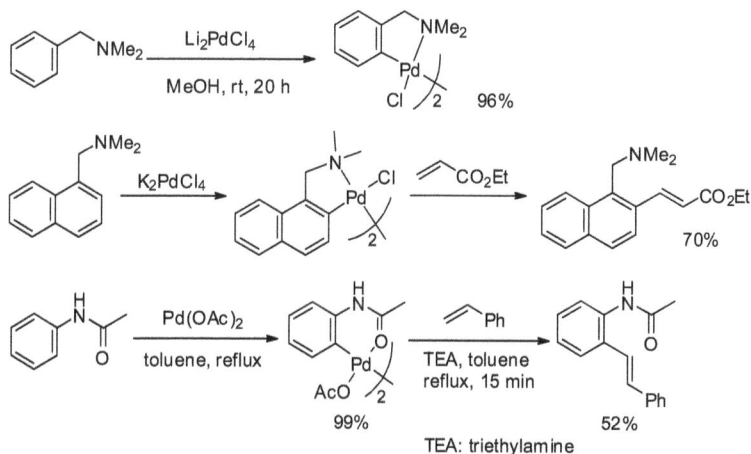

Scheme 2.28 Formation of cyclopalladates and their reactions with alkenes.

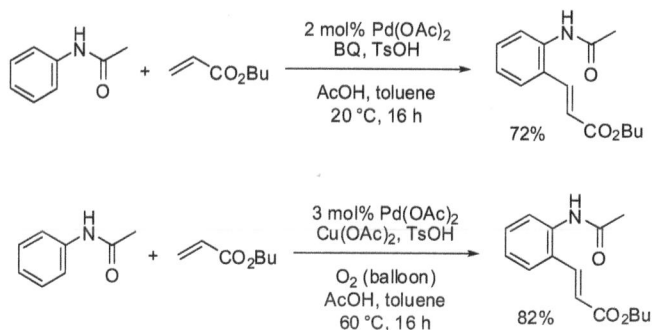

Scheme 2.29 *ortho*-Directing CDC reactions of acetanilide and butyl acrylate.

complex (50–65% yield), which undergoes reaction with styrene in the presence of triethylamine to give the corresponding 2-acetylaminostilbene in 52% yield.[53]

ortho-Selective CDC takes place catalytically in the reaction of anilides with alkenes in the presence of Pd(OAc)$_2$. The representative catalytic systems used in the *ortho*-coupling reactions are Pd(OAc)$_2$/BQ/TsOH[54] and Pd(OAc)$_2$/Cu(OAc)$_2$/TsOH/O$_2$ (balloon),[55] as shown in Scheme 2.29.

Substituted acetanilides, *N*-arylacetamides, undergo an *ortho*-coupling reaction to give the corresponding 2-alkenylated anilides in good yields. For example, the reaction of 3-acetanilde with butyl acrylate proceeds smoothly and selectively at 20 °C using 2 mol% Pd(OAc)$_2$ and BQ as the oxidant to yield (*E*)-3-(2-acetylamino-4-methylphenyl)propenoic acid butyl ester in a 91% yield.[54] Other examples are given in Scheme 2.30.[54,55] However, electron-deficient acetanilides bearing CF$_3$, Cl, or Br groups result in a lower yield of the coupling products. The case of *N*-(4-nitrophenyl)acetamide does not give the expected coupling product at all. The effect of TsOH is considered to be based on the increasing electrophilicity of Pd(II) due to the replacement of

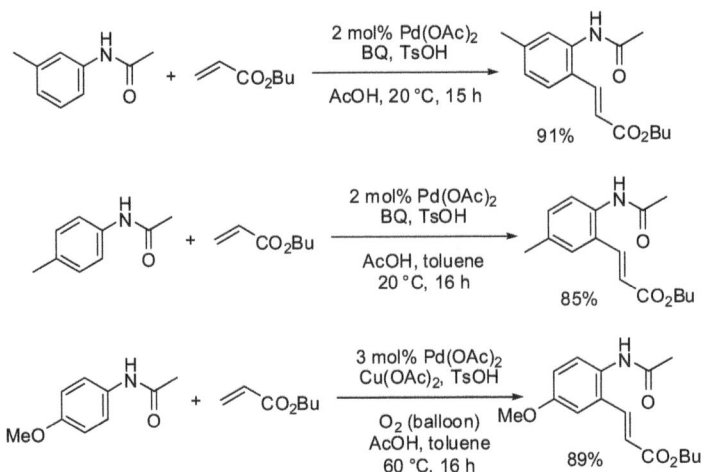

Scheme 2.30 *ortho*-Directing CDC reactions of *N*-arylacetamides and butyl acrylate.

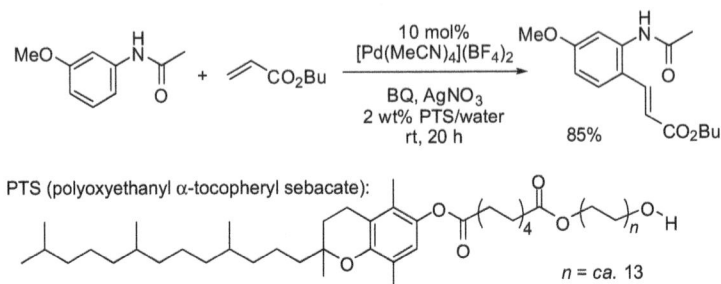

Scheme 2.31 *ortho*-Directing coupling reaction of *meta*-methoxyacetanilide and butyl acrylate in water.

AcO by TsO. In addition, replacement of BQ by PhCO$_3$^{t}Bu and Cu(OAc)$_2$ as oxidants improves the yield and substrate scope under milder conditions (room temperature).[56]

Pd-Catalyzed reactions in water containing the surfactant polyoxyethanyl α-tocopheryl sebacate (PTS) have been applied to the *ortho*-directing CDC reactions of *meta*-alkoxy-functionalized anilides and ureas with acrylates (Scheme 2.31).[57] The coupling reactions take place effectively when the cationic complex [Pd(MeCN)$_4$](BF$_4$)$_2$ is used as the catalyst together with BQ and AgNO$_3$ as the oxidant.

CDC reactions of *N,N*-dimethyl-*N'*-aryl ureas with butyl acrylate proceed effectively in acetone in the presence of a Pd(OAc)$_2$/BQ/TsOH catalytic system (Scheme 2.32).[58] It is considered that aryl ureas show higher reactivity than anilides due to their better co-ordinating ability.

Other nitrogen-containing groups, such as aminomethyl and aminoethyl groups also participate in *ortho*-directing CDC reactions. Examples are given in

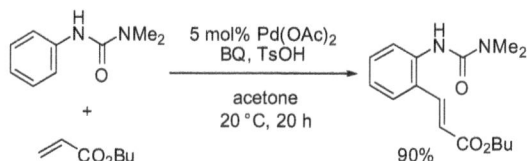

Scheme 2.32 *ortho*-Directing coupling reaction of *N,N*-dimethyl-*N*-phenyl urea and butyl acrylate.

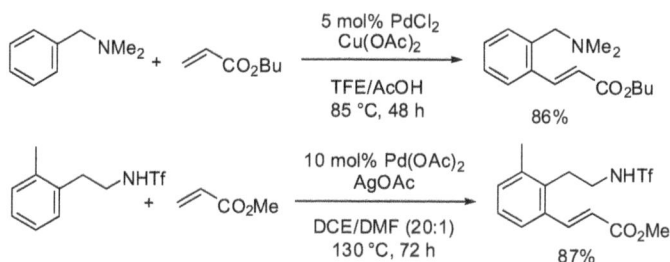

Scheme 2.33 *ortho*-Directing CDC reaction with an aminomethyl or aminoethyl group.

Scheme 2.33. The reaction of *N,N*-dimethylbenzylamine and alkenes is conducted in the presence of a PdCl$_2$/Cu(OAc)$_2$ catalytic system in a mixed solvent of trifluoroethanol (TFE) and AcOH (4 : 1) to give the coupling products in good yields.[59] Similar coupling reactions are applied to arenes substituted by a C2 or C3 alkyl tether terminated by an NHTf group.[60] Using a Pd(OAc)$_2$/AgOAc catalytic system in a mixed solvent of 1,2-dichloroethane (DCE) and DMF, *ortho*-alkenylated arenes are obtained in good yields. Although a variety of amino group co-ordination-directed CDC reactions have been developed, aniline itself does not show the coupling reaction, but provides the double insertion of ethylene leading to a heterocycle, 2-methylquinoline.[61]

Hydroxy groups also promote *ortho*-metalation but afford the heterocycle by subsequent cyclization in most cases.[62–64] Representative examples of hydroxyl-group-directing *ortho*-alkenylations are shown in Scheme 2.34. In the reaction with styrenes, the coupling products are obtained. The Pd(OAc)$_2$-catalyzed reaction of 2-hydroxybiphenyl and styrene in DMF gives the corresponding alkenylated product in 68% yield. In the case of arenes bearing 2-hydroxyethyl groups, the coupling reaction with styrenes proceeds in the presence of Pd(OAc)$_2$, AgOAc and Li$_2$CO$_3$ with an *N*-protected amino acid ligand.[65] A silanol group is shown to act as a directing group for the CDC reaction of arenes and alkenes. The advantage of the use of the silanol group is that it can be removed by treatment with tetrabutylammonium fluoride (TBAF).[66]

Ether-directed CDC reactions occur with the Pd(OAc)$_2$/Ac-Gly-OH/Ag$_2$CO$_3$ catalytic system in hexafluoroisopropanol (HFIP) as solvent.[67] The reaction of 2-methoxypropylbenzene with ethyl acrylate gives the *ortho*-alkenylated product in 70% yield (Scheme 2.35).

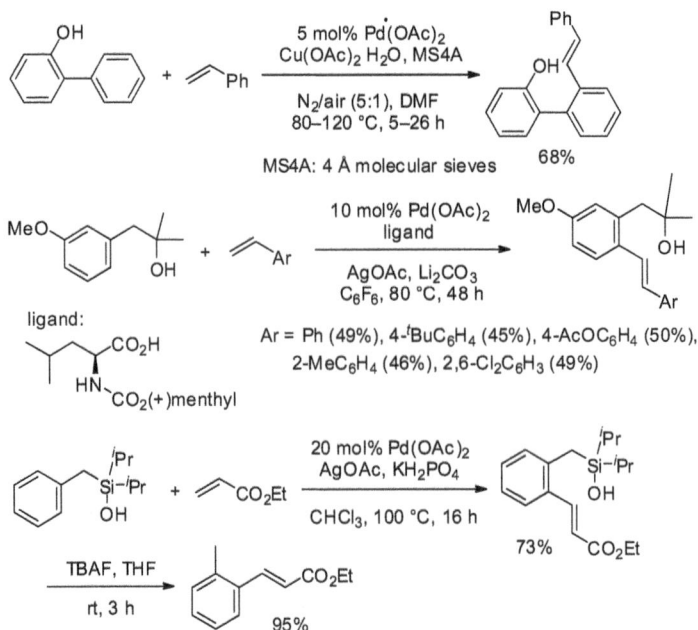

Scheme 2.34   Hydroxy-group-directing *ortho*-alkenylation.

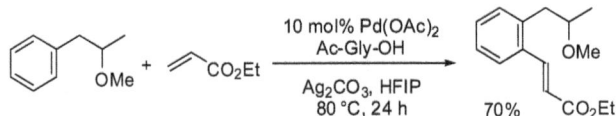

Scheme 2.35   Ether-directing *ortho*-alkenylation.

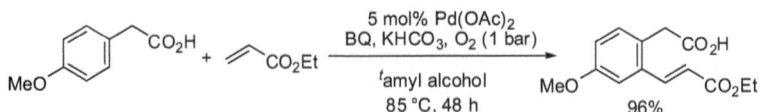

Scheme 2.36   Carboxyl-directing *ortho*-alkenylation.

The carboxyl group also assists *ortho*-metalation. In the case of benzoic acid, its derivatives undergo a CDC reaction with alkenes, but the subsequent reaction gives the cyclized product.[68] Phenylacetic and phenylpropionic acids also undergo CDC reactions.[69–71] The coupling reactions are conducted under basic conditions in *tert*-amyl alcohol using a Pd(OAc)$_2$/BQ/O$_2$ catalytic system (Scheme 2.36). Furthermore, the positional selectivity is increased by adding *N*-protected amino acids such as Boc-Leu-OH and Boc-Ile-OH. The effect of the added ligand is discussed in refs. 72 and 73.

A sophisticated nitrile-containing end-on template acts as a remote *meta*-directing group.[74] The template is considered to accommodate a macrocyclic

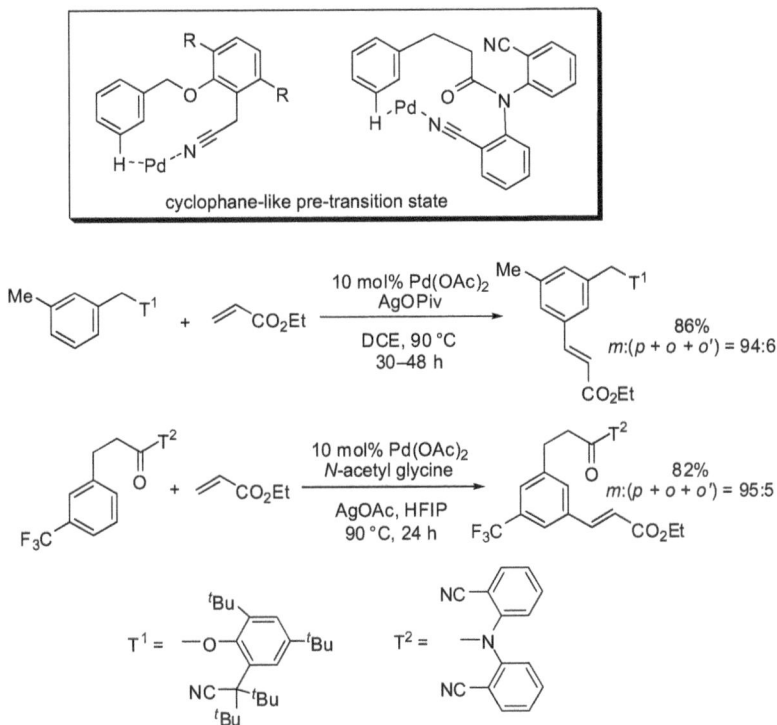

Scheme 2.37 Remote *meta*-selective alkenylation.

cyclophane-like pre-transition state. Using a Pd(OPiv)$_2$ (Piv: pivaloyl) and AgOPiv catalytic system, a high *meta*-selectivity is obtained, as shown in Scheme 2.37. In addition, the templates can be removed by Pd/C-mediated hydrogenolysis and hydrolysis using LiOH.

2.5 Conclusions and Outlook

This chapter demonstrated a dehydrogenative Heck-type reaction, that is, the Fujiwara–Moritani reaction. This reaction is a direct coupling reaction between an aromatic C–H bond and a vinylic C–H bond with the aid of a palladium catalyst, and provides aromatic alkenes by a simple and straightforward method. Since this reaction does not require pre-functionalization of hydrocarbons (arenes and alkenes), it is a direct, environmentally benign reaction. Furthermore, recent studies have enabled regioselective coupling reactions such as *ortho*- or *meta*-directing couplings. Mechanistically, the importance of the role of carboxylates in this reaction has been pointed out, and the features of the whole coupling reaction have been explained. Since this reaction was discovered in 1967, the utility of this coupling reaction has been comprehensively explored. In the future, this coupling reaction is expected to play a important role in organic synthesis.

References

1. (a) C. S. Yeung and V. M. Dong, *Chem. Rev.*, 2011, **111**, 1215; (b) M. Klussmann and D. Sureshkumar, *Synthesis*, 2011, 353; (c) C.-J. Li and W.-J. Yoo, *Top. Curr. Chem*, 2010, **292**, 281; (d) C.-J. Li, *Acc. Chem. Res.*, 2009, **42**, 335.

2. (a) I. Moritani and Y. Fujiwara, *Synthesis*, 1973, 524; (b) Y. Fujiwara, *Handbook of Organopalladium Chemistry for Organic Synthesis*, ed. E.-I. Negishi, John Wiley & Sons, New York, 2002, p. 2863; (c) C. Jia, T. Kitamura and Y. Fujiwara, *Acc. Chem. Res.*, 2001, **34**, 633.

3. I. Moritani and Y. Fujiwara, *Tetrahedron Lett.*, 1967, 1119.

4. (a) M. B. Smith, *March's Advanced Organic Chemistry*, Wiley, New York, 7th edn, 2013; (b) F. A. Carey and R. J. Sunderberg, *Advanced Organic Chemistry, Part B*, Springer, New York, 5th edn, 2007.

5. Y. Fujiwara, I. Moritani and M. Matsuda, *Tetrahedron*, 1968, **24**, 4819.

6. Y. Fujiwara, I. Moritani, M. Matsuda and S. Teranishi, *Tetrahedron Lett.*, 1968, 633.

7. Y. Fujiwara, I. Moritani, R. Asano, H. Tanaka and S. Teranishi, *Tetrahedron*, 1969, **25**, 4815.

8. Y. Fujiwara, I. Moritani, S. Danno, R. Asano and S. Teranishi, *J. Am. Chem. Soc.*, 1969, **91**, 7166.

9. Y. Fujiwara, I. Moritani, R. Asano and S. Teranishi, *Tetrahedron Lett.*, 1968, 6015.

10. I. Moritani, S. Danno, Y. Fujiwara and S. Teranishi, *Bull. Chem. Soc. Jpn.*, 1971, **44**, 578.

11. M. Yoshidomi, Y. Fujiwara and H. Taniguchi, *Nippon Kagaku Kaishi.*, 1985, 512.

12. K. Yamamura, *J. Org. Chem.*, 1978, **43**, 724.

13. S. Danno, I. Moritani and Y. Fujiwara, *Tetrahedron*, 1969, **25**, 4819.

14. T. Itahara, *J. Org. Chem.*, 1985, **50**, 5546.

15. (a) S. Czernecki and V. Dechavanne, *Can. J. Chem.*, 1983, **61**, 533; (b) V. Bellosta, S. Czernecki, D. Avenel, S. El Bahij and H. Gillier-Pandraud, *Can. J. Chem*, 1990, **68**, 1364.

16. K. Yamamura, *J. Chem. Soc., Perkin Trans. 1*, 1975, 988.

17. K. Mikami, M. Hatano and M. Terada, *Chem. Lett.*, 1999, 55.

18. Y. Fujiwara, R. Asano, I. Moritani and S. Teranishi, *Chem. Lett.*, 1975, 1061.

19. (a) R. Asano, I. Moritani, Y. Fujiwara and S. Teranishi, *Chem. Commun.*, 1970, 1293; (b) R. Asano, I. Moritani, A. Sonoda, Y. Fujiwara and S. Teranishi, *J. Chem. Soc. C*, 1971, 3691.

20. Y. Fujiwara, R. Asano, I. Moritani and S. Teranishi, *J. Org. Chem.*, 1976, **41**, 1681.

21. Y. Fjiwara, O. Maruyama, M. Yoshidomi and H. Taniguchi, *J. Org. Chem.*, 1981, **46**, 851.

22. O. Maruyama, Y. Fujiwara and H. Taniguchi, *Bull. Chem. Soc. Jpn.*, 1981, **46**, 2851.

23. T. Itahara, M. Ikeda and T. Sakakibara, *J. Chem. Soc., Perkin Trans. 1*, 1983, 1361.
24. T. Itahara, K. Kawasaki and F. Ouseto, *Synthesis*, 1984, 236.
25. T. Itahara and F. Ouseto, *Synthesis*, 1984, 488.
26. T. Itahara, K. Kawasaki and F. Ouseto, *Bull. Chem. Soc. Jpn.*, 1984, **57**, 3488.
27. A. Kasahara, T. Izumi, M. Yodono, R. Saito, T. Takeda and T. Sugawara, *Bull. Chem. Soc. Jpn.*, 1973, **46**, 1220.
28. R. Asano, I. Moritani, Y. Fujiwara and S. Teranishi, *Bull. Chem. Soc. Jpn.*, 1973, **46**, 663.
29. L. Ackermann, *Chem. Rev.*, 2011, **111**, 1315.
30. (a) K. M. Engle, D.-H. Wang and J.-Q. Yu, *J. Am. Chem. Soc.*, 2010, **132**, 14137; (b) S. Zhang, L. Shi and Y. Ding, *J. Am. Chem. Soc.*, 2011, **133**, 20218.
31. A. E. Shilov and G. B. Shul'pin, *Chem. Rev.*, 1997, **97**, 2879.
32. Y. Fuchita, K. Hiraki, Y. Kamogawa, M. Suenaga, K. Tohgoh and Y. Fujiwara, *Bull. Chem. Soc. Jpn.*, 1989, **62**, 1081.
33. (a) A. Vasseur, D. Harakat, J. Muzart and J. Le Bras, *J. Org. Chem.*, 2012, **77**, 5751; (b) A. Vasseur, D. Harakat, J. Muzart and J. Le Bras, *Adv. Synth. Catal*, 2013, **355**, 59.
34. Y. Fujiwara, I. Moritani, M. Matsuda and S. Teranishi, *Tetrahedron Lett.*, 1968, 3863.
35. T. Yamada, A. Sakakura, S. Sakaguchi, Y. Obora and Y. Ishii, *New J. Chem.*, 2008, **32**, 738.
36. M. Dams, D. E. De Vos, S. Celen and P. A. Jacobs, *Angew. Chem., Int. Ed.*, 2003, **42**, 3512.
37. M. Tani, S. Sakaguchi and Y. Ishii, *J. Org. Chem.*, 2004, **69**, 1221.
38. J. Tsuji and H. Nagashima, *Tetrahedron*, 1984, **40**, 2699.
39. C. Jia, W. Lu, T. Kitamura and Y. Fujiwara, *Org. Lett.*, 1999, **1**, 2097.
40. T. Yokota, M. Tani, S. Sakaguchi and Y. Ishii, *J. Am. Chem. Soc.*, 2003, **125**, 1476.
41. Y.-H. Zhang, B.-F. Shi and J.-Q. Yu, *J. Am. Chem. Soc.*, 2009, **131**, 5072.
42. A. Kubota, M. H. Emmert and M. S. Sanford, *Org. Lett.*, 2012, **14**, 1760.
43. (a) X. Zhang, S. Fan, C.-Y. He, X. Wan, Q.-Q. Min, J. Yang and Z.-X. Jiang, *J. Am. Chem. Soc.*, 2010, **132**, 4506; (b) S. Fan, C.-Y. He and X. Zhang, *Chem. Commun*, 2010, **46**, 4923.
44. J. Zhao, L. Huang, K. Cheng and Y. Zhang, *Tetrahedron Lett.*, 2009, **50**, 2758.
45. C. Aouf, E. Thiery, J. Le Bras and J. Muzart, *Org. Lett.*, 2009, **13**, 4096.
46. E. M. Beck, N. P. Grimster, R. Hatley and M. J. Gaunt, *J. Am. Chem. Soc.*, 2006, **128**, 2528.
47. E. M. Beck, R. Hatley and M. J. Gaunt, *Angew. Chem., Int. Ed.*, 2008, **47**, 3004.
48. N. P. Grimster, C. Gauntlett, C. R. A. Godfrey and M. J. Gaunt, *Angew. Chem., Int. Ed.*, 2005, **44**, 3125.
49. M. Ye, G.-L. Gao and J.-Q. Yu, *J. Am. Chem. Soc.*, 2011, **133**, 6964.

50. S. H. Cho, S. J. Hwang and S. Chang, *J. Am. Chem. Soc.*, 2008, **130**, 9254.
51. A. C. Cope and E. C. Friedrich, *J. Am. Chem. Soc.*, 1968, **90**, 909.
52. M. Julia, M. Duteil and J. Y. Lallemand, *J. Organomet. Chem.*, 1975, **102**, 239.
53. (a) H. Horino and N. Inoue, *Tetrahedron Lett.*, 1979, 2403; (b) H. Horino and N. Inoue, *J. Org. Chem.*, 1981, **46**, 4416.
54. M. D. K. Boele, G. P. F. van Strijdonck, A. H. M. de Vries, P. C. J. Kamer, J. G. de Vries and P. W. N. M. van Leeuwen, *J. Am. Chem. Soc.*, 2002, **124**, 1586.
55. J.-R. Wang, C.-T. Yang, L. Liu and Q.-X. Guo, *Tetrahedron Lett.*, 2007, **48**, 5449.
56. X. Liu and K. K. Hii, *J. Org. Chem.*, 2011, **76**, 8022.
57. T. Nishikata and B. H. Lipshutz, *Org. Lett.*, 2010, **12**, 1972.
58. W. Rauf, A. L. Thompson and J. M. Brown, *Chem. Commun.*, 2009, 3874.
59. G. Cai, Y. Fu, Y. Li, X. Wan and Z. Shi, *J. Am. Chem. Soc.*, 2007, **129**, 7666.
60. J.-J. Li, T.-S. Mei and J.-Q. Yu, *Angew. Chem., Int. Ed.*, 2008, **47**, 6452.
61. S. E. Diamond, A. Szalkiewicz and S. Mares, *J. Am. Chem. Soc.*, 1979, **101**, 490.
62. M. Miura, T. Tsuda and M. Nomura, *Chem. Lett.*, 1997, 1103.
63. M. Miura, T. Tsuda, T. Satoh, S. Pvsa-Art and M. Nomura, *J. Org. Chem.*, 1998, **63**, 5211.
64. S. Aoki, J. Oyamada and T. Kitamura, *Bull. Chem. Soc. Jpn.*, 2005, **78**, 468.
65. Y. Lu, D.-H. Wang, K. M. Engle and J.-Q. Yu, *J. Am. Chem. Soc.*, 2010, **132**, 5916.
66. C. Wang and H. Ge, *Chem.–Eur. J*, 2011, **17**, 14371.
67. G. Li, D. Leow, L. Wan and J.-Q. Yu, *Angew. Chem., Int. Ed.*, 2013, **52**, 1245.
68. M. Miura, T. Tsuda, T. Satoh, S. Pvsa-Art and M. Nomura, *J. Org. Chem.*, 1998, **63**, 5211.
69. D.-H. Wang, K. M. Engle, B.-F. Shi and J.-Q. Yu, *Science*, 2010, **327**, 315.
70. B.-F. Shi, Y.-H. Zhang, J. K. Lam, D.-H. Wang and J.-Q. Yu, *J. Am. Chem. Soc.*, 2010, **132**, 460.
71. K. M. Engle, D.-H. Wang and J.-Q. Yu, *Angew. Chem., Int. Ed.*, 2010, **49**, 6169.
72. K. M. Engle, D.-H. Wang and J.-Q. Yu, *J. Am. Chem. Soc.*, 2010, **132**, 14137.
73. R. D. Baxter, D. Sale, K. M. Engle, J.-Q. Yu and D. G. Blackmond, *J. Am. Chem. Soc.*, 2012, **134**, 4600.
74. D. Leow, G. Li, T.-S. Mei and J.-Q. Yu, *Nature*, 2012, **486**, 518.

CHAPTER 3

Copper-Catalyzed Cross-Dehydrogenative-Coupling Reactions

XIAOJIAN ZHENG AND ZHIPING LI*

Department of Chemistry, Renmin University of China, Beijing 100872, China
*Email: zhipingli@ruc.edu.cn

3.1 Introduction

Transition metal catalyzed C–H bond functionalization has greatly influenced traditional synthetic chemistry.[1] Copper-catalyzed cross-dehydrogenative-coupling (CDC) reactions have been developed as an attractive methodology for the construction of functional molecules from simple starting materials.[2] The transformation involves the connection of sp^3 C–H bonds to a variety of nucleophiles (X–H) to form sp^3 C–X bonds (where X = C, O, N, S, P, *etc.*) in the presence of a copper catalyst. The following section will focus on advances in copper-catalyzed CDC reactions. These reactions are classified by the type of C–H bonds, including C–H bonds adjacent to nitrogen, C–H bond adjacent to oxygen, allylic C–H bonds and benzylic C–H bonds. Asymmetric CDC reactions catalyzed by the combination of copper and an organocatalyst will also be considered.

RSC Green Chemistry No. 26
From C–H to C–C Bonds: Cross-Dehydrogenative-Coupling
Edited by Chao-Jun Li
© The Royal Society of Chemistry 2015
Published by the Royal Society of Chemistry, www.rsc.org

3.2 Copper-Catalyzed CDC Reactions Involving C–H Bonds Adjacent to Nitrogen

Much effort has been made to explore efficient CDC reactions with C–H bonds adjacent to nitrogen, because they greatly improve the atom economy. Copper often plays the key role in these kinds of transformations, and sp^3–sp^3, sp^3–sp and sp^3–sp^2 C–C bonds can be constructed (Scheme 3.1).

3.2.1 sp^3 C–H Bonds and sp C–H Bonds

In 2004, Li[3] and co-workers reported an effective CuBr-catalyzed direct alkynylation of an sp^3 C–H bond adjacent to a nitrogen atom *via* the CDC strategy (Scheme 3.2). The reaction was carried out without a solvent and under nitrogen atmosphere, and a variety of propargylic amines were efficiently synthesized. These compounds are of great pharmaceutical interest and are synthetic intermediates for various nitrogen-containing compounds.[4]

Using *N*-bromosuccinimide (NBS) as the free radical initiator combined with CuBr as a catalyst, Fu[5] and co-workers developed a CDC reaction of tertiary aliphatic amines and terminal alkynes to generate propargylic amines (Scheme 3.3). In this transformation, the alkynylation is selective for methyl substituents directly attached to the amine.

The mechanism of the above reactions involves the same process[3] (Scheme 3.4). First is the copper-catalyzed activation of the sp^3 C–H bond adjacent to the amine forming an imine-type intermediate (co-ordinated to copper) *via* a single-electron-transfer (SET) process. Meanwhile, copper also activates the terminal alkyne. Subsequently, the imine-type intermediate is attacked by the activated nucleophile copper acetylide to form the desired products and re-generate the copper salt. Importantly, the iminium ion

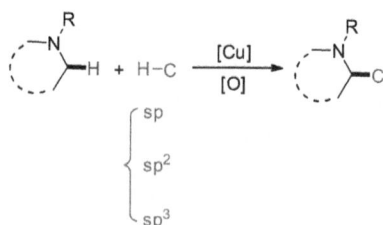

Scheme 3.1 Copper-catalyzed CDC reactions of C–H bonds adjacent to nitrogen.

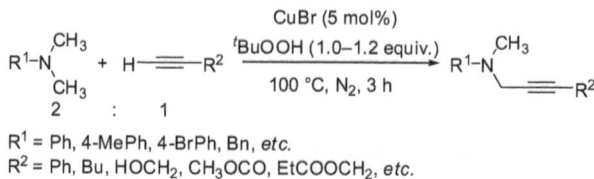

R^1 = Ph, 4-MePh, 4-BrPh, Bn, *etc.*
R^2 = Ph, Bu, HOCH$_2$, CH$_3$OCO, EtCOOCH$_2$, *etc.*

Scheme 3.2 Copper-catalyzed CDC reaction: alkynylation.

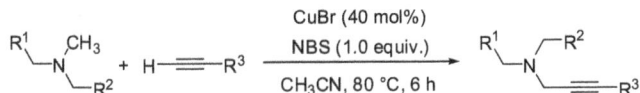

R[1] = Ph, CH$_3$(CH$_2$)$_{10}$CH$_2$, cyclohexane, *etc.*
R[2] = H, Me, *etc.*
R[3] = Ph, 4-MeOPh, CH$_3$(CH$_2$)$_4$CH$_2$, *etc.*

Scheme 3.3 Copper-catalyzed CDC reaction of tertiary aliphatic amines and terminal alkynes.

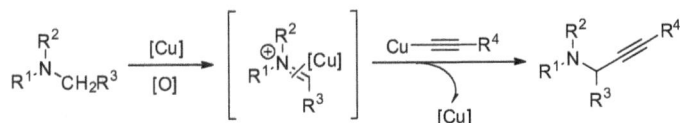

Scheme 3.4 Tentative mechanism for copper-catalyzed sp^3–sp C–C bond formation.

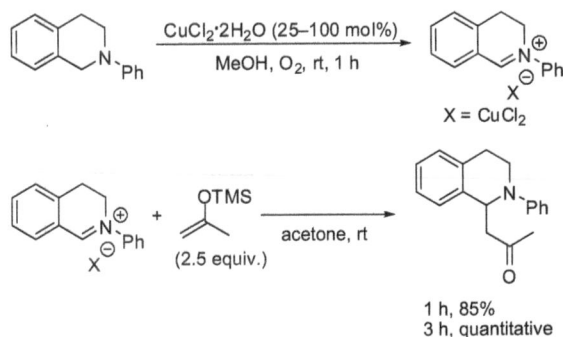

Scheme 3.5 Intermediate study in copper-catalyzed CDC reactions.

intermediates have been characterized by Klussmann[6] and co-workers (Scheme 3.5). To gain information on potential intermediates, *N*-phenyl-substituted tetrahydroisoquinoline (THIQ) was reacted with CuCl$_2$ · H$_2$O in methanol at room temperature under oxygen. After the full conversion of *N*-phenyl-substituted THIQ, a precipitate was separated. This precipitate was shown to be an iminium ion by X-ray diffraction, and the counterion was dichlorocuprate. Copper is in oxidation state (I) and bound ionically, not covalently. More importantly, the isolated iminium ion could be fully converted to the terminal product in the presence of nucleophiles in good yield.

3.2.2 sp^3 C–H Bonds and sp^2 C–H Bonds

Tetrahydroisoquinolines and indoles are common structures in natural products.[7] Based on previous work, Li[8] and co-workers developed a new type of CDC reaction, directly connecting THIQ sp^3 C–H bonds and indole sp² C–H bonds, leading to the indolation of THIQs (Scheme 3.6). The advantages of this method include high regioselectivity, using a free (NH) indole and use of a relatively cheap metal, copper, as a catalyst. Under similar reaction conditions, Huang[9] and co-workers extended the direct indolation of amines to *N,N*-dimethylanilines, affording the alkylated indoles in 52–78% yields (Scheme 3.7). Recently, similar work was also reported by Schnürch's group[10] (Scheme 3.8). Using Cu(NO₃)₂ · 3H₂O as the catalyst, they realized the direct functionalization of free THIQ with indoles and pyrroles, and a variety of unprotected *N*-containing products could be synthesized without the step of removing the *N*-protecting groups.

In 2011, Zhang[11] and co-workers developed an efficient CuBr-catalyzed oxidative cross-coupling of *N,N*-dimethylanilines and heteroarenes to give *N*-containing heterocyclic compounds under mild conditions. In this transformation, they employed the environmentally benign molecular oxygen as the oxidant instead of *tert*-butyl hydroperoxide (TBHP) (Scheme 3.9).

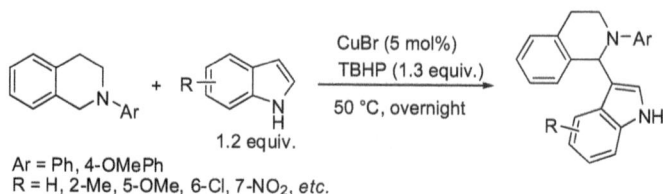

Ar = Ph, 4-OMePh
R = H, 2-Me, 5-OMe, 6-Cl, 7-NO₂, *etc.*

Scheme 3.6 Copper-catalyzed direct indolation of THIQs.

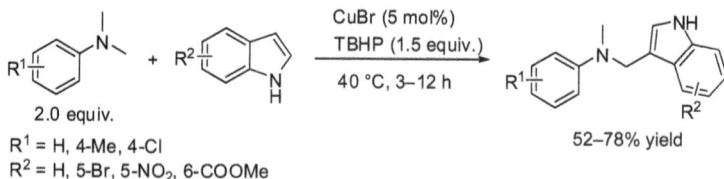

R¹ = H, 4-Me, 4-Cl
R² = H, 5-Br, 5-NO₂, 6-COOMe

Scheme 3.7 Copper-catalyzed direct indolation of *N,N*-dimethylanilines.

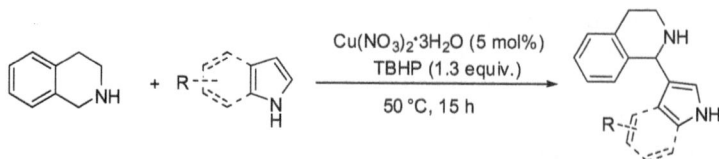

Scheme 3.8 Copper-catalyzed direct functionalization of free THIQ.

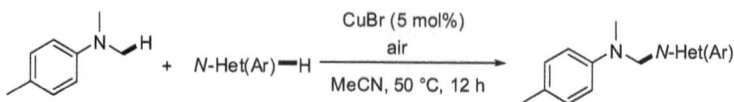

Scheme 3.9 Copper-catalyzed CDC reaction under air.

3.2.3 sp³ C–H Bonds

Further demonstration of the utility of CDCs was shown by the successful coupling of two challenging sp³ C–H bonds. In 2005, Li[12] and co-workers reported the first example of efficient CDC reaction between sp³ C–H bonds to construct β-nitroamines catalyzed by CuBr in the presence of TBHP (Scheme 3.10). Dialkyl malonates are another important synthon with relatively reactive sp³ C–H bonds, thus were also tested in copper-catalyzed CDC reactions with THIQs; β-diester amine derivatives could be obtained in high yields[13] (Scheme 3.11). The use of malononitrile as a pronucleophile generated β-dicyano-THIQ under the same reaction conditions.[13] Interestingly, a significant development in CDC reactions has been achieved by Li; it was found that both the nitroalkane and malonate reactions could be conducted in water using molecular oxygen instead of TBHP as an oxidant.[14] This offered a safer and environmentally benign process (Scheme 3.12).

The direct α-functionalization of natural peptides takes advantage of existing structure and provides rapid access to diverse new peptides. In 2008, Li[15] reported a new method for functionalizing peptido-amides through the CDC reaction at α-peptido C–H bonds (Scheme 3.13). To realize this process, 2.0 equiv. of Cu(OAc)₂ and 20 mol% of ligand are necessary, due to the formation of an active six-membered ring complex. In addition, the presence of a catalytic amount of base is beneficial to the reaction.

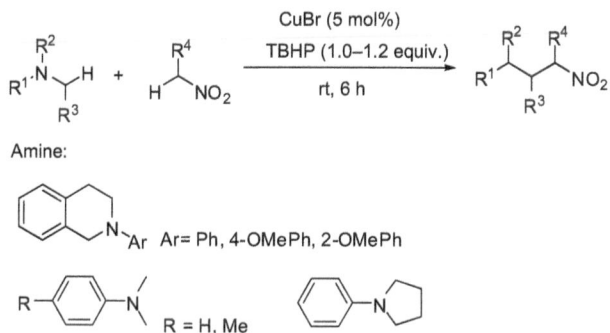

Scheme 3.10 Copper-catalyzed CDC reaction of tertiary amines and nitroalkanes.

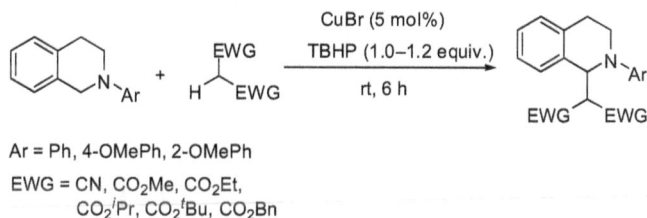

Scheme 3.11 Copper-catalyzed CDC reaction of tertiary amines and malonates. EWG = electron withdrawing groups.

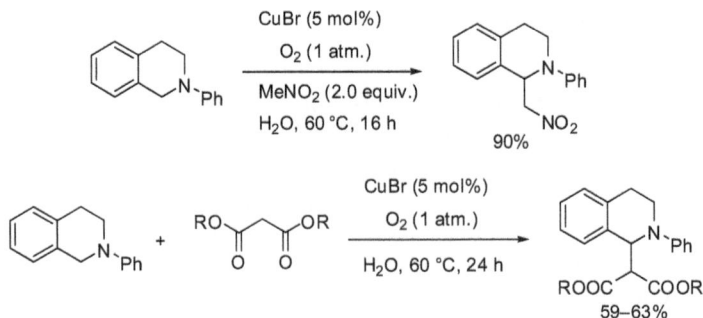

Scheme 3.12   Copper-catalyzed CDC reactions in water using molecular oxygen.

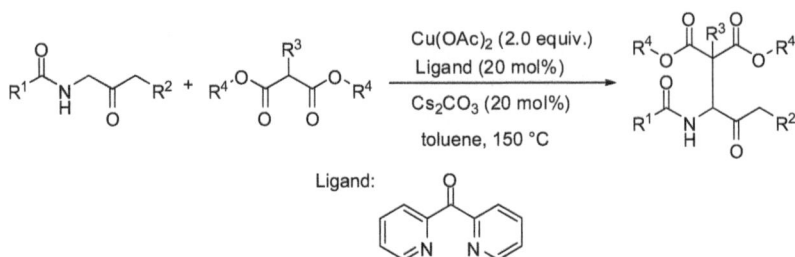

Scheme 3.13   Copper-catalyzed CDC reaction of peptido-amides and malonates.

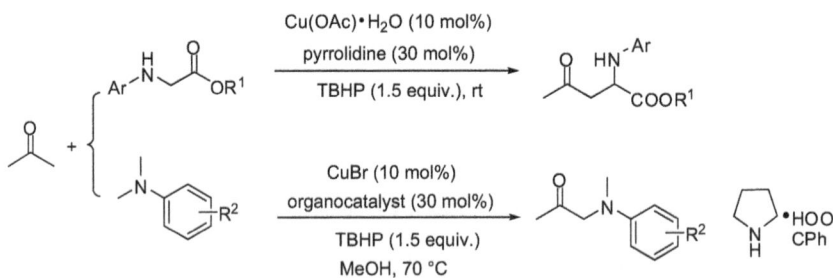

Scheme 3.14   Co-operative copper and organocatalyst-catalyzed CDC reaction.

Interestingly, Huang[9,16] and co-workers successfully applied a co-operative transition metal and organocatalyst to CDC reactions, achieving the direct coupling of secondary amines and ketones (Scheme 3.14). During this transformation, the organocatalyst is crucial due to the *in situ* enamine formed from the ketone, and pyrrolidine served as a nucleophile to attack the imine cation.

3.3   Copper-Catalyzed CDC Reactions Involving C–H Bonds Adjacent to Oxygen

Functionalization of the α-C–H bond of oxygen is more challenging than functionalizing the α-C–H bond of nitrogen due to the higher oxidation

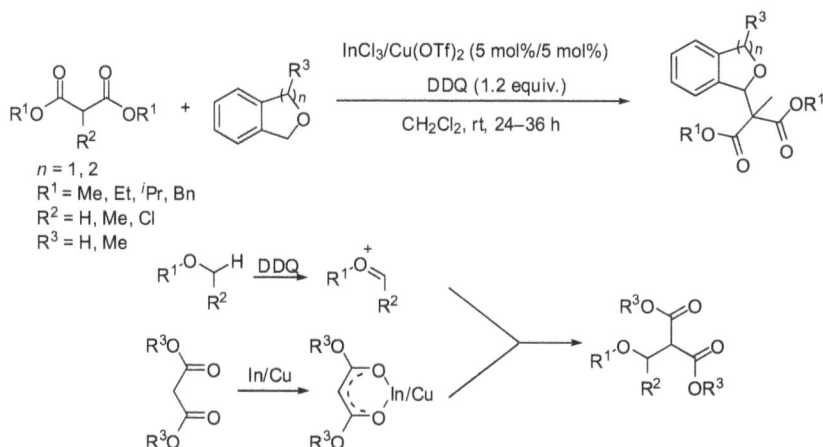

Scheme 3.15 Indium/copper co-catalyzed CDC reaction of α-C–H bonds adjacent to oxygen.

potential of ethers than amines. In 2006, Li[17] and co-workers demonstrated the oxidative alkylation of cyclic benzyl ethers with malonates, catalyzed by a combination of indium and copper catalysts in the presence of 2,3-dichloro-5,6-dicyanobenzoquinone (DDQ) (Scheme 3.15).

Benzyl ether reacted with DDQ to generate oxonium ions, the malonate was activated by the In/Cu catalyst, and the subsequent coupling of the two intermediates resulted in the desired product.

3.4 Copper-Catalyzed CDC Reactions Involving Other C–H Bonds

3.4.1 Allylic C–H Bonds

Transition metal (Pd, Ir, *etc.*) catalyzed allylic alkylation is one of the most important reactions for constructing C–C bonds in modern organic synthesis. The development of a catalytic allylic alkylation through a CDC process would be complementary to this. In 2006, Li[18] and co-workers reported the first catalytic allylic alkylation directly using allylic sp^3 C–H and methylenic C–H bonds *via* a CDC reaction (Scheme 3.16). A combination of copper and cobalt could furnish this transformation, and the ratio of the two metals (CuBr-to-CoCl$_2$ = 4 : 1) was found to be important to the reaction.

3.4.2 Benzylic C–H Bonds

In 2008, using an inexpensive copper salt as the catalyst, Powell[19] and co-workers developed the oxidative coupling of benzylic C–H bonds with 1,3-dicarbonyl compounds (Scheme 3.17). Kinetic isotope studies support a mechanism involving a benzylic H-atom abstraction.

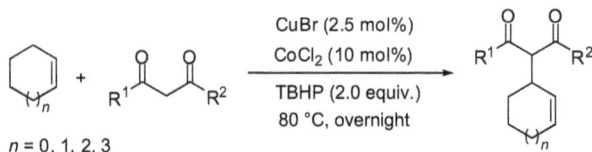

Scheme 3.16   Copper/cobalt co-catalyzed CDC reaction of allylic C–H bonds with methylenic C–H bonds.

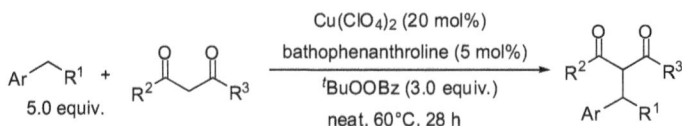

Scheme 3.17   Cu-Catalyzed CDC reaction of benzylic C–H bonds.

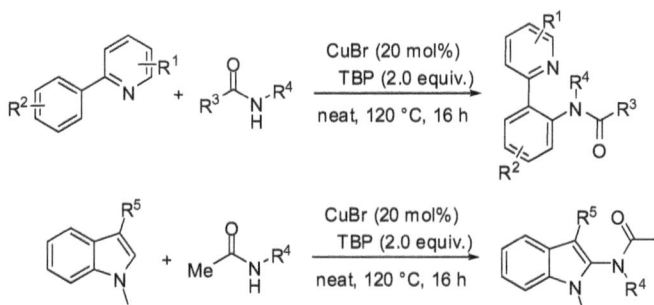

Scheme 3.18   Amidation of 2-arylpyridines and 1-methylindoles. TBP = di-tert-butyl peroxide.

3.4.3   C–N Bond Formation

Copper-catalyzed C–N bond formation can lead to structural moieties that are prevalent in the building blocks of active molecules in life sciences and in many material precursors. Consequently, a lot of effort has been made to construct C–N bonds. Recently, copper-catalyzed oxidative C–N bond formation *via* a CDC process has been developed. In 2010, Li[20] and co-workers presented a CuBr-catalyzed direct and highly site-selective amidation of 2-arylpyridines and 1-methylindoles through a CDC reaction (Scheme 3.18), which can serve as a novel approach to modify amides and directly produce various biologically useful arylamide derivatives.

3.4.4   C–P Bonds

α-Aminophosphonates and their corresponding α-aminophosphonic acids have received much interest in organic and medicinal chemistry because they are analogous to both natural and unnatural amino acids.[21] There have been several methods for the synthesis of dialkyl arylphosphonates.[22]

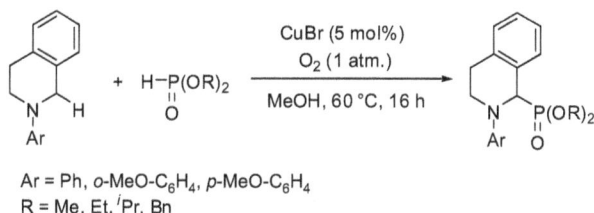

Ar = Ph, o-MeO-C$_6$H$_4$, p-MeO-C$_6$H$_4$
R = Me, Et, ^{i}Pr, Bn

Scheme 3.19 Cu-catalyzed CDC reaction to construct C–P bonds.

Among them, copper-catalyzed CDC reaction to construct C–P bonds has received more attention. Li[23] and co-workers developed a simple way to synthesize α-aminophosphonates by an efficient CDC reaction between sp^3 C–H and P–H bonds using a copper salt as a catalyst and molecular oxygen as the terminal oxidant (Scheme 3.19).

3.5 Asymmetric CDC Reactions

Enantioselective catalytic C–C bond formation has attracted much attention in both academic and industrial research. From the asymmetric synthetic concept point of view, a prochiral sp^2 carbon center is generally necessary as a precursor for constructing a chiral carbon center, while enantioselective catalytic C–C bond formation *via* C–H bond activation is still a challenge. Achieving this will have an important impact on asymmetric synthesis. The success of copper-catalyzed CDC reactions of C–H bonds opens up an opportunity to achieve catalytic asymmetric C–C bond formation based on the reaction of sp^3 C–H bonds.

Tetrahydroisoquinoline alkaloids with a stereocenter at C1 exist widely in nature and are compounds of extensive interest due to their biological and pharmacological properties. Various methodologies have been developed to construct this stereogenic center.[24] Employing the CDC reaction, Li[25] developed a novel catalytic asymmetric 1-alkynylation of THIQs in the presence of a chiral ligand, leading to optically active C1-substituted THIQ derivatives in moderate enantiomeric excess (ee) (Scheme 3.20). Recently, Chi[26] and co-workers also reported an enantioselective oxidative coupling reaction of THIQs with aldehydes under co-operative copper and amine catalysis (Scheme 3.21). Moreover, this co-operative catalytic system could also be applied to electron-donating-group substituted anilines and aliphatic amines.

Optically active α-amino acid derivatives are of great importance and have applications in the synthesis of biologically active peptides, natural products and organocatalysts. Based on the concept that catalytic oxidative α-sp^3 C–H bonds of secondary amines provide reactive imines, and considering that the nucleophilic addition to imines has been studied clearly, Wang[27] and co-workers studied the asymmetric CDC reactions of glycine esters with α-substituted ketoesters in the presence of Cu(OTf)$_2$ and a chiral bisoxazoline, obtaining a series of optically active α-alkyl α-amino acids and C1-alkylated THIQs (Scheme 3.22).

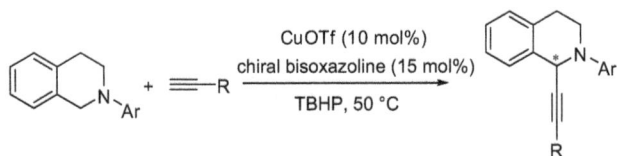

Scheme 3.20   Asymmetric synthesis of 1-alkynylated THIQs.

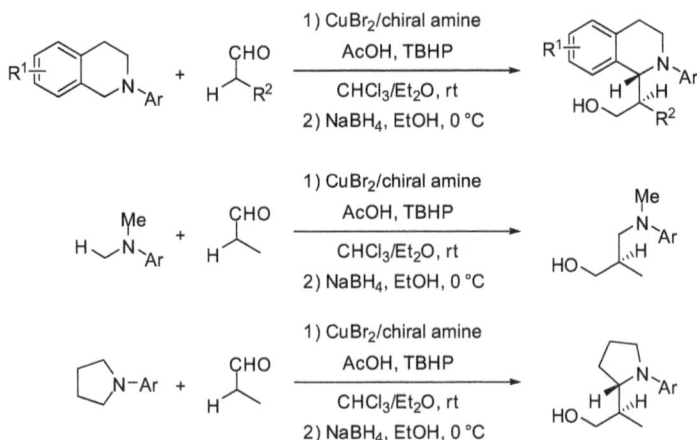

Scheme 3.21   Cu/amine co-catalyzed asymmetric oxidative coupling of tertiary amines with aldehydes.

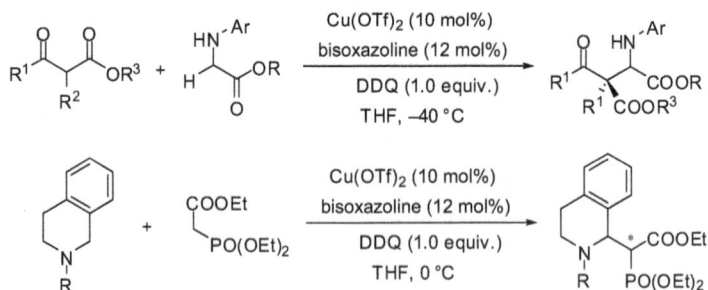

Scheme 3.22   Asymmetric synthesis of α-alkyl α-amino acids and C1-alkylated THIQs. THF = tetrahydrofuran.

3.6   Summary and Outlook

Since Li and co-workers reported the first example of a copper-catalyzed CDC reaction, numerous important advances have been made in the formation of C–C, C–N, and C–P bonds. This process represents a significant advance in view of green chemistry for several reasons: (1) simple organic substrates can be effectively coupled without the need for pre-functionalization; (2) there is a significant reduction in the amount of waste; and (3) such reactions present the most direct and efficient methods for C–C and C–X bond formation.

There is a significant amount of work still to be done in this exciting area, including: (1) expanding the substrates to other C–H bonds, not limited to C–H bonds adjacent to nitrogen or oxygen atoms and active C–H bonds such as benzylic or allylic C–H bonds; (2) asymmetric CDC reactions for the synthesis of optically active compounds remain a challenge issue; and (3) using environmentally benign molecular oxygen instead of TBHP or DDQ as oxidant is still desirable. These challenges are stimulating researchers' enthusiasm and the further development of CDCs is envisioned in the near future.

References

1. (a) G. Dyker, *Angew. Chem., Int. Ed.*, 1999, **38**, 1698; (b) V. Ritleng, C. Sirlin and M. Pfeffer, *Chem. Rev.*, 2002, **102**, 1731; (c) F. Kakiuchi and N. Chatani, *Adv. Synth. Catal.*, 2003, **345**, 1077; (d) D. Alberico, M. E. Scott and M. Lautens, *Chem. Rev.*, 2007, **107**, 174; (e) B.-J. Li, S.-D. Yang and Z.-J. Shi, *Synlett*, 2008, 949; (f) R. Giri, B.-F. Shi, K. M. Engle, N. Maugel and J.-Q. Yu, *Chem. Soc. Rev.*, 2009, **38**, 3242; (g) D. A. Colby, R. G. Bergman and J. A. Ellman, *Chem. Rev.*, 2010, **110**, 624; (h) T. W. Lyons and M. S. Sanford, *Chem. Rev.*, 2010, **110**, 1147; (i) C. S. Yeung and V. M. Dong, *Chem. Rev.*, 2011, **111**, 1215; (j) P. B. Arockiam, C. Bruneau and P. H. Dixnef, *Chem. Rev.*, 2012, **112**, 5879.
2. (a) Z. Li, D. S. Bohle and C.-J. Li, *Proc. Natl. Acad. Sci. U. S. A.*, 2006, **103**, 8928; (b) C.-J. Li and Z. Li, *Pure Appl. Chem.*, 2006, **78**, 935; (c) C.-J. Li, *Acc. Chem. Res.*, 2009, **42**, 335; (d) C. J. Scheuermann, *Chem. Asian J.*, 2010, **5**, 436; (e) C. Zhang, C. Tang and N. Jing, *Chem. Soc. Rev.*, 2012, **41**, 3464.
3. Z. Li and C.-J. Li, *J. Am. Chem. Soc.*, 2004, **126**, 11810.
4. H. Nakamura, T. Kamakura, M. Ishikura and J.-F. Biellman, *J. Am. Chem. Soc.*, 2004, **126**, 5958.
5. M. Niu, Z. Yin, H. Fu, Y. Jiang and Y. Zhao, *J. Org. Chem.*, 2008, **73**, 3961.
6. (a) E. Boess, D. Sureshkumar, A. Sud, C. Wirtz, C. Farès and M. Klussmann, *J. Am. Chem. Soc.*, 2011, **133**, 8106; (b) E. Boess, C. Schmita and M. Klussmann, *J. Am. Chem. Soc.*, 2012, **134**, 5317.
7. (a) K. W. Bentley, *Nat. Prod. Rep.*, 2004, **21**, 395; (b) K. W. Bentley, *Nat. Prod. Rep.*, 2005, **22**, 249; (c) M. Somei and F. Yamada, *Nat. Prod. Rep.*, 2005, **22**, 73.
8. Z. Li and C.-J. Li, *J. Am. Chem. Soc.*, 2005, **127**, 6968.
9. F. Yang, J. Li, J. Xie and Z.-Z. Huang, *Org. Lett.*, 2010, **12**, 5214.
10. M. Ghobrial, M. Schnürch and M. D. Mihovilovic, *J. Org. Chem.*, 2011, **76**, 8781.
11. L. Huang, T. Niu, J. Wu and Y. Zhang, *J. Org. Chem.*, 2011, **76**, 1759.
12. Z. Li and C.-J. Li, *J. Am. Chem. Soc.*, 2005, **127**, 3672.

13. Z. Li and C.-J. Li, *Eur. J. Org. Chem.*, 2005, 3173.
14. O. Baslé and C.-J. Li, *Green Chem.*, 2007, **9**, 1047.
15. L. Zhao and C.-J. Li, *Angew. Chem., Int. Ed.*, 2008, **47**, 7075.
16. J. Xie and Z.-Z. Huang, *Angew. Chem., Int. Ed.*, 2010, **49**, 10181.
17. Y. Zhang and C.-J. Li, *Angew. Chem., Int. Ed.*, 2006, **45**, 1949.
18. Z. Li and C.-J. Li, *J. Am. Chem. Soc.*, 2006, **128**, 56.
19. N. Borduas and D. A. Powell, *J. Org. Chem.*, 2008, **73**, 7822.
20. Q. Shuai, G. Deng, Z. Chua, D. S. Bole and C.-J. Li, *Adv. Synth. Catal.*, 2010, **352**, 632.
21. (a) S. C. Fields, *Tetrahedron*, 1999, **55**, 1237; (b) E. K. Fields, *J. Am. Chem. Soc.*, 1952, **74**, 1528; (c) C. Yuan and S. Chen, *Synthesis*, 1992, 1124; (d) D. Redmore, *J. Org. Chem.*, 1978, **43**, 992; (e) T. Yokomatsu, Y. Yoshida and S. Shibuya, *J. Org. Chem.*, 1994, **59**, 7930; (f) J. Hiratake and J. Oda, *Biosci., Biotechnol., Biochem.*, 1997, **61**, 211.
22. (a) S. Lachat and H. Kunz, *Synthesis*, 1992, 90; (b) C. Qian and T. Huang, *J. Org. Chem.*, 1998, **63**, 4125; (c) K. Manabe and S. Kobayashi, *Chem. Commun.*, 2000, 669; (d) M. R. Saidi and N. Azizi, *Synlett*, 2002, 1347; (e) T. Akiyama, M. Sanada and K. Fuchibe, *Synlett*, 2003, 1463.
23. O. Baslé and C.-J. Li, *Chem. Commun.*, 2009, 4124.
24. (a) M. Chrzanowska and M. D. Rozwadowska, *Chem. Rev.*, 2004, **104**, 3341; (b) K. W. Bentley, *Nat. Prod. Rep.*, 2004, **21**, 395.
25. (a) Z. Li and C.-J. Li, *Org. Lett.*, 2004, **6**, 4997; (b) Z. Li, P. D. Macleod and C.-J. Li, *Tetrahedron: Asymmetry*, 2006, **17**, 590.
26. J. Zhang, B. Tiwari, C. Xing, X. Chen and Y. R. Chi, *Angew. Chem., Int. Ed.*, 2012, **51**, 3649.
27. G. Zhang, Y. Zhang and R. Wang, *Angew. Chem., Int. Ed.*, 2011, **50**, 10429.

CHAPTER 4

Iron-Catalyzed Cross-Dehydrogenative-Coupling Reactions

CHRISTOPHE DARCEL,* JEAN-BAPTISTE SORTAIS AND
SAMUEL QUINTERO DUQUE

UMR 6226 "Institut des Sciences Chimiques de Rennes" – Team
"Organometallics: Materials and Catalysis" – Centre for Catalysis and
Green Chemistry, Campus de Beaulieu, 35042 Rennes, France
*Email: christophe.darcel@univ-rennes1.fr

4.1 Introduction

The building of C–C bonds is one of the fundamental tasks in molecular synthesis. The last few decades have witnessed significant advances in the field of transition metal catalyzed cross-coupling reactions as powerful methods for the formation of C–C bonds, as exemplified by the Suzuki, Negishi and Heck reactions.[1] However these reactions often require pre-functionalized starting materials to generate the desired product. Even if such methods have evolved as a general tool for the synthesis of fine chemicals, there are some major drawbacks in terms of green chemistry; in many of these reactions, catalysts derived from heavy or rare metals are used, and their toxicity and elevated prices constitute a severe disadvantage for large-scale applications. Furthermore, such methodologies require a transmetalation step, which generates a stoichiometric amount of metal salt as waste. An elegant alternative methodology to construct C–C bonds is the

RSC Green Chemistry No. 26
From C–H to C–C Bonds: Cross-Dehydrogenative-Coupling
Edited by Chao-Jun Li
© The Royal Society of Chemistry 2015
Published by the Royal Society of Chemistry, www.rsc.org

$$C \xrightarrow{a} H + H \xrightarrow{b} C \xrightarrow[\text{Oxidant (stoichiometric)}]{\text{[Fe] (cat.)}} C \xrightarrow{a} \xrightarrow{b} C + \text{"} (H \xrightarrow{} H) \text{"} \quad \begin{array}{ll} C^a\text{: } sp^3 & C^b\text{: } sp^3 \\ sp^2 & sp^2 \\ sp \end{array}$$

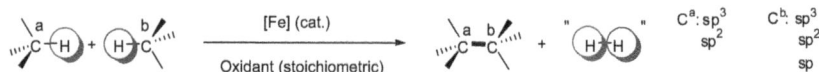

Scheme 4.1 Principle of CDC reactions with formal liberation of H_2.

direct coupling between two C–H bonds. Such a catalytic methodology has already attracted much attention using noble transition metals as catalysts.[2–4] During the last decade, the group of C. J. Li has made a huge and remarkable pioneering contribution to the construction of new C–C bonds, starting directly with C–H bonds under oxidative conditions using transition metal catalysts from the first row in the periodic table, such as copper or iron.[5] Even if this reaction is termed "cross-dehydrogenative-coupling" (CDC), it is noteworthy that H_2 is not usually the by-product of these transformations, since the thermodynamics of making a C–C bond with loss of H_2 are typically unfavorable, thus an external driving force is required, namely, an appropriate sacrificial oxidant.

This chapter describes the spectacular and fast development of CDC reactions using iron catalysts involving the oxidative coupling of various sp, sp^2 and sp^3 C–H bonds (Scheme 4.1).

4.2 Iron-Catalyzed CDC Reactions Between sp^3 C–H Bonds Leading to sp^3 C–C Bond Formation

In a seminal contribution,[6] C.-J. Li reported in 2007 the first iron-catalyzed CDC reaction between alkanes and activated methylene derivatives such as β-ketoesters or β-diketones. A simple $FeCl_2$ catalytic system (20 mol%) can promote the coupling reaction in the presence of 2 equiv. of di-*tert*-butyl peroxide ($^tBuOO^tBu$) as the oxidant in an alkane as the solvent (Scheme 4.2). The CDC reaction is efficient with cycloalkanes and β-ketoesters (48–88% yields). It is worth mentioning that linear alkanes such as *n*-hexane produce moderate yields (42%) with a mixture of two regio-isomers in a 1.2 : 1 ratio. By contrast, the reaction with β-diketones is a more difficult task, and only low yields (10–15%) are obtained.

In a similar manner, using the same catalytic system and conditions [$FeCl_2$ (20 mol%), $^tBuOO^tBu$ (2 equiv.), 80 °C, 8–36 h], diarylmethane derivatives can react efficiently with β-ketoesters, β-ketoamides and β-diketones (Scheme 4.3).[7]

Interestingly, such a methodology can be extended to an asymmetric version using a co-operative bimetallic system: various xanthene derivatives bearing a quaternary stereogenic center were obtained with enantiomeric excesses (ee) up to 99%. Indeed, using 10 mol% of a system composed of $Fe(BF_4)_2 \cdot 6H_2O$, $NiBr_2$ and the (L)-proline-derived N,N-dioxide ligands **4-A/B** (in a ratio 0.2 : 1 : 1.2), moderate-to-good yields (34–90%) and excellent enantioselectivities (97–99% ee) were obtained.[8] The catalytic system based on the ligand **4-A** is well adapted to perform CDC reactions with indanone derivatives, whereas the ligand **4-B** seems adequate for the reaction with 1-tetralone compounds (Scheme 4.4).

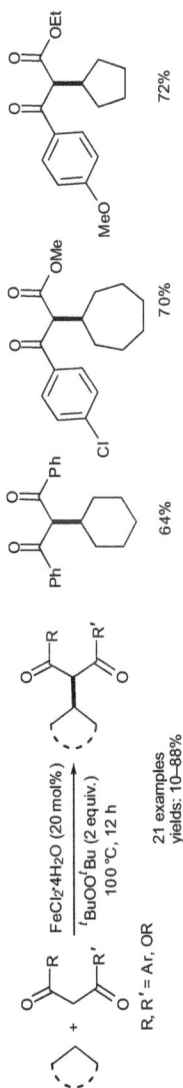

Scheme 4.2 CDC reaction between alkanes and activated methylene derivatives.

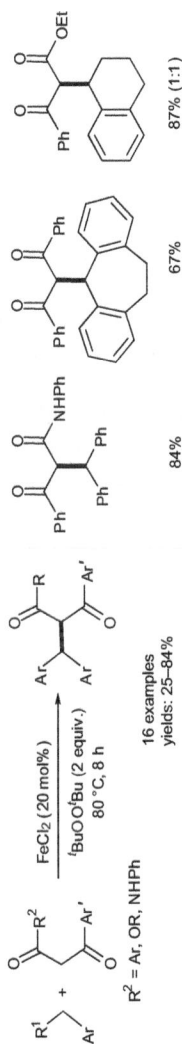

Scheme 4.3 CDC reaction between diarylmethanes and activated methylene derivatives.

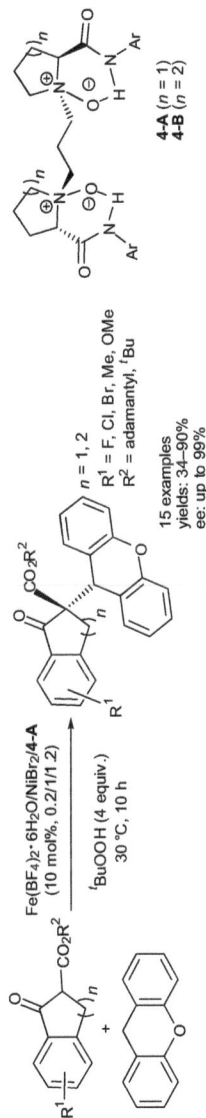

Scheme 4.4 Enantioselective version of the CDC reaction between β-ketoesters and xanthenes.

Notably, this methodology can be used for the regioselective CDC reaction of toluene derivatives with activated methylene compounds.[9] Using the system [FeCl$_2$ (10 mol%), (^{t}BuOO^{t}Bu) (2 equiv.), toluene, 120 °C for 24 h], 41% of the alkylated 1,3-dicarbonyl compound was obtained. The best catalytic system was based on Fe(OAc)$_2$ (10 mol%) in combination with 4 equiv. of ^{t}BuOO^{t}Bu (Scheme 4.5).

It is noteworthy that the addition of a radical scavenger such as 2,2,6,6-tetramethylpiperidinyl-1-oxyl (TEMPO) completely suppressed the reactivity, showing that a radical mechanism is most likely involved in the transformation. Furthermore, a significant amount of 1,2-diphenylethane was obtained under such standard conditions, which also supports the formation of a benzyl radical **6-D** during the reaction (Scheme 4.6). Firstly, *tert*-butyl peroxide decomposes to give a *tert*-butyl peroxide radical and oxidizes Fe(II) **6-A** into Fe(III) **6-B**, which reacts with the 1,3-dicarbonyl derivative leading to an Fe(III) enolate **6-C**. Meanwhile, a benzyl radical **6-D** obtained by abstraction of H$^\bullet$ can then react with **6-C**.

Notably, oxygen can be used as the oxidant to perform such a reaction.[10] To perform the CDC reaction of 1,3-dicarbonyl derivatives with diphenylmethane, copper has to be used as a co-catalyst in the presence of *N*-hydroxyphthalimide (NHPI). The optimized reaction conditions consist of the use of 5 mol% of FeCl$_2$, 5 mol% of CuCl and 20 mol% of NHPI in the presence of 5 equiv. of diphenylmethane at 105 °C overnight under an oxygen atmosphere. The role of CuCl and NHPI as co-catalysts in the presence of

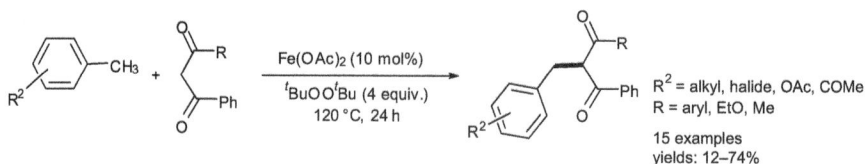

Scheme 4.5   CDC reaction between toluene derivatives and activated methylene compounds.

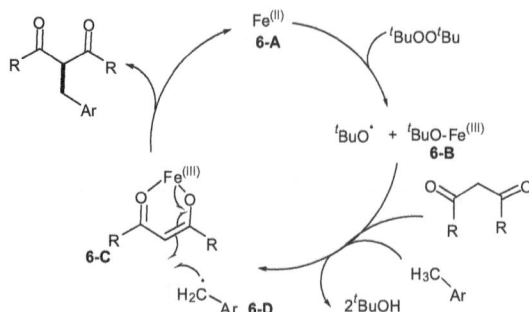

Scheme 4.6   Plausible mechanism for the selective CDC reaction between toluene derivatives and activated methylene compounds.

O_2 is to promote the formation of the benzyl radical which then reacts with the Fe(III) enolate giving the CDC adducts.

In the same way, the direct functionalization of sp^3 C–H bonds adjacent to heteroatoms such as oxygen or nitrogen can be also successfully performed using Fe-catalyzed CDC methodology. Indeed, *via* a single-electron-transfer (SET) process, sp^3 C–H bonds in α-position to an oxygen or a nitrogen can lead to activated oxonium or iminium species which can react with carbon nucleophiles. In 2008, Z. Li reported a convenient CDC reaction with cyclic or acyclic ethers and 1,3-dicarbonyl derivatives.[11a] $Fe_2(CO)_9$ was revealed to be the best catalytic system (5–20 mol%) in association with 3 equiv. of $^tBuOO^tBu$ in refluxing ether derivatives for 1 h. The CDC reaction is efficient with β-diketones, β-ketoesters and β-ketoamides, leading to the corresponding alkylated products with yields of up to 98% (Scheme 4.7). It is noteworthy that the reaction works also with sulfide and amine derivatives such as tetrahydrothiophene and *N,N*-dimethylaniline, respectively. The same reaction can be extended using TEMPO oxoammonium tetrafluoroborate salts as the oxidant (1.2 equiv.) and $Fe(OTf)_2$ (10 mol%) as the catalyst in CH_2Cl_2 at rt.[11b]

β-Diketones can also react with propargylic ethers using 20 mol% of $FeCl_2$ in the presence of 1.2 equiv. of 2,3-dichloro-5,6-dicyanobenzoquinone (DDQ) in CH_2Cl_2 at 30 °C for 15 h[12] (Scheme 4.8). Notably, when using *tert*-butyl-hydroperoxide (TBHP) or oxygen as the oxidant instead of DDQ, only a trace amount of the CDC coupling product was detected. Furthermore, when using $FeCl_3$ as the catalyst, a lower yield (48%) was obtained.

Subsequently, Z. Li extended the CDC reaction with amines bearing an sp^3 C–H bond in the α-position: using the reaction between *N,N*-dimethylaniline and β-dicarbonyl derivatives in the presence of 2.5 mol% of $Fe_2(CO)_9$ and TBHP (2 equiv.), unexpected dialkylation of the methylene group of the amine partner occurred, with good yields, producing methylene-bridged bis-1,3-dicarbonyl products (Scheme 4.9).[13]

Interestingly, they were able to perform this transformation sequentially starting from a large excess of *N,N*-dimethylaniline (10 equiv.) and the ketoester **10-A** under standard conditions; this reaction was stopped at the formation of the oxidative coupling adduct **10-B** which could be obtained in 47% yield (together with 37% of **10-C**). The reaction of 1 equiv. of **10-B** with one extra equivalent of **10-A** led to the methylene bis-dicarbonyl compound **10-C** in 91% yield, indicating that **10-B** can be an intermediate of the reaction (Scheme 4.10).

This crucial result suggested the plausible mechanism described in Scheme 4.11. Two pathways are proposed for the transformation of the product resulting from the CDC reaction of **11-A**: (1) either a direct S_N2 reaction; or (2) a consecutive sequence with a Cope-type elimination then a Michael addition.

Using a similar methodology, the group of C.-J. Li succeeded in performing a CDC reaction between tertiary amines and nitroalkanes where an oxygen atmosphere was used as the oxidant. Interestingly, the catalyst used was magnetically recoverable Fe_3O_4 nanoparticles (10 mol%). Under such conditions, 1,2,3,4-tetrahydroisoquinoline (THIQ) compounds were

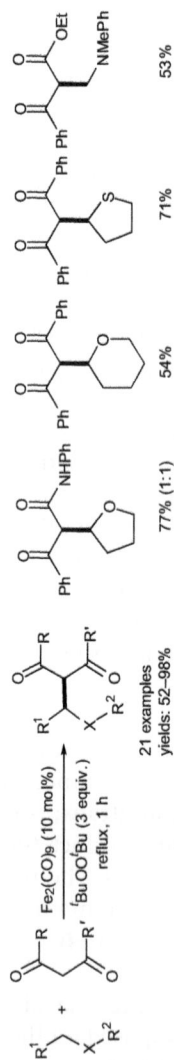

Scheme 4.7 CDC reaction of activated methylene compounds with 1,3-dicarbonyl derivatives.

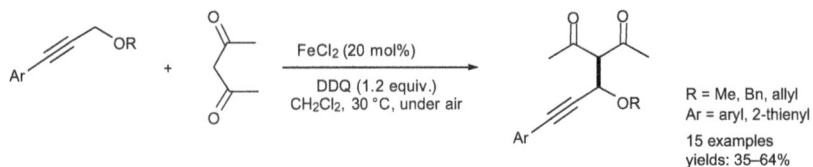

Scheme 4.8 CDC Reaction of β-diketones with propargylic ethers.

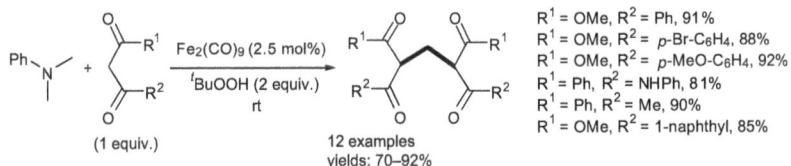

R^1 = OMe, R^2 = Ph, 91%
R^1 = OMe, R^2 = *p*-Br-C$_6$H$_4$, 88%
R^1 = OMe, R^2 = *p*-MeO-C$_6$H$_4$, 92%
R^1 = Ph, R^2 = NHPh, 81%
R^1 = Ph, R^2 = Me, 90%
R^1 = OMe, R^2 = 1-naphthyl, 85%

Scheme 4.9 CDC Reaction of amines bearing an sp^3 C–H bond in the α-position, with β-dicarbonyl compounds.

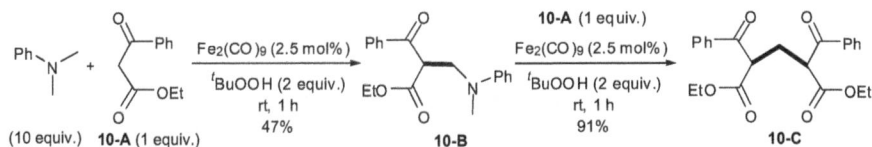

Scheme 4.10 Proposed pathway to obtaining methylene bis-dicarbonyl compounds.

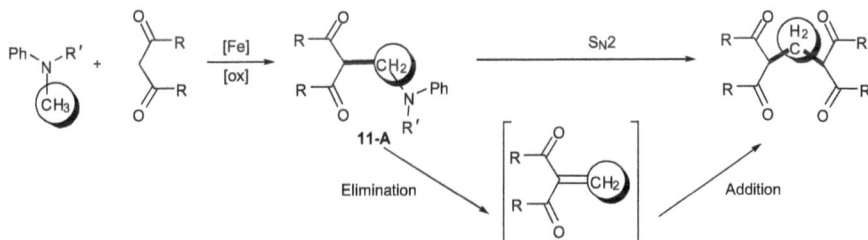

Scheme 4.11 Possible mechanism for the formation of methylene bis-dicarbonyl compounds.

obtained in moderate-to-good yields using a nitroalkane as the solvent (Scheme 4.12).[14] Notably, the Fe$_3$O$_4$ nanoparticle catalyst can be recovered by simple removal of the magnetic stirring bar and re-used up to nine times without significant loss of catalytic activity. Furthermore, the CDC reaction can be also performed with acetone as the coupling partner instead of nitroalkanes, leading to the corresponding alkylated compounds. The notable interest in this procedure is the use of oxygen as a clean oxidant and re-usable Fe$_3$O$_4$ nanoparticles as an efficient catalyst.

Recently, an efficient methodology for the vinylation of benzylic methyl groups using a CDC reaction was reported.[15] With *N,N*-dimethylacetamide

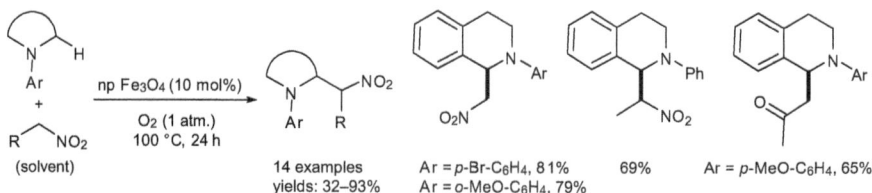

Scheme 4.12 CDC Reaction for the synthesis of 1,2,3,4-THIQs, np = nanoparticles.

(DMAc) as the solvent, 2-methyl-aza-arenes such as quinoxaline derivatives were transformed into 2-vinyl-aza-arenes using $FeCl_3 \cdot 6H_2O$ (2 mol%) and with $K_2S_2O_8$ (2 equiv.) as the oxidant at 110 °C under air (Scheme 4.13).

From a mechanistic point of view, experimental evidence such as the inhibition of the reaction in the presence of a radical scavenger, or the detection of the CDC adduct in mass spectroscopy, suggests a CDC mechanism leading to the intermediate **14-B** from DMAc, (Scheme 4.14) which reacts with the enamine **14-C**. The obtained adduct **14-D** then undergoes elimination to produce the 2-vinyl-2-aza-arene derivatives.

4.3 Iron-Catalyzed CDC Reactions Between sp^3 C–H and sp^2 C–H Bonds Leading to sp^3–sp^2 C–C Bond Formation

In arene chemistry, two of the main ways to build an sp^3–sp^2 C–C bond are namely; (1) Friedel–Crafts type reactions; and (2) transition metal cross-coupling methodologies. One of the main drawbacks of both methodologies is the use of stoichiometric amounts of metal precursors or Lewis acids. In terms of greener chemistry, both CDC reactions and iron catalysts are beneficial for sp^3–sp^2 C–C bond formation, and this was reported in pioneering contributions by Minisci.[16] In 2009, Z.-J. Li described the arylation of diarylmethane derivatives *via* a CDC methodology. Indeed, using $FeCl_2$ (10 mol%) as the catalyst in the presence of 2.5 equiv. of DDQ, various electron-rich arenes can be alkylated by diarylmethane compounds in 1,2-dichloroethane (DCE) at 100 °C for 36 h. Notably, DDQ was found to be a more effective oxidant than peroxide reagents (Scheme 4.15).[17]

The high regioselectivity of the reaction is mainly controlled by the electronic properties of the arenes. Double alkylation may occur when using more electron-rich arenes such as 1,2,3-trimethoxybenzene. The mechanism of this reaction is based on an iron-promoted SET oxidation leading to a benzyl radical **16-B** and then to a benzyl cation **16-C**, which reacts with electron-rich arenes *via* a Friedel–Crafts mechanism, the reduced hydroquinone **16-E** abstracting a proton, leading to the alkylated arenes (Scheme 4.16).

The same group has shown that 1-arylvinylacetates are able to react as nucleophiles with diarylmethanes in the presence of di-*tert*-butyl peroxide (DTBP) as the oxidant, leading to substituted acetophenone derivatives[18] (Scheme 4.17).

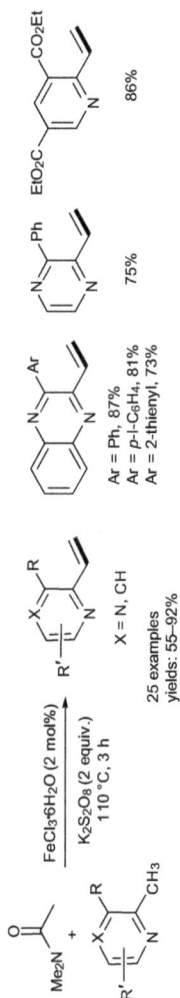

Scheme 4.13 Synthesis of 2-vinyl-aza-arenes from 2-methyl-aza-arenes *via* a CDC reaction.

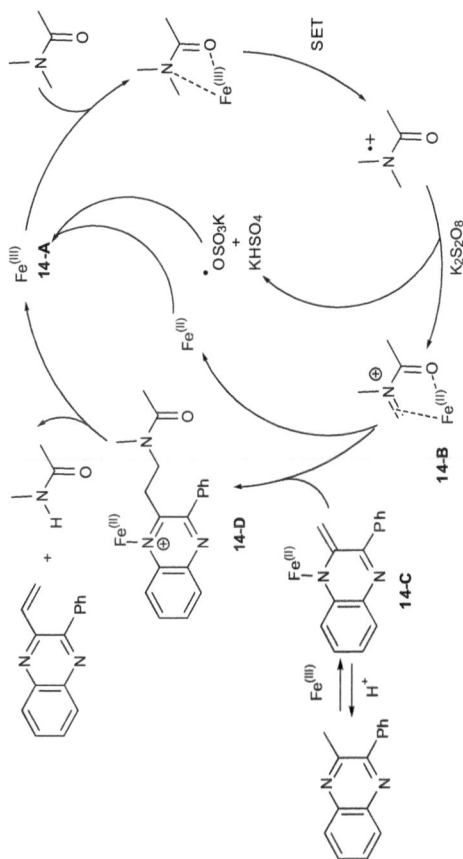

Scheme 4.14 Proposed mechanism for the vinylation of benzylic methyl groups.

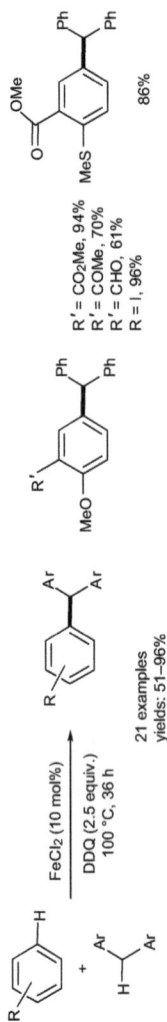

Scheme 4.15 Arylation of diarylmethane derivatives.

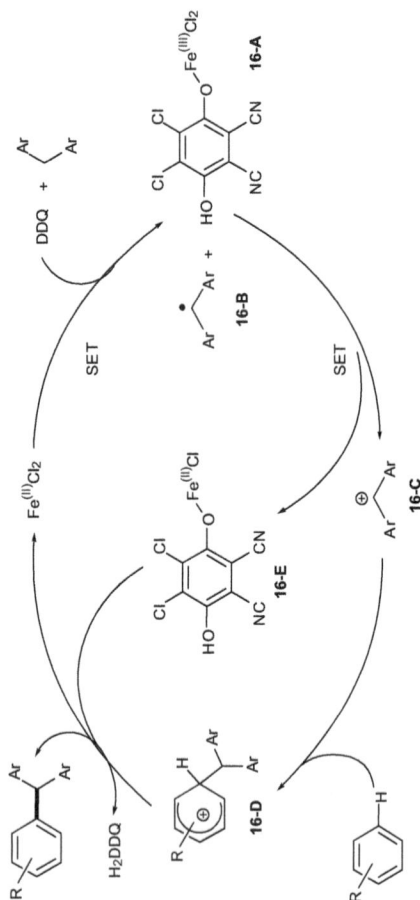

Scheme 4.16 Plausible mechanism for the iron-catalyzed CDC arylation of diarylmethanes.

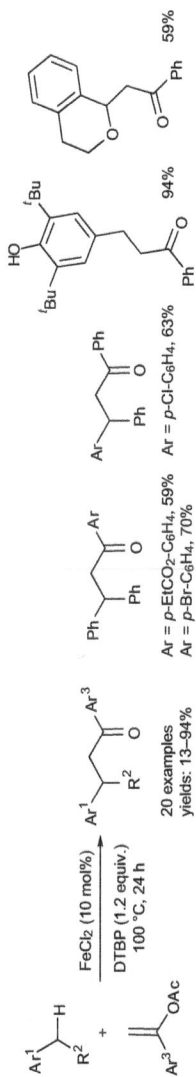

Scheme 4.17 Synthesis of substituted acetophenones by CDC reaction between 1-aryl vinylacetates and diarylmethanes.

The observed by-products (*tert*-butyl ether and the compound resulting from the dimerization of the diphenylmethane) seem to support a radical mechanism (Scheme 4.18). Firstly, the benzylic radical intermediate and the $^tBuO^{\bullet}$ radical **18-A**, generated *via* an SET process, undergo an electrophilic addition to styryl acetate leading to an iron-coordinated radical intermediate **18-B**. The latter, after decomposition of the intermediate **18-C**, leads to the desired product, AcO^{t}Bu, and re-generates the iron catalytic species.

With sp^3 C–H bonds adjacent to heteroatoms such as nitrogen, an sp^3–sp^2 C–C bond can also be formed by CDC reaction. In 2009, the group of Itami described the direct functionalization of methylamines (2.5 equiv.) with various heterocycles (1 equiv.) in the presence of FeCl$_2 \cdot$4H$_2$O (10 mol%), with pyridine-*N*-oxide (2 equiv.) as the oxidant and two additives [KI (0.2 equiv.) and 2,2′-Bipyridine (bipy) (0.1 equiv.)] at 130 °C for 24 h (Scheme 4.19).[19]

In 2010, the group of Schnürch reported the direct functionalization of *N*-protected THIQs by reaction with indoles in the presence of 5 mol% of Fe(NO$_3$)$_3 \cdot$9H$_2$O and 1.3 equiv. of ^{t}BuOOH at 50 °C. Interestingly, the type of *N*-protecting group seems to have a crucial influence in the efficiency of the reaction, with the Boc group affording the best results (Scheme 4.20).[20–21]

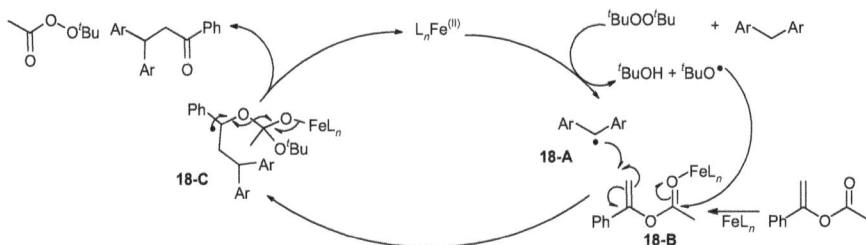

Scheme 4.18 Possible mechanism for reaction of 1-aryl vinylacetates with diarylmethanes.

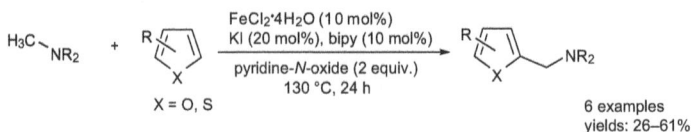

Scheme 4.19 CDC Reaction between methylamines and heterocycles.

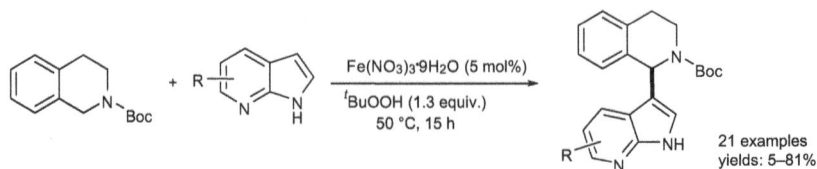

Scheme 4.20 CDC reaction of *N*-protected THIQs with indoles.

N-Protected THIQs and isochromans can also react under such conditions with electron-rich arenes (mainly with mono- and dimethoxyarenes) leading to the corresponding arylated compounds (12–81% yields). The same reaction can be performed using a silica-supported iron terpyridine complex [(SBA-15-[2,2′:6′,2″-terpyridine (terpy)])Fe(NCMe)$_3$][(ClO$_4$)$_2$] as the catalyst (3 mol%) with ^{t}BuOOH (3 equiv.).[22] Notably, only the silica-supported terpy ligand can be recycled (for up to five consecutive runs), 3 mol% of additional Fe(ClO$_4$)$_2$ · 6H$_2$O must be added to re-generate the active species for each run.

4.4 Iron-Catalyzed CDC Reactions Between sp^3 and sp C–H Bonds Leading to sp^3–sp C–C Bond Formation

The formation of sp^3–sp C–C bonds is important in molecular synthesis because internal alkynes are widely present in natural products and drugs. Several methodologies have been developed to prepare such internal alkynes including; (1) nucleophilic substitution of an alkyl halide by an acetylide; and (2) Pd/Cu Sonogashira or Pd Heck alkynylation reactions. CDC reactions can also be performed using terminal alkynes as nucleophiles using iron as the catalyst. In 2009, Vogel described that FeCl$_2$ (10 mol%) catalyzed the oxidative cross-coupling of *N,N*-dimethylarylamines with terminal alkynes in the presence of (^{t}BuO)$_2$ (as the most effective oxidant) at 100 °C for 24 h, leading to the corresponding propargylamines (24–93% yields) (Scheme 4.21).[23] Interestingly, starting from *N,N*-dimethyl-*N*-alkylamines, a regioselective reaction of the terminal alkynes took place at the methyl position, which is explained by steric requirements.

Similar to the CDC reaction leading to sp^3–sp^2 C–C bonds, the direct functionalization of sp^3 C–H bonds with terminal alkynes adjacent to heteroatoms can be also performed. In early 2012, Jiao reported the CDC reaction of terminal alkynes with benzylic ethers such as isochromans, using 5 mol% of Fe(OTf)$_2$ · 2MeCN and 1.2 equiv. of DDQ (as the most effective oxidant) in chlorobenzene at 90 °C for 24 h (Scheme 4.22).[24] Under such

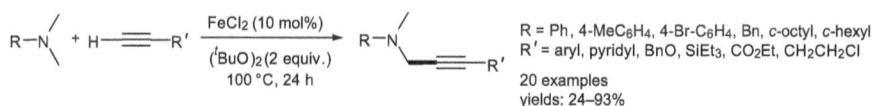

$$R-N \begin{array}{c}/\\\backslash\end{array} + H{-}{\equiv}{-}R' \xrightarrow[\substack{(^tBuO)_2(2\ equiv.) \\ 100\ °C,\ 24\ h}]{FeCl_2\ (10\ mol\%)} R-N \begin{array}{c}/\\\backslash\end{array}{\equiv}{-}R'$$

R = Ph, 4-MeC$_6$H$_4$, 4-Br-C$_6$H$_4$, Bn, *c*-octyl, *c*-hexyl
R′ = aryl, pyridyl, BnO, SiEt$_3$, CO$_2$Et, CH$_2$CH$_2$Cl
20 examples
yields: 24–93%

Scheme 4.21 Synthesis of propargylamines from terminal alkynes and *N,N*-dimethylarylamines.

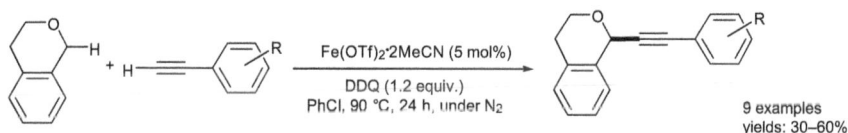

$$\text{(isochroman)} + H{-}{\equiv}{-}\text{(aryl-R)} \xrightarrow[\substack{DDQ\ (1.2\ equiv.) \\ PhCl,\ 90\ °C,\ 24\ h,\ under\ N_2}]{Fe(OTf)_2\cdot2MeCN\ (5\ mol\%)} \text{(product)}$$

9 examples
yields: 30–60%

Scheme 4.22 CDC reaction of terminal alkynes with benzylic ethers.

conditions, other benzylic ethers, such as benzyl methyl ether, have shown only very low reactivity.

4.5   Iron-Catalyzed CDC Reactions Between sp^2 C–H Bonds Leading to sp^2 C–C Bond Formation

In biaryl synthesis, although transition metal catalyzed aryl–aryl cross-coupling reactions have been widely exemplified using Suzuki-type reactions, CDC reactions between sp^2 C–H bonds leading to sp^2 C–C bonds are scarce. In 2010, Katsuki reported an enantioselective CDC reaction between two naphthol moieties catalyzed by a chiral iron(salan) complex under aerobic oxidative conditions (Scheme 4.23).[25] Using 4 mol% of the iron(salan) complex **23-A** as the catalyst in toluene at 60 °C for 48 h, two different 2-naphthols were coupled and the cross-coupled derivatives were isolated with moderate yields (44–70%) and good ee (87–95%). It must be pointed out that the homocoupling derivatives are also obtained in low yields.

A regioselective CDC reaction between *N,N*-dimethylaniline derivatives and 2-naphthol can be also performed under oxidative conditions (TBHP, 2 equiv.) in the presence of FeCl$_3 \cdot$6H$_2$O (20 mol%) in toluene at 0 °C for 4–5 h (Scheme 4.24).[26] The coupling reaction took place regioselectively at the *ortho*-position of both aniline and 2-naphthol, leading to the corresponding dialkylamino- and hydroxy-substituted biaryls. Remarkably, no homo-coupling products and no cross-coupling with the methyl group of the *N,N*-dimethylaniline derivatives was detected. The mechanism is rationalized by the electrophilic attack of 2-naphthol on the radical cationic species, obtained by an SET process on the aniline derivatives.

Indole derivatives can be prepared *via* an iron-catalyzed intramolecular CDC reaction starting from vinylaniline derivatives in the presence of 10 mol% of FeCl$_3$, 1.5 equiv. of Cu(OAc)$_2 \cdot$CuCl$_2$ as the oxidant, and 3 equiv. of K$_2$CO$_3$ in *N,N*-dimethylformamide (DMF) at 120 °C (Scheme 4.25).[27]

Intramolecular CDC reactions can also be efficient for the preparation of fluorenones and xanthones starting from *ortho*-formylbiphenyls and *ortho*-formylbiphenylethers, respectively. Notably, the ferrocene can be used in low catalytic amounts (0.1 to 1 mol%) in the presence of 2.2 equiv. of TBHP as the oxidant (Scheme 4.26).[28]

4.6   Iron-Catalyzed CDC Reactions as a Tool to Generate Molecular Complexity: Cascade Reactions

4.6.1   Michael Addition Cascade

β-1,3-Dicarbonyl aldehydes were synthesized from 1,3-dicarbonyl derivatives and two molecules of tertiary amines in the presence of a catalytic amount of

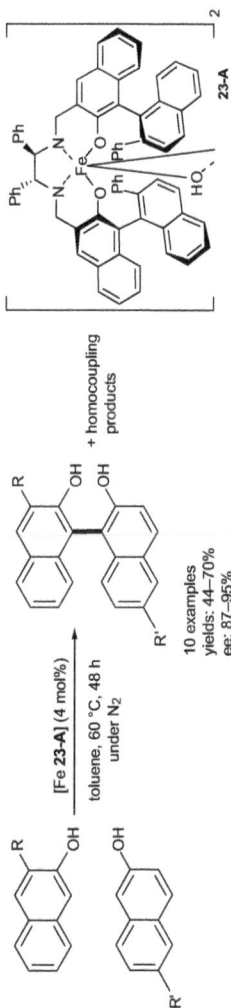

Scheme 4.23 Enantioselective CDC reaction towards binaphthols.

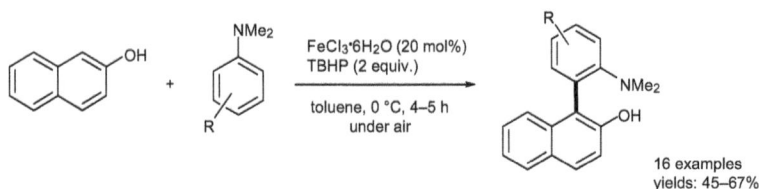

Scheme 4.24   Synthesis of dialkylamino- and hydroxy-substituted biaryls by CDC reaction.

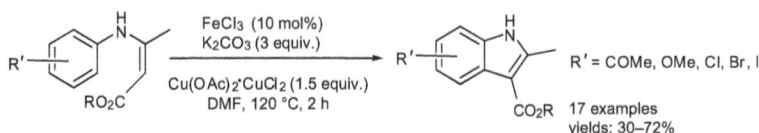

Scheme 4.25   Indole derivatives synthesis *via* iron-catalyzed intramolecular CDC reaction.

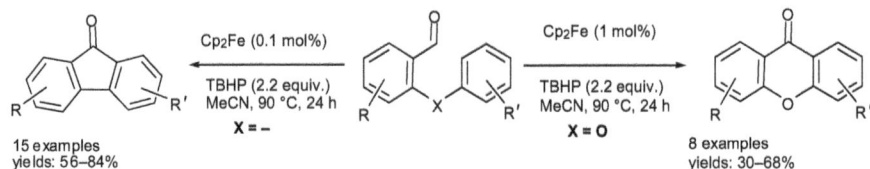

Scheme 4.26   Intramolecular CDC reactions for the synthesis of fluorenones and xanthones.

$Fe_2(CO)_9$ (2.5 mol%) and 3 equiv. of ^{t}BuOOH at 25 °C for 10 min (Scheme 4.27).[29]

The first step of the mechanism is based on a CDC reaction leading to an iminium salt **28-A** starting from alkylamines. Then two scenarios can be imagined. Firstly, the hydrolysis of **28-A** leads to the corresponding aldehyde which can undergo aldol condensation giving the α,β-unsaturated aldehyde **28-C**. The latter, by Michael addition with the β-ketoester gives the desired product (Scheme 4.28). The second scenario, which cannot be excluded, is a pathway involving a direct aldol condensation of **28-A** with **28-B**, followed by a Michael addition.

4.6.2   Friedel–Crafts Reaction Cascade

In 2009, Z. Li extended the CDC reaction to sp^3 C–H bonds adjacent to the oxygen of furans, with indoles, leading to symmetrical and unsymmetrical 1,1-bis-indolylmethane derivatives (Scheme 4.29).[30] Notably, the reaction can selectively produce either symmetrical or unsymmetrical 1,1-bis-indolylmethane compounds. To obtain unsymmetrical derivatives, a sequential addition of the indole reagent has to be performed. Interestingly, the nature

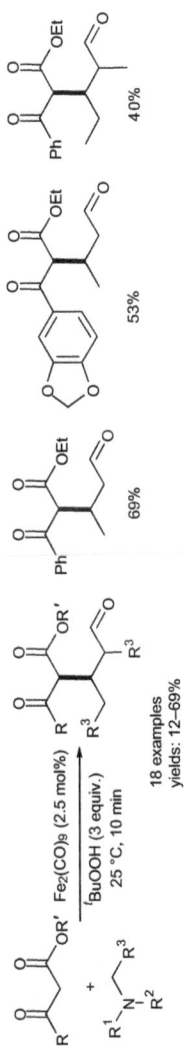

Scheme 4.27 Synthesis of β-1,3-dicarbonyl aldehydes by a CDC–Michael addition cascade.

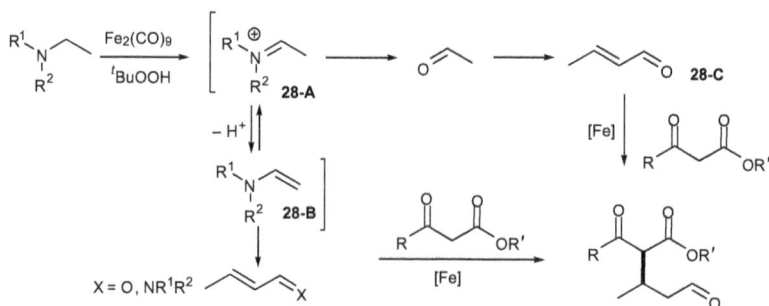

Scheme 4.28   Possible mechanism for the synthesis of β-1,3-dicarbonyl aldehydes.

of the substituent on the indole partner plays a crucial role in the chemo-selection; while the nature of the substituent has a moderate influence on the first indolation reaction, the electron-rich indoles are the most effective for performing the second one.

In order to rationalize the mechanism of this reaction, firstly a CDC reaction involves the furane derivative, then a radical intermediate **30-A** is produced and its addition to the first equiv. of indole leads to a radical species **30-B**. A second SET process gives the substituted furane **30-C**. Then, the double indolation can be explained by a Friedel–Crafts type alkylation catalyzed by iron, which plays the role of a Lewis acid (Scheme 4.30).

Using the same methodology (a CDC/Friedel–Crafts reaction), efficient production of quinolines was described using 10 mol% of FeCl$_3$ in the presence of 2 equiv. of (^{t}BuO)$_2$ in DCE at 80 °C for 12 h: a large variety of glycine derivatives (1 equiv.) and terminal aryl alkynes (1.2 equiv.) can react, leading to substituted quinolines (62–88% yields) (Scheme 4.31).[31]

From a mechanistic point of view, after a CDC reaction and the loss of a proton (see § 4.3), the imine intermediate **32-B** is produced and can then react with terminal alkynes to give the alkynylated species **32-C**. Then, *via* a Friedel–Crafts reaction of the electron-rich aryl ring with the internal triple bond, followed by an iron-mediated re-aromatization, the desired quinolines are obtained (Scheme 4.32).

Similar substituted quinolines can be efficiently obtained starting from *N*-alkylaniline (including glycine) derivatives by reaction with alkenes in the presence of FeCl$_3$ (10 mol%) and TEMPO oxoammonium tetrafluoroborate salt (2 equiv.) as the oxidant in dichloromethane (DCM) at 60 °C. Interestingly, the same quinolines can be obtained starting from anilines, ethyl glyoxalate and alkenes, in a one-pot multi-component reaction, under similar conditions.[32]

4.6.3   Cyclization Cascade

In 2012, Deng reported an FeCl$_3$-catalyzed sp^3–sp^2 C–C bond formation *via* a cross-dehydrogenative homocoupling reaction of two units of (*E*)-1,2-diaryl-prop-1-enes under oxidative conditions, leading to tetra-arylcyclopentenes.

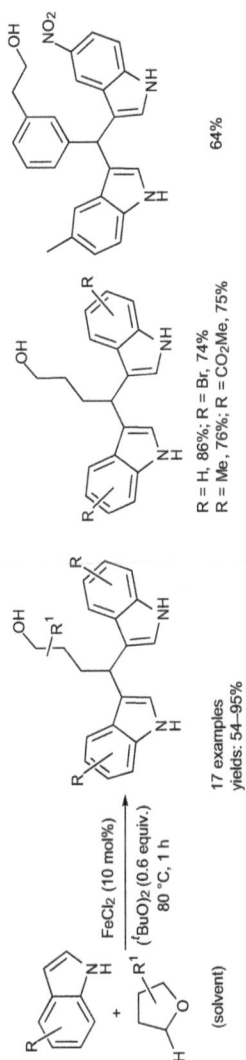

Scheme 4.29 Synthesis of symmetrical and unsymmetrical 1,1-bis-indolylmethane derivatives.

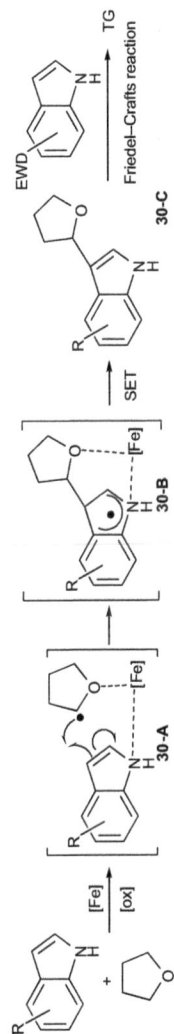

Scheme 4.30 Plausible mechanism for the synthesis of 1,1-bis-indolylmethane derivatives. TG = Target Goal.

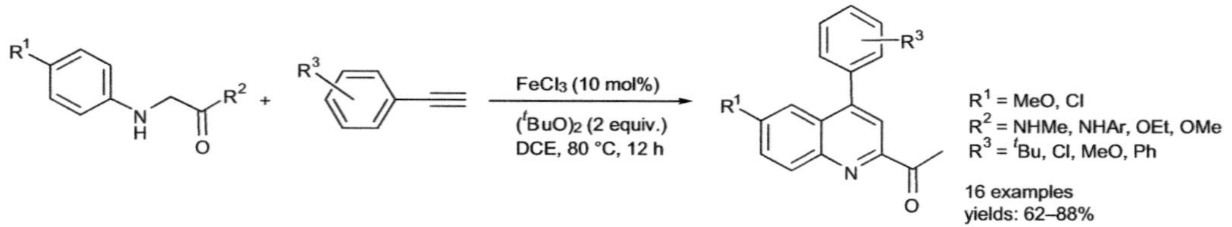
R1
R2
R3
FeCl3 (10 mol%)
(tBuO)2 (2 equiv.)
DCE, 80 °C, 12 h
N
H
O
R1 = MeO, Cl
R2 = NHMe, NHAr, OEt, OMe
R3 = tBu, Cl, MeO, Ph
16 examples
yields: 62–88%

Scheme 4.31 Synthesis of substituted quinolines.

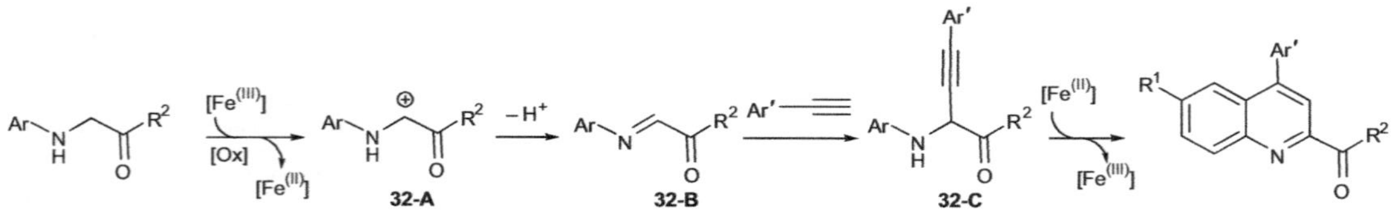
Ar
N
H
O
R2
[Fe(III)]
[Ox]
[Fe(II)]
32-A
– H+
32-B
Ar′
32-C
[Fe(II)]
[Fe(III)]
R1

Scheme 4.32 Proposed mechanism for the formation of substituted quinolines.

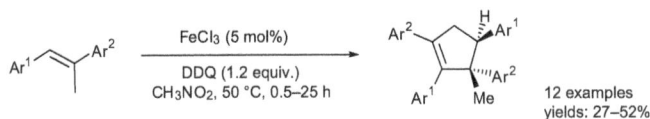

Scheme 4.33 Synthesis of tetra-arylcyclopentenes starting from (*E*)-1,2-diarylprop-1-enes.

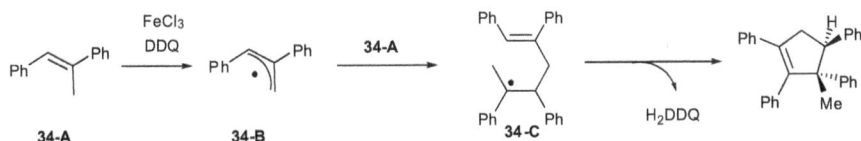

Scheme 4.34 Tentative mechanism for the synthesis of tetra-arylcyclopentenes.

Scheme 4.35 Synthesis of polysubstituted naphthalene compounds.

The reaction used 5 mol% of FeCl$_3$ in the presence of 1.2 equiv. of DDQ in nitromethane at 50 °C for 0.5–25 h (Scheme 4.33).[33]

Notably, the reaction proceeds with aryl groups having an activating effect, and no reaction was observed with strong electron-withdrawing groups (EWGs) such as NO$_2$ on the starting material. In the first step, a CDC reaction takes place leading to the allylic radical intermediate **34-B** which can react with a second equivalent of 1,2-diarylprop-1-ene leading to the species **34-C**. A subsequent tandem cyclization/oxidation gives the desired product (Scheme 4.34).

In a continuation of this work, the authors also reported the cross-coupling between 1,3-diarylprop-1-enes and styrenes using FeCl$_3$ (10 mol%) as a catalyst with DDQ (2.5 equiv.). In nitromethane at 50 °C, 1,3-diarylprop-1-enes react with styrene derivatives to give polysubstituted naphthalene compounds (yields: 30–80%) (Scheme 4.35).[34]

It is noteworthy that the substitution on the diarylpropene has a crucial influence on the reactivity, and that methyl and halo substituents favor the reaction. Nevertheless, it proceeds regioselectively with unsymmetrical di-substituted diarylpropenes, as only one regio-isomer was obtained with the aryl groups on the more substituted carbon of the alkene, fusing to lead to the naphthalene moiety. The mechanism is similar to the previous one involving a CDC reaction of diarylpropene with styrene, with a subsequent benzoannulation.

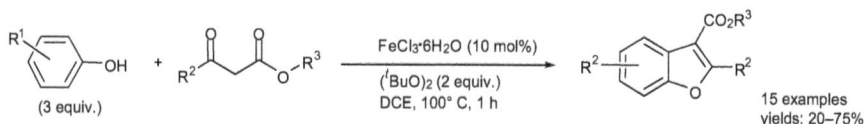

Scheme 4.36   Preparation of polysubstituted benzofurans.

In 2009, Z. Li reported the preparation of polysubstituted benzofurans *via* a reaction between phenols and β-ketoesters. Using 10 mol% of $FeCl_3 \cdot 6H_2O$ and 2 equiv. of $(^tBuO)_2$ in DCE, 3 equiv. of phenol react with 1 equiv. of β-ketoester affording the resulting benzofurans in 20–75% yields (Scheme 4.36).[35] The nature of the substituents on the phenol has a significant influence on the activity, with electron-donating *p*-substituents favoring the reaction.

In 2013, Lei[36] and Pappo[37] simultaneously reported an efficient and general way to prepare dihydrobenzofurans from phenols and alkenes (Scheme 4.37). The reaction was performed in the presence of 10–20 mol% of $FeCl_3$ and DDQ (1.2 equiv.) in toluene at room temperature or DTBP (2 equiv.) in DCE at 80 °C. It is noteworthy that the reactions are highly chemoselective, and can also work with α-alkylstyrenes and alkyl-based olefins.

Based on experimental evidence *i.e.*, inhibition of the reaction with radical scavengers, detection of a putative species in Electron Paramagnetic Resonance (EPR) corresponding to a HDDQ radical, and the active role of $FeCl_3$ in the oxidative radical coupling as a Lewis acid, a plausible mechanism was proposed (Scheme 4.38). In the first step, DDQ oxidizes the phenol leading to the phenol radical ArO• **38-A** and the HDDQ• radical. As a Lewis acid, $FeCl_3$ permits the generation of the carbon radical **38-B** thus controlling the regioselectivity of the addition of the alkenes, leading to the species **38-C** which finally reacts with the HDDQ• radical to give the desired product.

Finally, Pappo has shown great interest in CDC reactions for the synthesis of sophisticated molecules bearing quaternary carbon centers within polycyclic hemi-acetal or spirolactone derivatives, demonstrating that CDC reactions can also lead to complexity. Indeed, using the association of $FeCl_3$ (10 mol%) and 1,10-phenanthroline (5 mol%) as the catalyst system in the presence of $(^tBuO)_2$, phenols and α-substituted-β-ketoesters can be coupled, leading to polycyclic hemi-acetal derivatives with 13–95% yields. The crucial role of the 1,10-phenanthroline in some cases must be emphasized, as it permits selective transformations, reducing the Lewis acidity of the iron salt by co-ordination, and avoiding side-reactions such as naphthol dimerization or Friedel–Crafts alkylation. On the other hand, when the stability of the obtained hemi-acetal is low, a subsequent transesterification step leads to polyspirolactones (Scheme 4.39).[38] The key step is the formation of the intermediate **39-A** *via* a CDC reaction.

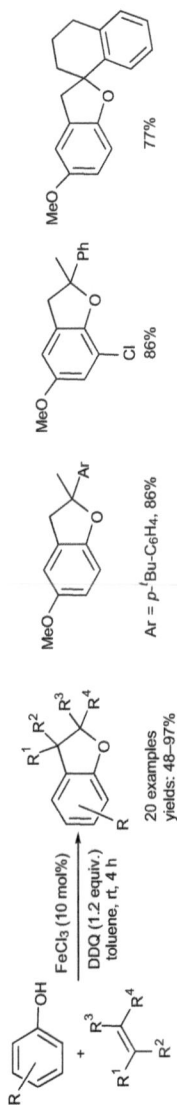

Scheme 4.37 Preparation of dihydrobenzofurans from phenols and alkenes.

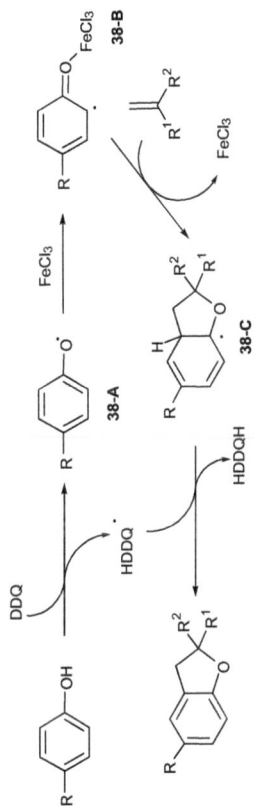

Scheme 4.38 Proposed mechanism for the formation of dihydrobenzofurans.

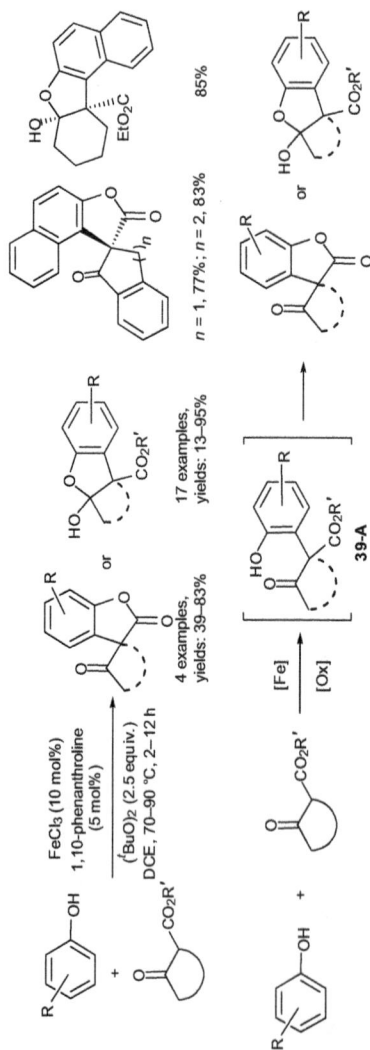

Scheme 4.39 Polycyclic hemi-acetals or spirolactones from phenols and β-ketoesters.

4.7 Conclusions and Outlook

Iron-catalyzed cross-dehydrogenative-coupling is a very attractive reaction for building C–C bonds efficiently starting from two C–H bonds under oxidative conditions. This methodology is highly attractive as it permits the building of such C–C bonds by avoiding the use of main transition metal organometallic reagents and halides or pseudo-halides as coupling partners, evading the generation of metal wastes in a stoichiometric amount. This is a breakthrough in terms of green chemistry! In the presence of simple and cheap catalysts such as iron salts and oxidants such as hydrogen peroxide, oxygen, *tert*-butyl hydroperoxide, TEMPO and DDQ, the functionalization of various sp^3 C–H bonds by other C–H bonds can be performed without pre-activation. Furthermore, this methodology also permits the preparation of more sophisticated heterocyclic and/or polycyclic compounds in a one-pot sequence.

As we can see over the five last years, from the pioneering reports of C.-J. Li to the impressive molecular complexity made by Pappo, iron-catalyzed CDC reactions have a promising future in molecular synthesis!

References

1. (a) E. Negishi and S. Baba, *J. Am. Chem. Soc.*, 1976, **98**, 6729; (b) J. K. Stille and D. Milstein, *J. Am. Chem. Soc.*, 1978, **100**, 3636; (c) M. Kumada, *Pure Appl. Chem.*, 1980, **52**, 669; (d) A. Suzuki and N. Miyaura, *J. Chem. Soc.*, 1979, 8.
2. C.-L. Sun, B.-J. Li and Z.-J. Shi, *Chem. Rev.*, 2011, **111**, 1293.
3. C. Liu, H. Zhang, W. Shi and A. Lei, *Chem. Rev.*, 2011, **111**, 1780.
4. C. S. Yeug and V. M. Dong, *Chem. Rev.*, 2011, **111**, 1215.
5. C.-J. Li, *Acc. Chem. Res.*, 2009, **42**, 335.
6. Y. Zhang and C.-J. Li, *Eur. J. Org. Chem.*, 2007, **28**, 4654.
7. Z. Li, L. Cao and C.-J. Li, *Angew. Chem., Int. Ed.*, 2007, **46**, 6505.
8. W. Cao, X. Liu, R. Peng, P. He, L. Lin and X. Feng, *Chem. Commun.*, 2013, **49**, 3470.
9. S. G. Pan, J. H. Liu, Y. M. Li and Z. P. Li, *Chin. Sci. Bull.*, 2012, **57**, 2382.
10. C. A. Correia and C.-J. Li, *Tetrahedron Lett.*, 2010, **51**, 1172.
11. (a) Z. Li, R. Yu and H. Li, *Angew. Chem., Int. Ed.*, 2008, **47**, 7497; (b) H. Richter and O. García Mancheño, *Eur. J. Org. Chem.*, 2010, 4460.
12. Y. Xie, M. Yu and Y. Zhang, *Synthesis*, 2011, 2803.
13. H. Li, Z. He, X. Guo, W. Li, X. Zhoa and Z. Li, *Org. Lett.*, 2009, **11**, 4176.
14. T. Zeng, G. Song, A. Moores and C.-J. Li, *Synlett*, 2010, 2002.
15. S.-J. Lou, D.-Q. Xu, D.-F. Shen, Y.-F. Wang, Y.-K. Liu and Z.-Y. Xu, *Chem. Commun.*, 2012, **48**, 11993.
16. F. Minisci, *Synthesis*, 1973, 1.
17. Y. Z. Li, B. J. Li, X.-Y. Lu, S. Lin and Z.-J. Shi, *Angew. Chem., Int. Ed.*, 2009, **48**, 3817.
18. C.-X. Song, G.-X. Cai, T. R. Farell, Z.-P. Jiang, H. Li, L.-B. Gan and Z.-J. Shi, *Chem. Commun.*, 2009, 6002.

19. M. Ohta, M. P. Quick, J. Yamaguchi, B. Wünsch and K. Itami, *Chem. Asian. J.*, 2009, **4**, 1416.
20. M. Ghobrial, K. Harhammer, M. D. Mihovilovic and M. Schnürch, *Chem. Commun.*, 2010, **46**, 8836.
21. M. Ghobrial, M. Schnürch and M. D. Mihovilovic, *J. Org. Chem.*, 2011, **76**, 8781.
22. P. Liu, C.-Y. Zhou, S. Xiang and C.-M. Che, *Chem. Commun.*, 2010, **46**, 2739.
23. C. M. R. Volla and P. Vogel, *Org. Lett.*, 2009, **11**, 1701.
24. S.-K. Xiang, B. Zhang, L.-H. Zhang, Y. X. Cui and N. Jiao, *Sci. China Chem*, 2012, **55**, 50.
25. H. Egami, K. Matsumoto, T. Oguma, T. Kunisu and T. Katsuki, *J. Am. Chem. Soc.*, 2010, **132**, 13633.
26. M. Chandrasekharam, B. Chiranjeevi, K. S. V. Gupta and B. Sridhar, *J. Org. Chem.*, 2011, **76**, 10229.
27. Z. H. Guan, Z. Y. Yan, Z. H. Ren, X. Y. Liu and Y. M. Liang, *Chem. Commun.*, 2010, **46**, 2823.
28. S. Wertz, D. Leifert and A. Studer, *Org. Lett.*, 2013, **15**, 928.
29. W. Liu, J. Liu, D. Ogawa, Y. Nishihara, X. Guo and Z. Li, *Org. Lett.*, 2011, **13**, 6272.
30. X. Guo, S. Pan, J. Liu and Z. Li, *J. Org. Chem.*, 2009, **74**, 8848.
31. P. Liu, Z. Wang, J. Lin and X. Hu, *Eur. J. Org. Chem.*, 2012, 1583.
32. (a) H. Richter and O. García Mancheño, *Org. Lett.*, 2011, **13**, 6066; (b) R. Rohlmann, T. Stopka, H. Richter and O. García Mancheño, *J. Org. Chem.*, 2013, **78**, 6050.
33. Y. Li, L. Cao, X. Luo and W.-P. Deng, *China J. Chem*, 2012, **30**, 2834.
34. H. Liu, L. Cao, J. Sun, J. S. Fossey and W.-P. Deng, *Chem. Commun.*, 2012, **48**, 2674.
35. X. Guo, R. Yu, H. Li and Z. Li, *J. Am. Chem. Soc.*, 2009, **131**, 17387.
36. Z. Huang, L. Jin, Y. Feng, P. Peng, H. Yi and A. Lei, *Angew. Chem., Int. Ed.*, 2013, **52**, 7151.
37. U. A. Kshirsagar, C. Regev, R. Parnes and D. Pappo, *Org. Lett.*, 2013, **15**, 3174.
38. R. Parnes, U. A. Kshirsagar, A. Werbeloff, C. Regev and D. Pappo, *Org. Lett.*, 2012, **14**, 3324.

CHAPTER 5

Cross-Dehydrogenative-Coupling Reactions Involving Allyl, Benzyl and Alkyl C–H Bonds

GUO-JUN DENG,* FUHONG XIAO AND LUO YANG*

College of Chemistry, Xiangtan University, Xiangtan 411105, China
*Email: gjdeng@xtu.edu.cn; yangluo@xtu.edu.cn

5.1 Introduction

Carbon–carbon bond forming reactions are central to organic chemistry and are a very important method for converting simple molecules into more complex compounds. The direct use of two C–H bonds to form C–C bonds is highly desirable since it has the potential to streamline the synthetic scheme by eliminating the need for the preparation and isolation of activated substrates prior to the coupling event. Since the pioneering work of Murai[1] and Fujiwara[2] on C–C bond-forming reactions through the catalytic cleavage of C–H bonds, this research area has developed rapidly and is becoming an increasingly viable alternative.[3] Most of the methodology developed centers on the functionalization of two sp^2 C–H bonds, especially for electron-rich heterocycles and directing-group-containing arenes. For both kinetic and thermodynamic reasons, the metal-catalyzed C–H activation of unactivated sp^3 C–H bonds (without any functional groups) is more difficult than that of sp^2 hybridized C–H bonds. Although unactivated sp^3 C–H bonds can be

RSC Green Chemistry No. 26
From C–H to C–C Bonds: Cross-Dehydrogenative-Coupling
Edited by Chao-Jun Li

$$\text{Csp}^3\text{-H} \quad + \quad \begin{array}{c} \text{Csp}-\text{H} \\[1em] \text{Csp}^2\text{-H} \\[1em] \text{Csp}^3\text{-H} \end{array} \xrightarrow[\text{[O]}]{\text{cat.}} \begin{array}{c} \text{Csp}^3\text{-Csp} \\[1em] \text{Csp}^3\text{-Csp}^2 \\[1em] \text{Csp}^3\text{-Csp}^3 \end{array}$$

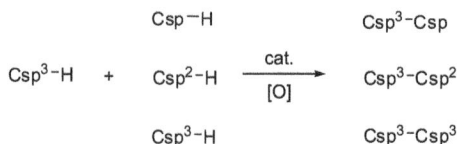

Scheme 5.1   Cross-dehydrogenative-coupling reactions involving sp³ C–H bonds.

functionalized by highly reactive species such as radicals, carbenes, and superacids, these transformations are generally not selective and do not tolerate other functionalities. In 2004, the Li group first specifically proposed, and has subsequently been exploring, the possibility of developing a methodology to construct functional molecules by using two sp³ C–H bonds.[4] Following that, various efficient methods were developed for sp³ C–H bond activation and subsequent coupling with sp, sp² and sp³ C–H bonds under relatively mild reaction conditions with good selectivity (Scheme 5.1).[5] However, most sp³ C–H bond activations involve the α-C–H bond of heteroatoms such as nitrogen and oxygen due to the easy formation of iminium ions and oxonium ions under oxidative reaction conditions. The selective activation of sp³ C–H bonds without any functional groups is very challenging. In this chapter, an overview of the recent developments, especially in the past several years, in oxidative cross-coupling reactions will be presented, with a focus on C–C bond formation between sp³ C–H bonds adjacent to carbon atoms and other C–H bonds.

5.2   CDC Reactions Involving Allylic C–H Bonds

Palladium-catalyzed allylic alkylation (the Tsuji–Trost reaction) is an important approach for constructing C–C bonds in modern organic synthesis [Scheme 5.2, route (1)].[6] As a general protocol, a leaving group such as carboxylate is always required at the allylic position, which is activated by a palladium catalyst during the reaction with a pronucleophile. The direct utilization of an allylic C–H bond rather than an allylic functional group would avoid the need to synthesize the allylic functional group, thus leading to a reduced number of synthetic steps [Scheme 5.2, route (2)].

5.2.1   Alkylation (sp³–sp³ Coupling)

Li and co-workers developed a CDC allylic alkylation between allylic sp³ C–H and methylenic sp³ C–H bonds. By using a combination of CuBr and CoCl₂ as the catalyst and a stoichiometric amount of TBHP (*tert*-butyl hydroperoxide) as the oxidant, various 1,3-dicarbonyl compounds reacted smoothly with cyclohexene and gave the desired products in moderate yields (Scheme 5.3). Sensitive substituents such as chloro and bromo were tolerated under the optimized reaction conditions. Other cyclic alkenes including diallylic systems were also transformed into the desired products when

$$\text{LG = OAc, Cl, OC(=O)OR', etc.}$$

(1)

(2)

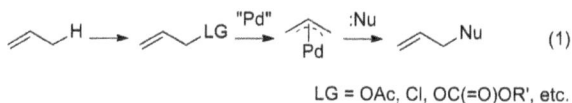

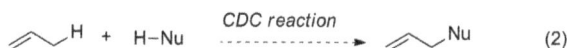

Scheme 5.2   The Tsuji–Trost reaction (1) and an allylic CDC reaction (2).

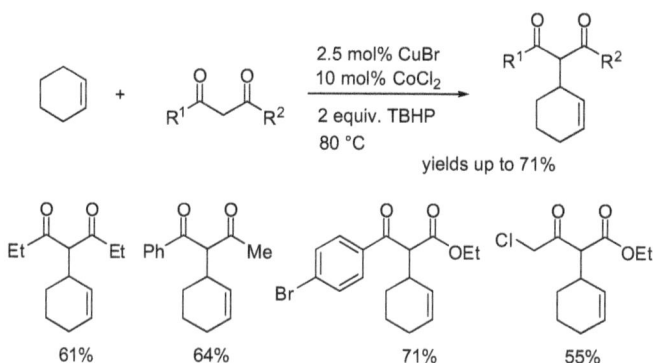

Scheme 5.3   CDC allylic alkylation of diketones.

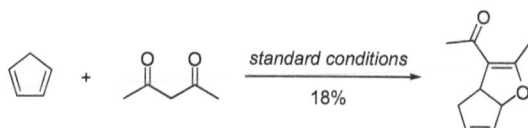

Scheme 5.4   Tandem allylic CDC alkylation and cyclization.

reacted with diketones.[7] Interestingly, if cyclopentadiene was used, the major product obtained was the dihydrofuran derivative, due to further cyclization of the alkylation product *in situ* (Scheme 5.4).

The Shi group reported a complementary Pd-catalyzed allylic alkylation using benzoquinone as the sacrificial oxidant under an oxygen atmosphere. The reaction showed good selectivity; in intermolecular tandem C–H bond functionalization reactions, the linear product dominated. During the intramolecular counterpart reactions, cyclizations yielding five- or six-membered rings proceeded smoothly with excellent diastereoselectivity. Although palladium salts were used as catalysts, aryl bromides and aryl chlorides were well tolerated under the reaction conditions. The addition of a radical inhibitor did not affect the reactivity, which provides evidence against single-electron-transfer (SET) steps (Scheme 5.5).[8]

The White group concurrently reported a similar allylic alkylation using the same ligand **BS**. With methylnitroacetate as the coupling partner, a wide range of allylarenes underwent direct allylic alkylation and afforded the linear isomer products with good yields. Functional groups such as esters, acetyl groups and halogens were tolerated under the optimized reaction

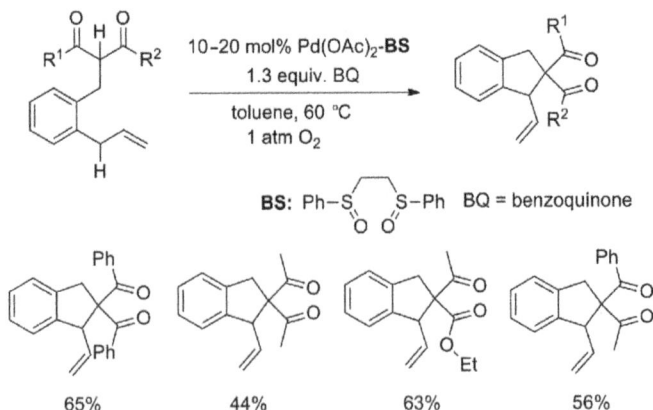

Scheme 5.5 Intramolecular direct allylic alkylation.

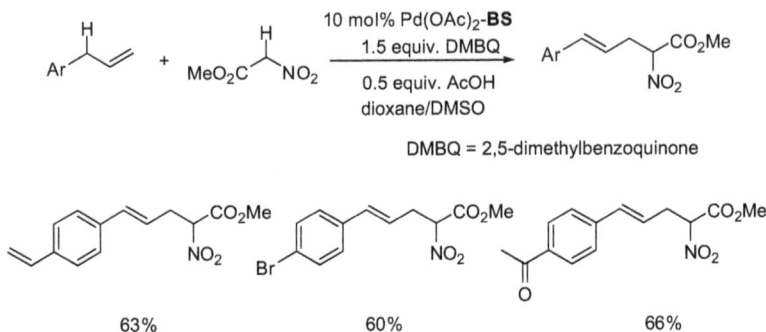

Scheme 5.6 Pd-Catalyzed allylic alkylation of methylnitroacetate.

conditions. DMSO was required for the catalyst and a Pd π-allylic complex resulting from C–H activation was isolated for the first time, and its reactivity examined in stoichiometric experiments (Scheme 5.6).[9]

Bao and co-workers demonstrated a CDC reaction between 1,3-diaryl-propenes and 1,3-diketones in the absence of metal catalyst using DDQ (2,3-dichloro-5,6-dicyano-1,4-benzoquinone) as the oxidant. The reaction proceeded smoothly under very mild conditions and gave the corresponding products in good-to-excellent yields (Scheme 5.7).[10]

5.2.2 Arylation (sp³–sp² Coupling)

The Bao group further developed a palladium-catalyzed allylation reaction of indoles using DDQ as the oxidant. The CDC reaction occurred smoothly even at 0 °C and gave the 1,3-diphenylallylindoles in moderate to good yields. The allylation took place selectively at the 3-position of indoles and no N-allylation by-products were detected (Scheme 5.8).[11] The authors propose that the reaction proceeds *via* a DDQ-mediated oxidation of the allylic substrates into

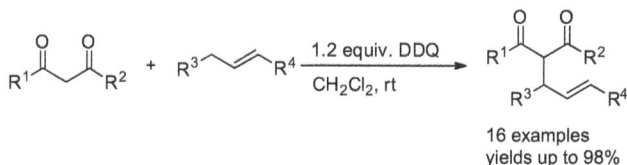

Scheme 5.7 CDC alkylation of allylic sp^3 C–H bonds.

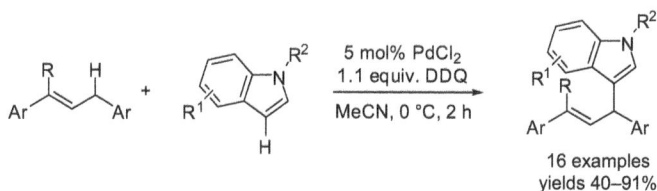

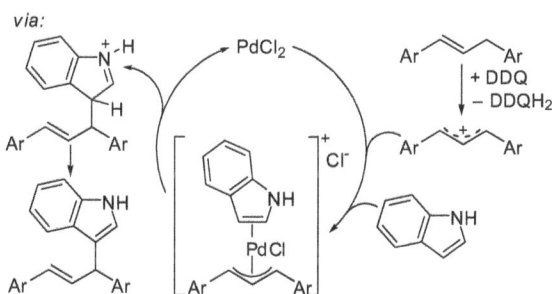

Scheme 5.8 CDC arylation of allylic sp^3 C–H bonds.

Pd-stabilized allyl cations that subsequently react with the indoles in a Friedel–Crafts type reaction.

5.2.3 Tandem CDC Processes

Meanwhile, Deng and co-workers disclosed a new methodology for rapid construction of biologically and synthetically important polysubstituted naphthalene derivatives from 1,2-aryl-propenes and styrenes (Scheme 5.9). FeCl$_3$ efficiently catalyzed the tandem CDC reaction and subsequent benzoannulation gave the corresponding products in good yields. A radical mechanism was proposed and the addition of 2.0 equiv. of TEMPO (2,2,6,6-tetramethyl-1-piperidinyloxy) completely inhibited the reaction.[12]

5.3 CDC Reactions Involving Benzyl C–H Bonds

5.3.1 Alkylation (sp^3–sp^3 Coupling)

During recent years, great progress has been made on the CDC reaction of benzylic C–H bonds adjacent to nitrogen or oxygen atoms. The formation of iminium or oxonium ions plays an important role in these coupling

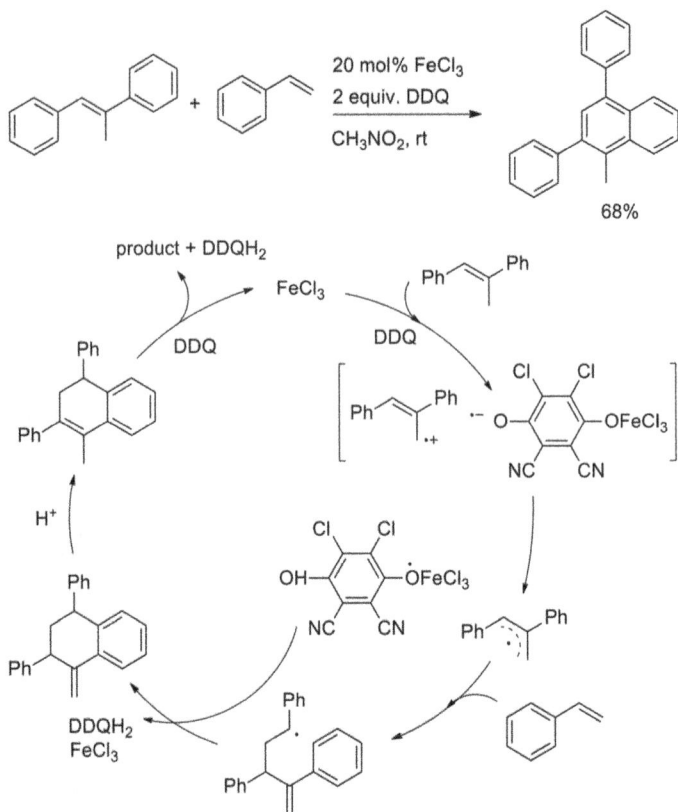

Scheme 5.9 Synthesis of naphthalenes *via* a tandem CDC process.

reactions. Thus, the CDC reaction of benzylic C–H bonds lacking an adjacent heteroatom is a more challenging task.[13] The numerous advantages of iron make it highly attractive as a catalyst for CDC reactions.[14] Li and co-workers developed an FeCl$_2$-catalyzed oxidative activation of a benzylic C–H bond lacking an adjacent heteroatom, which was followed by a cross-coupling reaction with diketones to form C–C bonds (Scheme 5.10).[15] Good-to-excellent yields were obtained when the reaction was performed at 80 °C by using *tert*-butyl peroxide (*^{t}*BuOO*^{t}*Bu or TBP) as the oxidant. The reaction was also found to proceed efficiently at room temperature, and the desired product was isolated in up to an 87% yield. Furthermore, Li and co-workers found that simple benzylic C–H bonds could be alkylated with 1,3-dicarbonyl compounds using oxygen as the terminal oxidant, in the presence of catalytic amounts of FeCl$_2$, CuCl and NHPI (*N*-hydroxyphthalimide).[16]

Xanthene and acridane derivatives are pharmaceutically active compounds, as well as natural compounds, and their synthesis has attracted much attention. Recently, Klussmann and co-workers reported the oxidative coupling of xanthene and other activated benzylic compounds with carbon-based nucleophiles such as ketones. The reaction proceeded smoothly under

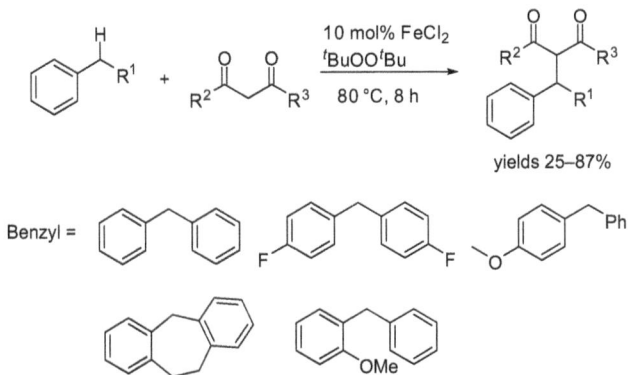

Scheme 5.10   FeCl$_2$-Catalyzed alkylation of benzylic C–H bonds.

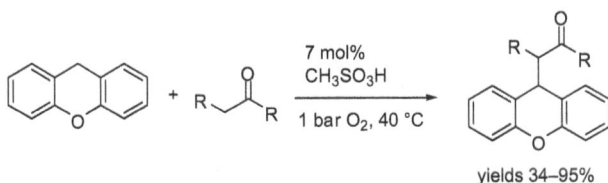

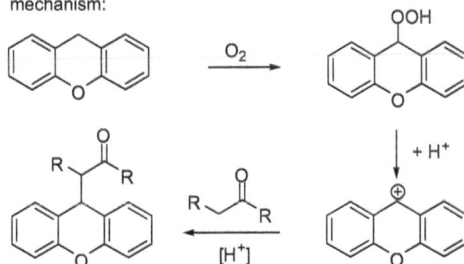

Scheme 5.11   Metal-free CDC reaction of benzylic C–H bonds.

ambient conditions by using elemental oxygen and a catalytic amount of methanesulfonic acid. Good yields could be obtained even the reactions were carried out at room temperature in the absence of a solvent. The proposed reaction mechanism involves auto-oxidative formation of a xanthene hydroperoxide intermediate and a subsequent acid-catalyzed S$_N$1 type reaction with nucleophiles (Scheme 5.11).[17]

1,3-Dicarbonyl compounds are able to co-ordinate to various Lewis acids to form nucleophiles and have been employed in various transformations. Gong and co-workers developed an enantioselective oxidative cross-coupling reaction of 3-indolylmethyl C–H bonds with 1,3-dicarbonyls using a chiral Lewis acid catalyst.[18] The combination of the Lewis acid Cu(OTf)$_2$ with a chiral bis(oxazoline) ligand exhibited the best reactivity and enantioselectivity. The reaction proceeded smoothly at 0 °C using DDQ as the oxidant

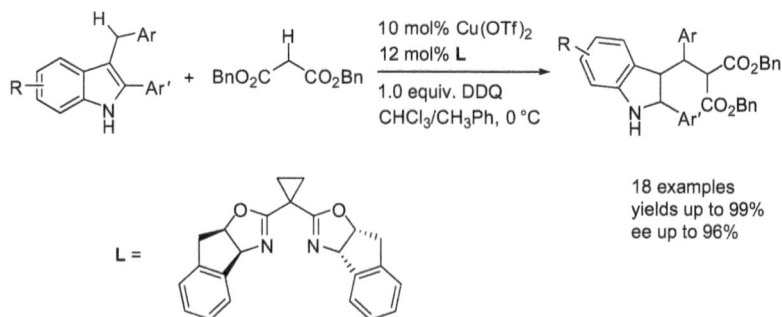

Scheme 5.12   Enantioselective alkylation of 3-indolylmethyl C–H bonds.

and afforded the corresponding alkylated products in up to 99% yield with high enantioselectivities (Scheme 5.12). Electron spin resonance (ESR) studies of the reaction intermediate did not show any signal, which indicated that a cationic rather than radical species was generated as the key intermediate of the coupling reaction. The presence of a chiral copper complex enhanced the oxidizing ability by co-ordinating to the oxygen of DDQ to facilitate the dehydrogenation of the indole substrate.

5.3.2   Arylation (sp³–sp² Coupling)

Lu and co-workers found that a homocoupling of *para*-xylene afforded biaryl- or diarylmethane using the catalytic system Pd(OAc)$_2$/TFA/K$_2$S$_2$O$_8$. In this coupling reaction, the aryl and benzylic C–H bonds of *para*-xylene could be selectively activated just by tuning the concentration of TFA [Scheme 5.13(a)].[19] Similarly, Zhang and co-workers developed a copper-catalyzed CDC reaction of *N-para*-tolylamides through successive C–H activations. Cu(OTf)$_2$ showed very good activity with Selectfluor® as the oxidant. This strategy could also be used for benzoxazine derivative preparation. Various 4*H*-3,1-benzoxazines were prepared in good-to-excellent yields through successive intermolecular CDC reaction of aromatic C–H and benzylic methyl C–H bonds, and subsequent intramolecular C–O bond formation. A catalytic amount of water was thought to play an important role for *in situ* generation of the key copper hydroxy complex catalyst. Sensitive functional groups such as halogens and trifluoromethyl groups were well tolerated under the optimized conditions. This method afforded an efficient route for rapid construction of heterocyclic compounds from readily available starting materials [Scheme 5.13(b)].[20]

Shi and co-workers disclosed the use of DDQ for the direct arylation of diphenylmethane derivatives.[21] A variety of electron-rich aromatic substrates were successfully coupled with the benzylic C–H bonds of diphenylmethane catalyzed by FeCl$_2$, and gave the corresponding products in good-to-excellent yields. The methoxy group of aromatic substrates played an important role in regioselectivity. Orthoester-, acetoxy-, carboxy-, carbonyl-, and

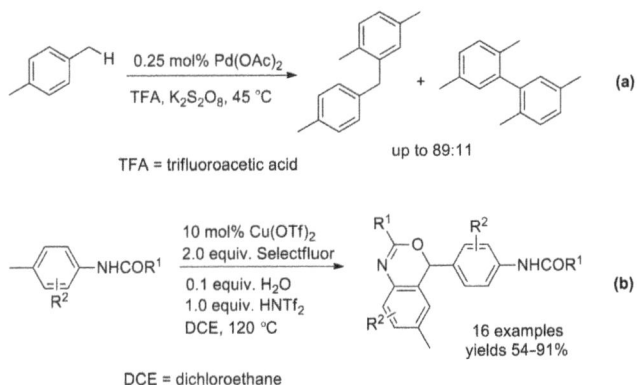

TFA = trifluoroacetic acid

up to 89:11   (a)

DCE = dichloroethane

16 examples
yields 54–91%   (b)

Scheme 5.13   CDC arylation.

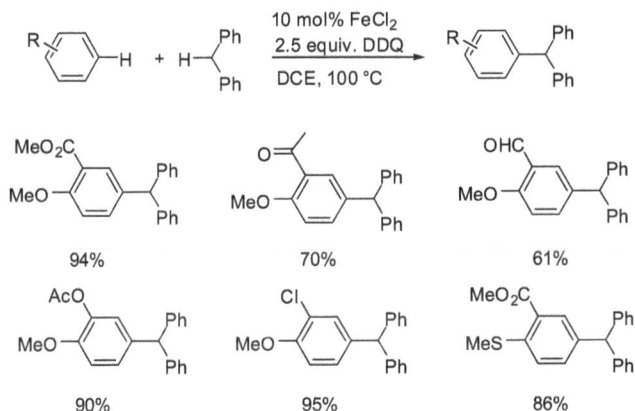

94%

70%

61%

90%

95%

86%

Scheme 5.14   Iron-catalyzed selective CDC reaction of benzylic C–H bonds with arenes.

halide-substituted anisoles gave only one regio-isomer. The regioselectivity could also be well controlled by a methylthio group instead of a methoxy group (Scheme 5.14). The group of Patel extended the scope of Pd-catalyzed chelation-assisted *ortho*-alkylation to the synthesis of benzophenone derivatives. Various 2-phenylpyridines were coupled with toluenes in the first step and followed by oxidation of the resulting diphenyl methanes into the corresponding ketones (Scheme 5.15).[22]

5.3.3   Alkynylation (sp³–sp Coupling)

Li and co-workers found that the direct alkynylation of diphenylmethane could also be achieved in the presence of a copper catalyst using DDQ as the oxidant (Scheme 5.16). Various alkynes could react with diphenylmethane

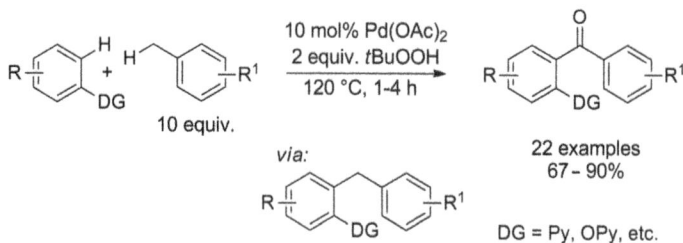

Scheme 5.15 Palladium-catalyzed CDC ketone formation.

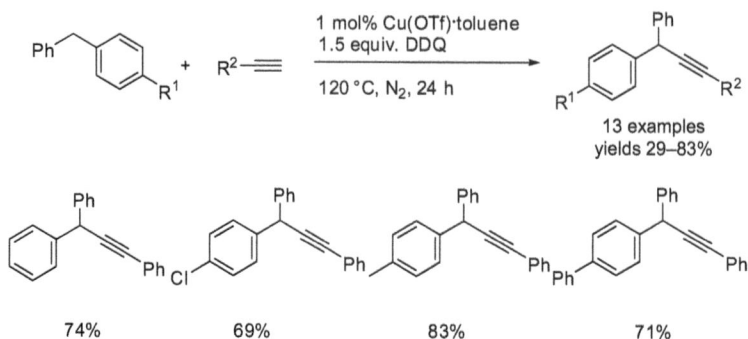

Scheme 5.16 CDC reaction of alkynes and diphenylmethane derivatives.

derivatives effectively to give the desired products in good yields using CuOTf as catalyst.[23] Under similar conditions, FeCl$_2$ did not demonstrate any catalytic activity, and InCl catalyzed the cross-coupling with reduced yields.

5.4 CDC Reactions Involving Alkane C–H Bonds

5.4.1 Alkane Alkylation (sp^3–sp^3 Coupling)

In recent years, various CDC reactions involving sp^3 C–H bonds have been successfully developed. However, it is a great challenge to use simple alkanes (unactivated and without any functional groups) to form C–C bonds by such an oxidative approach. Alkanes are major constituents of petroleum and natural gas, but they are difficult to convert directly to more valuable chemicals since they possess strong C–C and C–H bonds and lack Lewis acidic and basic sites of reactivity. Historically, the Fenton chemistry[24] and the Gif processes[25] established the conversion of aliphatic C–H bonds into C–O bonds under mild conditions by using peroxides as the oxidant and catalyzed by various iron catalysts.[26] Li and co-workers successfully used this process for C–C bond formation. It was found that by using 10 mol% FeCl$_2$ · 4H$_2$O as the catalyst and TBP as the oxidant at 100 °C for 12 h under an atmosphere of nitrogen, various activated methylene substrates reacted with simple cyclohexane, cyclopentane, cycloheptane, cyclooctane and

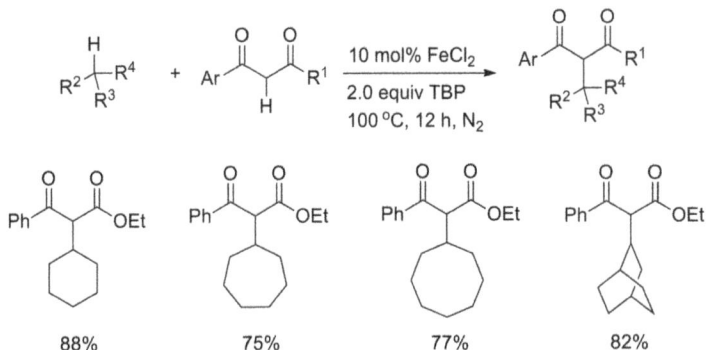

88%   75%   77%   82%

Scheme 5.17   Simple alkane CDC alkylation.

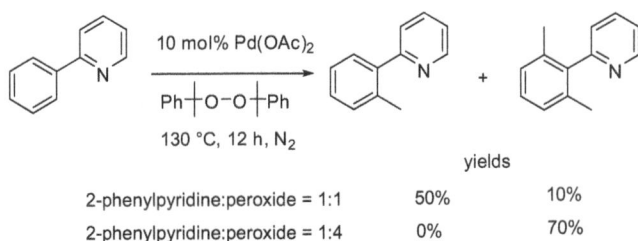

	yields	
2-phenylpyridine:peroxide = 1:1	50%	10%
2-phenylpyridine:peroxide = 1:4	0%	70%

Scheme 5.18   Methylation of 2-phenylpyridine with dicumyl peroxide.

adamantane to give the corresponding alkylation products in good yields in most cases (Scheme 5.17).[27] Other iron salts such as FeBr$_2$ and FeCl$_3$ were also effective for this kind of transformation.

5.4.2   Alkane Arylation (sp^3–sp^2 Coupling)

The Li group further extended this strategy for simple alkane arylation. When 2-phenylpyridine treated with dicumyl peroxide and combined with 10 mol% Pd(OAc)$_2$ was used as the catalyst at 130 °C under an atmosphere of nitrogen, methylated products were obtained in moderate-to-good yields. The ratio of mono- to and bis-methylated products depended on the ratio of the reactants (Scheme 5.18).[28] The methyl group came from the dicumyl peroxide. Other peroxides and palladium catalysts could also be used, albeit generating lower yields of the methylation products. With benzo[*h*]quinoline as the substrate, 76% of the mono-methylated product was obtained. Acet-anilide substrates are also effective in this transformation, generating the mono-methylation product in moderate yields. Further studies showed that the direct alkylation of simple alkanes could be observed by adding excess alkanes to the reaction mixture. However, the reaction of 2-phenylpyridines with cycloalkanes in the presence of TBP under palladium-catalyzed reaction conditions only gave a trace amount of the desired product. After systematic investigation of the reaction conditions, Li found that [Ru(*p*-cymene)Cl$_2$]$_2$

was an efficient catalyst for this transformation. The desired products could be obtained in good yields by using 10 mol% [Ru(*p*-cymene)Cl$_2$]$_2$ as the catalyst and TBHP as the hydrogen acceptor, at 135 °C for 16 h under an atmosphere of air (Scheme 5.19).[29]

The mechanism of the reaction was proposed to involve a ruthenium-catalyzed aryl C–H activation followed by an H–alkyl exchange. Reductive elimination of this intermediate generated the arene–cycloalkane coupling product and re-generated the active ruthenium catalyst (Scheme 5.20). Deuterium isotope experiments showed a large negative kinetic isotope effect, which suggests that the ruthenium-catalyzed aryl C–H activation is a fast equilibrium and the H–alkyl exchange is the rate-limiting step.

Air-sensitive phosphine ligands are rarely used for CDC reactions under strong oxidative conditions. However, the Li group developed a

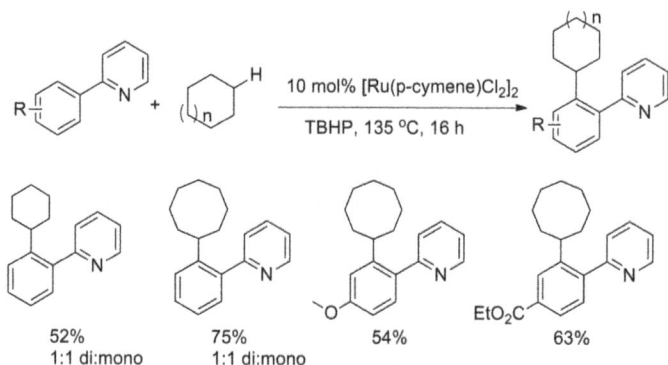

Scheme 5.19   CDC reaction between phenylpyridines and cycloalkanes.

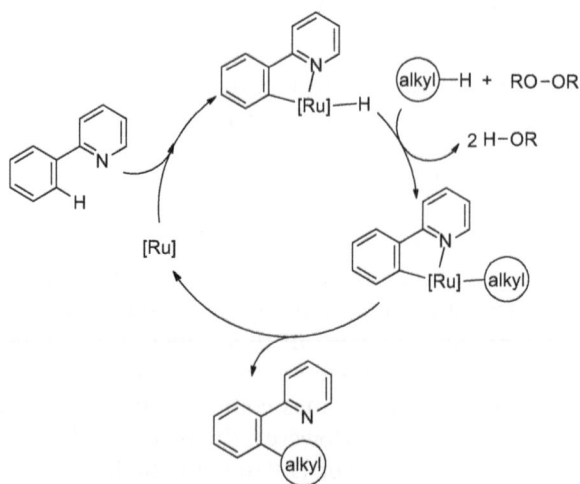

Scheme 5.20   Proposed reaction mechanism.

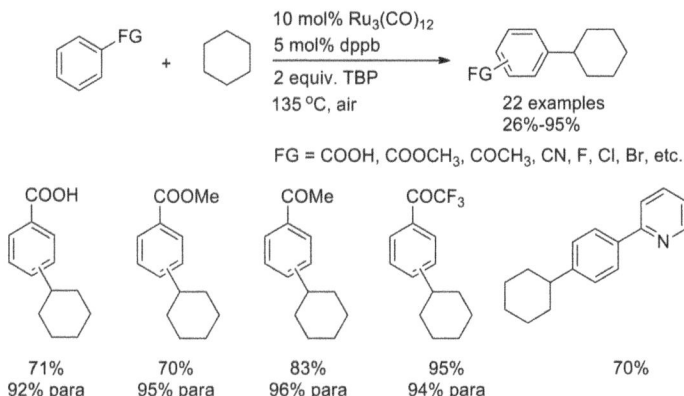

Scheme 5.21 *para*-Selective cross-coupling of benzene derivatives with cyclohexane.

ruthenium-catalyzed *para*-selective oxidative cross-coupling reaction of arenes and cycloalkanes in the presence of phosphine ligands. The combination of $Ru_3(CO)_{12}$ with DPPB (bis(diphenylphosphino)butane) could efficiently catalyze the coupling reaction of benzene derivatives and cycloalkanes and gave the products in high yields with good selectivity (Scheme 5.21). Carboxylic groups, esters and acetyl groups are good directing groups to achieve high selectivity. The commonly used 2-phenylpyridine was also tested in this reaction and the coupling took place exclusively at the *para*- position (>99%).[30] The exact reaction mechanism is not very clear. However, the kinetic isotope study revealed that the reaction most likely proceeds *via* a radical mechanism. This reaction overwhelmed the effect of strongly *ortho*-directing chelating substituents, and benzoic acid could be used directly.

Under similar conditions, however, unactivated pyridines or quinolines could not react with simple alkanes even when increasing the TBP to 3 equiv. Alternatively, a Lewis acid catalyst was added to increase the acidity of the C2–H bond on these heteroaromatic rings to increase the reactivity. Quinoline smoothly reacted with cycloalkanes in the presence of TBP catalyzed by $Sc(OTf)_3$ and gave the bis-alkylation product in good yields (Scheme 5.22).[31] Two equiv. of peroxide was necessary to achieve high reaction yields. No catalyst was required when the more active substrate pyridine-*N*-oxide was used as the substrate. The additional *N*-oxide moiety enhanced the reactivity significantly. Pyridine-*N*-oxide reacted with cyclohexanes in the presence of TBP and gave di- and tri-alkylation products in 70% total yield (Scheme 5.23).[32] The reaction also took place with norbornane and 1,4-dioxane to give structurally interesting molecules in moderate yields.

Similarly, purines were able to couple with simple alkanes in the presence of peroxides. Guo and co-workers found that the reaction occurred on C8 to give C8-alkylpurines by C–C bond formation that was only promoted by TBP, while it occurred on the amino group to give N6-alkylated purines by C–N

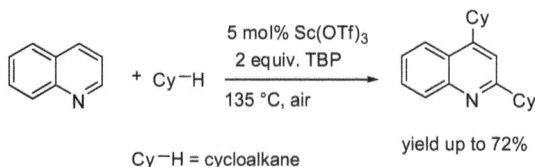

Scheme 5.22   CDC reaction between quinoline and cyclohexanes.

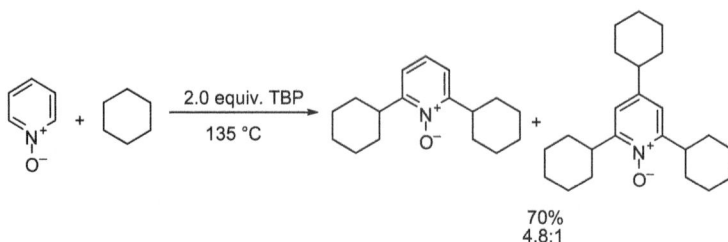

Scheme 5.23   CDC reaction between pyridine-*N*-oxide and cyclohexane.

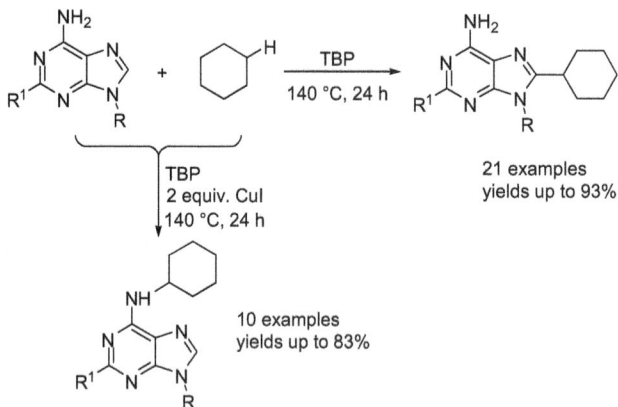

Scheme 5.24   CuI-Controlled alkylation of purines.

bond formation when 2 equiv. of CuI were added (Scheme 5.24).[33] A reaction mechanism involving a free-radical process was proposed. The cyclohexyl free radical reacted at the C8 position of the purine by radical addition–oxidation to afford the C8-alkylpurine. When CuI was added to the reaction, the cyclohexyl radical was converted into the cyclohexyl cation which then reacted with the amino group on account of its higher electrophilicity.

Antonchick and co-workers found that [bis(trifluoroacetoxy)iodo]benzene (PIFA) and NaN$_3$ could be used as an alternative radical initiator in the absence of a transition metal catalyst. Various nitrogen-containing heteroarenes were reacted with simple alkanes in the presence of PIFA and NaN$_3$ to selectively afford the corresponding alkylated products (Scheme 5.25).[34] Acyclic alkanes afforded moderate-to-high yields of products. Exclusive 2° site functionalization was observed, and reaction of *n*-butane led to the corresponding product in 72% yield when 50 equiv. of *n*-butane was used.

The authors proposed a mechanism that starts with the generation of a (trifluoroacetoxy)iodo benzene radical and an azide radical (Scheme 5.26). The azide radical abstracts a hydrogen atom from the cyclic alkane and the resulting hydrazoic acid reacts with a second PIFA molecule with release of trifluoroacetic acid that protonates the heterocycle and activates it for the addition of the alkyl radical. The resulting heterocyclic radical cation is oxidized into the corresponding product by the (trifluoroacetoxy)iodo benzene radical.

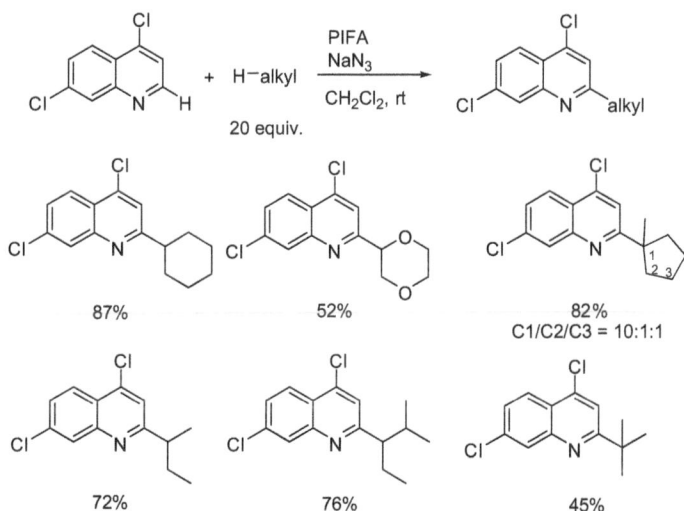

Scheme 5.25   CDC reaction of alkanes with heteroarenes.

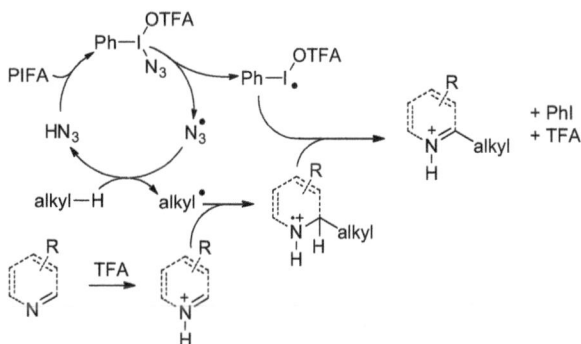

Scheme 5.26   Proposed mechanism of PIFA-mediated alkylation.

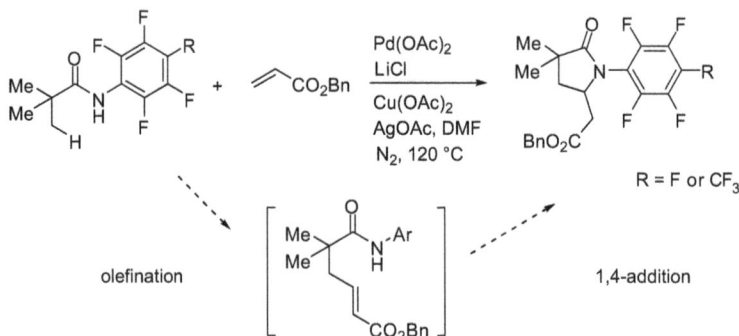

Scheme 5.27   Olefination of sp^3 C–H bonds.

5.4.3   Alkane Olefination (sp^3–sp^2 Coupling)

The transition metal catalyzed direct olefination of arene C–H bonds has emerged as a powerful method for alkene synthesis. However, this strategy is rarely employed for less reactive alkane sp^3 C–H bonds. After the realization of various versatile aryl C–H olefinations,[35] Yu and co-workers investigated the palladium-catalyzed olefination of *N*-arylpivalamides (Scheme 5.27).[36] The amide functional group acted as a directing group and played an important role in this transformation. An electron-withdrawing group on the amide significantly improved the reaction yield. The combined use of Cu(OAc)$_2$ and AgOAc was necessary to get satisfactory reaction yields. After olefination, the amide products underwent 1,4-conjugate addition to give the corresponding lactam compounds. The reaction conditions could also be applied to effect the olefination of cyclopropyl methylene C–H bonds and substrates containing α-hydrogen atoms. This strategy provided an efficient alternative for lactam preparation in one pot *via* C–H olefination and addition.

An intramolecular aerobic Pd-catalyzed synthesis of various 2,3-dihydroindolizinium salts was provided by the group of Sanford (Scheme 5.28).[37] Nitrogen heterocycles served as directing groups, and air was used as the terminal oxidant. The product underwent reversible intramolecular Michael addition, which could protect the mono-alkenylated product from over-functionalization. Hydrogenation of the Michael adducts provided access to bicyclic nitrogen-containing scaffolds that are prevalent in alkaloid natural products.

5.4.4   Alkane Cyclization (sp^3–sp^2 Coupling)

Several efficient procedures have been developed that allow the cross-dehydrogenative synthesis of heterocycles *via* intramolecular cyclization of acetanilide radicals. Fagnou and co-workers developed a palladium-catalyzed intramolecular coupling of arenes and alkanes in air. A new sp^2–sp^3 C–C

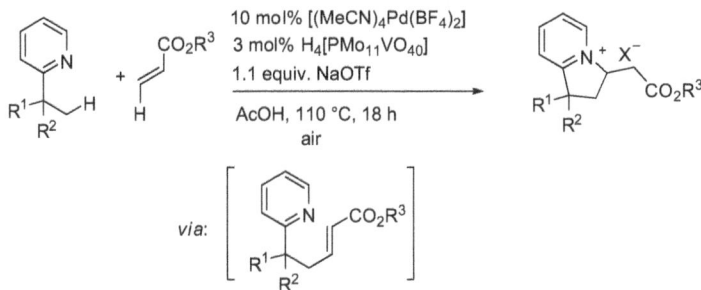

Scheme 5.28 2,3-Dihydroindolizinium salts preparation *via* CDC cyclization.

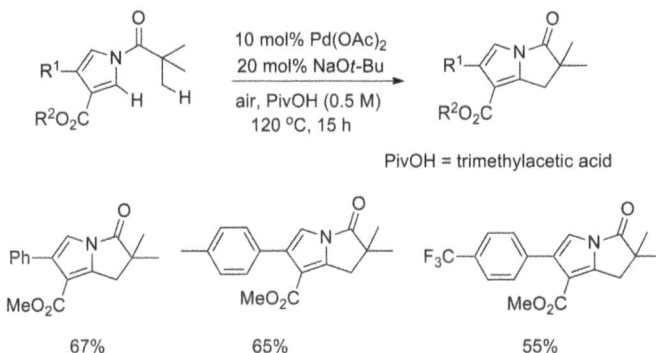

PivOH = trimethylacetic acid

| 67% | 65% | 55% |

Scheme 5.29 Intramolecular CDC reaction of substituted pyrroles.

bond between an azole ring and an unactivated methyl substituent was created. The reactions employed a simple, readily available palladium catalyst, exhibited high regioselectivity with respect to both the azole and the alkane moieties, and could be performed in an open flask using air as the terminal oxidant (Scheme 5.29).[38] Similarly, fused thiophene-cyclopentanes could be synthesized *via* palladium-catalyzed dehydrogenative sp^2–sp^3 C–H coupling under oxidative reaction conditions. The reaction showed good chemo- and diastereoselectvity by using 20 mol% of Pd(OAc)$_2$ and 3 equiv. of Ag$_2$CO$_3$ as the oxidant. However, a direct comparison showed that it is much less efficient than the traditional two-step sequence composed of electrophilic halogenation and Pd(0)-catalyzed sp^3 C–H arylation (Scheme 5.30).[39] In most cases, lower than 40% yields were obtained, although the diastereoselectivity was very high.

Pihko and co-workers developed a palladium-catalyzed dehydrogenative β-functionalization of β-ketoesters with indoles. The reactions proceeded smoothly and gave the corresponding products in good yields and high regioselectivities (C3-selective for the indole partner and β-selective for the β-ketoester) at room temperature using Pd(TFA)$_2$ as the catalyst and peroxide as the oxidant (Scheme 5.31).[40] The authors proposed two possible mechanisms after systematic investigation: a Saegusa-type mechanism ("late

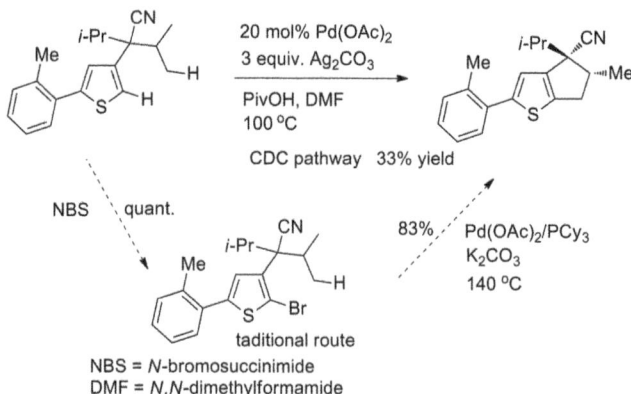

Scheme 5.30   Comparison of the CDC method and the traditional two-step sequence.

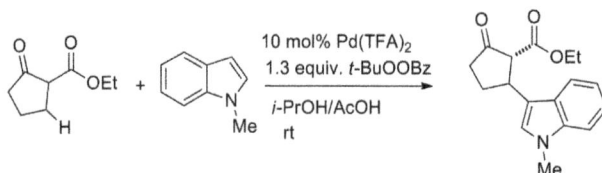

Scheme 5.31   Dehydrogenative β′-functionalization of β-ketoesters with indoles.

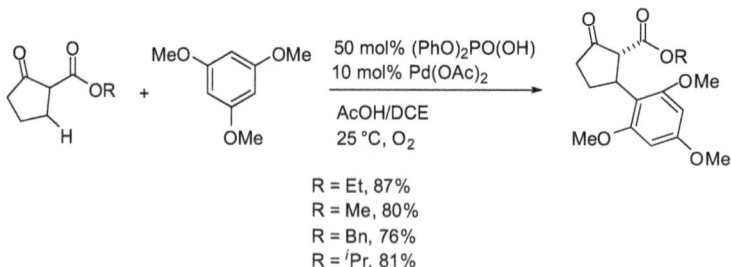

Scheme 5.32   Dehydrogenative β′-arylation of β-ketoesters with arenes.

indole") and an indole-assisted dehydrogenation mechanism ("early indole"). The same group also developed a palladium-catalyzed CDC reaction using electron-rich arenes as coupling partners (Scheme 5.32).[41] The rate of arylation in the presence of 50 mol% (PhO)$_2$P(O)OH was approximately 14-fold faster when compared to the rate without the acid additive. Molecular oxygen was used as the sole oxidant for this kind of transformation. A possible mechanism involving acid-assisted palladation of the arene and subsequent engagement of the β-ketoester was proposed on the basis of control experiments.

5.5   Conclusions and Outlook

In recent years, building a carbon–carbon linkage directly from two simple C–H bonds has emerged as an attractive and green method. Transition metal catalyzed functionalization of sp^3 C–H bonds adjacent to heteroatoms is quite common and many examples have been recently described. However, the activation of non-activated sp^3 C–H bonds especially with simple alkanes always requires very harsh reaction conditions, such as high reaction temperatures, strong oxidants and excess reactants. Symmetrical alkanes are preferred for CDC reactions since regioselectivity control is challenging under such harsh reaction conditions. The development of effective catalysts (metal or non-metal) for selective activation and functionalization of non-activated alkane C–H bonds under mild reaction conditions, especially using molecular oxygen as the oxidant, is highly desirable.

Acknowledgements

The present and past co-workers in Li's laboratory, whose names are given in the list of references, are highly acknowledged for their hard work. We also thank the Canada Research Chair (Tier I) foundation (to C.-J. Li), the CFI, NSERC, FQRNT, the US NSF CAREER Award, the CIC (Merck Frosst/Boh-ringer Ingelheim/AstraZeneca). The National Natural Science Foundation of China (21172185), the Hunan Provincial Natural Science Foundation of China (11JJ1003), the New Century Excellent Talents in University from Ministry of Education of China (NCET-11-0974) are also acknowledged.

References

1. S. Murai, F. Kakiuchi, S. Sekine, Y. Tanaka, A. Kamatani, M. Sonoda and N. Chatani, *Nature*, 1993, **366**, 529.
2. C. Jia, D. Piao, J. Oyamada, W. Lu, T. Kitamura and Y. Fujiwara, *Science*, 2000, **287**, 1992.
3. (a) *Activation and Functionalization of C–H Bonds*, ed. K. I. Goldberg and A. S. Goldman, American Chemical Society, Washingtom, DC, 2004, vol. 885; (b) *C–H Activation*, ed. J. Q. Yu and Z. J. Shi, Springer, Heidelberg, 2010; (c) T. W. Lyons and M. S. Sanford, *Chem. Rev.*, 2010, **110**, 1147; (d) X. Chen, K. M. Engle, D. H. Wang and J. Q. Yu, *Angew. Chem., Int. Ed.*, 2009, **48**, 5094; (e) J. A. Ashenhurst, *Chem. Soc. Rev.*, 2010, **39**, 540; (f) C. Johansson and T. J. Colacot, *Angew. Chem., Int. Ed.*, 2010, **49**, 676; (g) L. C. Campeau, D. R. Stuart and K. Fagnou, *Aldrich Acta*, 2007, **40**, 35.
4. C. J. Li, *Acc. Chem. Res.*, 2009, **42**, 335.
5. G. J. Deng and C.-J. Li, in *Organic Chemistry - Breakthroughs and Perspectives*, ed. K. L. Ding and L. X. Dai, Wiley-VCH, Weinheim, 2012, ch. 19.

6. (a) J. Tsuji, *Transition Metal Reagents and Catalysts: Innovations in Organic Synthesis*, Wiley, New York, 2000, ch. 4; (b) B. M. Trost and M. L. Grawley, *Chem. Rev.*, 2003, **103**, 2921.

7. Z. Li and C.-J. Li, *J. Am. Chem. Soc.*, 2006, **128**, 56.

8. S. Lin, C. Song, G. Cai, W. Wang and Z. J. Shi, *J. Am. Chem. Soc.*, 2008, **130**, 12901.

9. A. Young and M. C. White, *J. Am. Chem. Soc.*, 2008, **130**, 14090.

10. D. Cheng and W. Bao, *Adv. Synth. Catal.*, 2008, **350**, 1263.

11. H. Mo and W. Bao, *Adv. Synth. Catal.*, 2009, **351**, 2845.

12. H. Liu, L. Cao, J. Sun, J. Fossey and W. P. Deng, *Chem. Commun.*, 2012, **48**, 2674.

13. (a) N. Borduas and D. A. Powell, *J. Org. Chem.*, 2008, **73**, 7822; (b) C. X. Song, G. X. Cai, T. R. Farrell, Z. P. Jiang, H. Li, L. B. Gan and Z. J. Shi, , *Chem. Commun.*, 2009, 6002.

14. (a) C. Bolm, J. Legros, J. Le Paih and L. Zani, *Chem. Rev.*, 2004, **104**, 6217; (b) A. Furstner and R. Martin, *Chem. Lett.*, 2005, **34**, 624.

15. Z. P. Li, L. Cao and C.-J. Li, *Angew. Chem., Int. Ed.*, 2007, **46**, 6505.

16. C. A. Correia and C.-J. Li, *Tetrahedron Lett*, 2010, **51**, 1172.

17. Á. Pintér, A. Sud, D. Sureshkumar and M. Klussmann, *Angew. Chem., Int. Ed.*, 2010, **49**, 5004.

18. C. Guo, J. Song, S. W. Luo and L. Z. Gong, *Angew. Chem., Int. Ed.*, 2010, **49**, 5558.

19. Y. Rong, R. Li and W. J. Lu, *Organometallics*, 2007, **26**, 4376.

20. T. Xiong, Y. Li, X. Bi, Y. Lv and Q. Zhang, *Angew. Chem., Int. Ed.*, 2011, **50**, 7140.

21. Z. J. Shi, S. Li, X. Y. Lu, B. J. Lu and Y. Z. Li, *Angew. Chem., Int. Ed.*, 2009, **48**, 3817.

22. S. Guin, S. K. Rout, A. Banerjee, S. Nandi and B. K. Patel, *Org. Lett.*, 2012, **14**, 5294.

23. C. A. Correia and C.-J. Li, *Adv. Synth. Catal.*, 2010, **352**, 1446.

24. (a) D. T. Sawyer, A. Sobkowiak and T. Matsushita, *Acc. Chem. Res.*, 1996, **29**, 409; (b) C. Walling, *Acc. Chem. Res.*, 1998, **31**, 155.

25. D. H. R. Barton and D. Doller, *Pure Appl. Chem.*, 1991, **63**, 1567.

26. C. Bolm, J. Legros, J. L. Paih and L. Zani, *Chem. Rev.*, 2004, **104**, 6217.

27. Y. H. Zhang and C.-J. Li, *Eur. J. Org. Chem.*, 2007, 4654.

28. Y. Zhang, J. Feng and C.-J. Li, *J. Am. Chem. Soc.*, 2008, **130**, 2900.

29. G. Deng, L. Zhao and C.-J. Li, *Angew. Chem., Int. Ed.*, 2008, **47**, 6278.

30. X. Y. Guo and C.-J. Li, *Org. Lett.*, 2011, **13**, 4977.

31. G. J. Deng and C.-J. Li, *Org. Lett.*, 2009, **11**, 1171.

32. G. J. Deng, K. Ueda, S. Yanagisawa, K. Itami and C.-J. Li, *Chem.-Eur. J*, 2009, **15**, 333.

33. R. Xia, H. Y. Niu, G. R. Qu and H. M. Guo, *Org. Lett.*, 2012, **14**, 5546.

34. A. P. Antonchick and L. Burgmann, *Angew. Chem., Int. Ed.*, 2013, **52**, 3267.

35. D. H. Wang, K. M. Engle, B. F. Shi and J. Q. Yu, *Science*, 2010, **327**, 315.
36. M. Wasa, K. M. Engel and J. Q. Yu, *J. Am. Chem. Soc.*, 2010, **132**, 3680.
37. K. J. Stowers, K. C. Fortner and M. S. Sanford, *J. Am. Chem. Soc.*, 2011, **133**, 6541.
38. B. Liégault and K. Fagnou, *Organometallics*, 2008, **47**, 4841.
39. C. Pierre and O. Baudoin, *Tetrahedron*, 2013, **69**, 4473.
40. M. V. Leskinen, K. T. Yip, A. Valkonen and P. M. Pihko, *J. Am. Chem. Soc.*, 2012, **134**, 5750.
41. K. T. Yip, R. Y. Nimje, M. V. Leskinen and P. M. Pihko, *Chem.–Eur. J*, 2012, **18**, 12590.

CHAPTER 6

Aryl–Aryl Coupling via Cross-Dehydrogenative-Coupling Reactions

BRENTON DEBOEF* AND ASHLEY L. PORTER

Department of Chemistry, University of Rhode Island, Kingston,
Rhode Island 02881, USA
*Email: bdeboef@chm.uri.edu

6.1 Introduction

Currently there is a need for environmentally friendly ways to incorporate carbon–carbon bonds into organic molecules. Specifically, biaryl C–C bonds are a high priority, since 75% of marketed pharmaceuticals contain aryl or heterocyclic groups.[1] The traditional methods for making biaryl C–C bonds include Suzuki, Negeshi, Stille and Kumada couplings,[2] all of which involve the coupling of an aryl-halide and an aryl–metal species. These methods appear green, since they use small quantities of metal catalysts and require no external oxidant. However, they suffer from wasteful and costly pre-functionalization steps and poor atom economy.[3]

The ideal scenario would be to oxidatively couple two aryl C–H bonds and produce a new biaryl C–C bond. This coupling method, herein known as cross-dehydrogenative-coupling (CDC), requires no pre-functionalization steps and generates fewer by-products as waste (Scheme 6.1).

Compared to traditional coupling techniques, CDC provides a greener synthetic route by improving the overall step economy of the

RSC Green Chemistry No. 26
From C–H to C–C Bonds: Cross-Dehydrogenative-Coupling
Edited by Chao-Jun Li
© The Royal Society of Chemistry 2015
Published by the Royal Society of Chemistry, www.rsc.org

Cross-dehydrogenative-coupling

Traditional Suzuki-type coupling

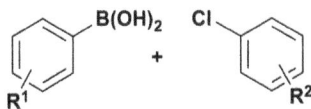

Scheme 6.1   Biaryl synthesis *via* CDC and traditional coupling techniques.

reaction, since functionalization is not a pre-requisite in CDC. The number of additional steps required to install functional groups prior to coupling is often time consuming, expensive, and wasteful.

In 1912, the first oxidative biaryl coupling was reported by Scholl, in which simple arenes were coupled using a stoichiometric amount of $AlCl_3$.[4,5] Since the introduction of the Scholl reaction, the number of publications concerning biaryl CDC reactions has continued to grow, attracting much interest in both academic and industrial settings. However, these reactions come with many challenges that must be overcome before they can be implemented in industry.

One of the main challenges with the CDC methodology is finding ways to readily activate C–H bonds, since they are short, strong, and hard to activate *via* oxidative addition with organometallic complexes. Traditional coupling methods, such as the Suzuki reaction, are successful because pre-functionalization with a halogen or boronic acid creates species that are amenable to metalation *via* oxidative addition or transmetalation. Another challenge is using inexpensive and readily available oxidants. Coupling between two C–H bonds could ideally produce hydrogen gas as a by-product; however, that is often thermodynamically unfavorable, thus, requiring an external oxidant to drive the reaction forward.[6] Finally, controlling the selectively of C–H activation has proved difficult. Unwanted dimer by-products often plague CDC reactions, and finding conditions that minimize homocoupling is difficult. Regioselectivity is the other problem due to the omnipresence of C–H bonds. Often this is resolved by the use of a pre-installed directing group that can co-ordinate with a metal catalyst to facilitate insertion into a nearby C–H bond.

Despite the difficulty of biaryl synthesis *via* CDC, much progress has been made over the past century, with most of the advances being realized in the past decade. This chapter will focus on some of the exciting highlights and recent advancements of these green reactions

6.2   Palladium-Catalyzed CDC Systems

Although many different metal catalysts have been used in oxidative coupling reactions, the most common ones are late transition metals, including Pd, Rh, Ru and Cu.[2] Undoubtedly, Pd is used in the majority of examples, probably due to its high versatility and activity.[7] The first biaryl CDC reaction catalyzed by Pd was reported by van Helden and Verberg, in which biphenyl was formed in high yields.[8] This initiated a wave of studies using Pd catalysts in CDC reactions.

Because Pd(0) has the potential to easily form a colloid and fall out of solution, a ligand or solvent system that can co-ordinate Pd(0) can be used to ameliorate this problem.[9] However, the phosphine ligands that are commonly employed in traditional cross-coupling chemistry are not stable in the oxidizing environments required for most CDC reactions, so simple palladium salts are often used, such as $Pd(OAc)_2$ or $PdCl_2$.[7] Hartwig and co-workers observed that ligand-less anionic palladium species more readily cleave C–H bonds compared to hindered phosphine-ligated palladium catalysts.[10] Thus providing evidence as to why most Pd-catalyzed CDC reactions work efficiently without ligands. Pd-Catalyzed CDC reactions typically run in an acidic medium for up to several days[11] at a temperature over 100 °C. An external oxidant is needed in these reactions. The greenest option is to use O_2; though, other less green alternatives including peroxysulfates and metal salts of Cu(II) or Ag(I) are more common.

6.2.1   Molecular Oxygen as an Oxidant and Ligand-Controlled Regioselectivity

Several groups have successfully utilized molecular oxygen as the sole oxidant in aerobic Pd-catalyzed biaryl CDC reactions. These reactions are probably the greenest CDC reactions that exist right now due to the fact that using O_2 or air as the oxidant is inexpensive and makes water as a final by-product. Most of these O_2-oxidized reactions take place in an acidic environment and at temperatures ranging anywhere from 55–120 °C.

In 1973, Yoshimoto and Itatani successfully carried out a biaryl Pd-catalyzed CDC reaction with pressurized O_2 as the oxidant.[12] They coupled biphenylethers intramolecularly, but also got an equal amount of intermolecular dimers as side-products. The experimental conditions were harsh (50 kg cm^{-2} of 1 : 1 oxygen and nitrogen and heated to 150 °C) and were not regioselective. In 1999, Akermark and co-workers published a Pd catalyzed CDC reaction that intramolecularly coupled arylaminoquinones.[13] This reaction worked in only 1 atm. of O_2; however, when coupling biphenylethers or biphenylamines, the reaction required additional Sn(II) acetate additives to prevent bulk Pd from falling out of solution.

Later, the groups of Fagnou, Ohno and Kandekar developed improved intramolecular aerobic CDC examples in which all experimental conditions included a $Pd(OAc)_2$ catalyst, a base, pivalic acid (PivOH) or acetic acid, and oxygen (Scheme 6.2). These reactions were successful on biphenylethers[14] and biphenylamines[14,15] with up to 99% yield and *N*-benzoylindoles[14,16] and *N*-benzoylpyrroles[17] with up to 92% yield.

Several examples of aerobic intermolecular coupling *via* CDC of aryls have also been published, using similar conditions.[18,19] Perhaps the most interesting example to date was by Stahl and co-workers in 2011. They aerobically coupled indoles and arenes *via* intermolecular CDC using Pd catalysts and diazafluorene ligands.[20] While optimizing the reaction, they found that both the anionic ligands in the Pd(II) salts and the neutral diazafluorene ligands could control the regioselectivity (Scheme 6.3). The combination of

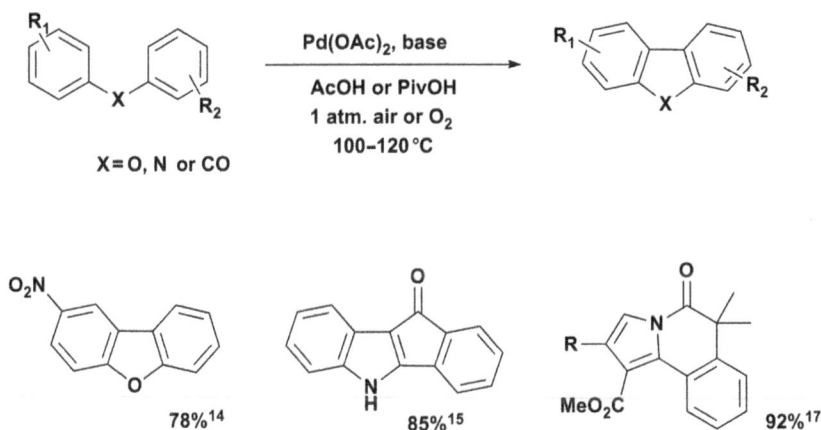

Scheme 6.2   Pd-Catalyzed aerobic intramolecular CDC reactions.[14–17]

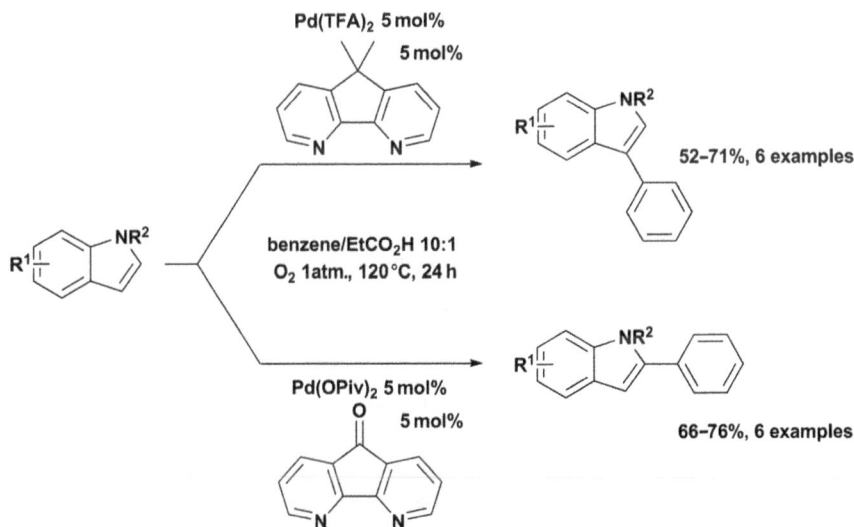

Scheme 6.3   Arylation of indoles *via* CDC with ligand-controlled regioselectivity.[20]

the pivalate anion with a diazafluorenone ligand was selective for C2 coupling, whereas the combination of the trifluoroacetate anion with a dimethyldiazafluorene ligand was selective for C3 coupling. In both catalyst systems the presence of electron-donating groups (EDGs) deteriorated the regioselectivity; however the presence of electron-withdrawing groups (EWGs) improved the regioselectivity with the C2-selective catalyst system.

6.2.2   Oxidant-Controlled Regioselectivity

Another aspect that has been shown to readily influence site selectivity in CDC reactions is the choice of oxidant (Scheme 6.4).

In 2007, DeBoef and co-workers showed that regioselectivity could be manipulated in the arylation of benzofurans with different oxidants.[21,22] They found upon examining the reaction in acidic conditions, that CDC reactions employing the heteropolyacid $H_4PMo_{11}VO_{40}$ (HPMV) gave almost exclusive C2-selectivity. The use of $Cu(OAc)_2$ also gave moderate C2-selectivity, and the use of AgOAc gave no selectivity. That same year Fagnou and

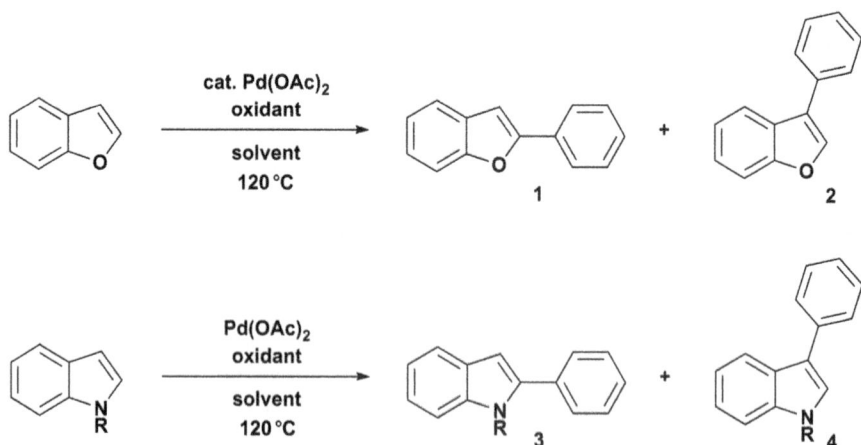

	Benzofuran 1:2	Indole 3:4
HPMV / acid	99:1	decomp.
Cu(OAc)$_2$ / acid	40:1	1:8.9
Ag(OAc) / acid	non-selective	8.7:1
Cu(OAc)$_2$ / dioxane	1:6	1:4
Ag(OAc) / dioxane	1:6	3.6:1

Scheme 6.4   Scope of oxidant-controlled regioselectivity.[21–27]

co-workers showed that, when arylating indoles, regioselectivity could be controlled by using different oxidants.[23,24] They found that using $Cu(OAc)_2$ gave C3-selectivity and using AgOAc gave C2-selectivity.

Further studies have come from the DeBoef group on oxidant-controlled regioselectivity, in which both benzofuran and indole arylations were examined in different solvent parameters.[25-27] They surmised that solvent had little overall effect on regioselectivity compared to oxidant choice,[27] and both DeBoef and Fagnou reasoned that site selectivity is most likely due to the formation of polymetallic clusters during catalysis.[23-25]

6.2.3 Catalytic Oxidants with O_2 as a Terminal Oxidant

Using O_2 as the only oxidant is, to date, the greenest condition for the synthesis of biaryls *via* CDC; however, several studies have shown that catalytic amounts of a metal oxidant can be used along with O_2 as a terminal oxidant (Scheme 6.5).[12,22,28-30]

These "co-catalyst" oxidants are sometimes needed to help prevent bulk Pd from falling out of solution. They are often strong oxidants that can readily re-oxidize Pd(0) back to Pd(II) before it falls out of solution.[13] Transition metal salts, such as $Sn(OAc)_2$,[13] Ag_2O,[29] $Cu(OTf)_2$[30] and $Mn(OAc)_3 \cdot H_2O$,[31] have been used, but perhaps one of the most unique examples was by DeBoef and co-workers in which they used HPMV as a co-catalyst oxidant to afford high yields in the arylation of benzofurans *via* CDC reactions (Scheme 6.6).[22]

Heteropolyacids have been shown to be quite efficient as catalytic oxidants due to their highly reversible oxidizability.[31] During DeBoef's study of the arylation of benzofurans, these oxidants contributed to high yields, and would tolerate substrates containing EDGs on the aryl moiety, which increased the reaction rate. Alternatively, the presence of EWGs shut down the reaction. Subsequent mechanistic studies suggest that HPMV/O_2-oxidized reactions proceed *via* a Pd(II)/Pd(IV) mechanism.[27]

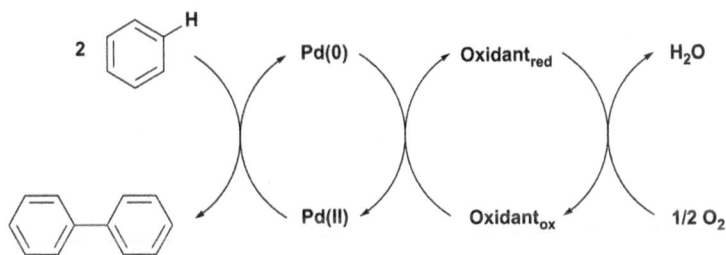

Scheme 6.5 Redox cycle for a Pd-catalyzed CDC with a catalytic metal oxidant and O_2 as a terminal oxidant.

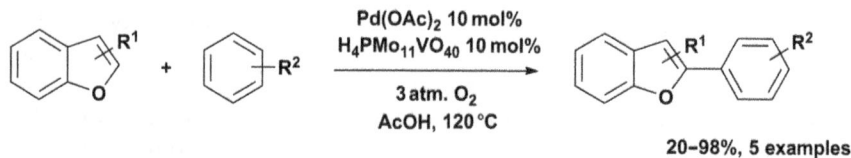

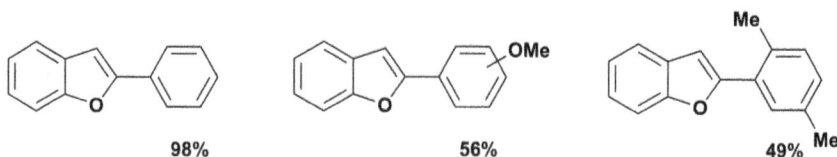

Scheme 6.6 Aerobic arylation of benzofurans *via* Pd-catalyzed CDC with HPMV as the oxidant.[22]

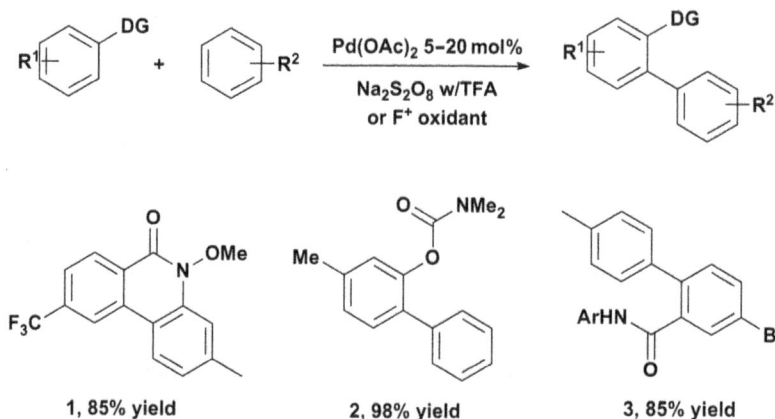

Scheme 6.7 Pd-Catalyzed CDC reactions utilizing organic peroxide and F⁺ oxidants.[32–39]

6.2.4 Organic Peroxide and F⁺ Oxidants

Another green oxidant option for Pd-catalyzed biaryl CDC reactions is to use organic peroxides or F⁺ oxidants such as sodium persulfate ($Na_2S_2O_8$) or *N*-fluorobenzenesulfonamide (NFSI) (Scheme 6.7).

All of these are strong oxidants and are simultaneously inexpensive and environmentally friendly, since they do not contain transition metals.[32] Given that these peroxides and F⁺ reagents are even stronger oxidants than most metal salts, it has been proposed that they are capable of causing the CDC to go through a Pd(II)/Pd(IV) catalytic cycle rather than the usual Pd(0)/Pd(II) cycle.[34–36] A directing group (DG in Scheme 6.7) is used in most Pd-catalyzed biaryl CDC reactions with organic oxidants, which is most likely to help control site selectivity.

The oxidant, sodium persulfate, has been used in several Pd-catalyzed biaryl CDC reactions by such groups as Dong, Cheng and Lu, all of which also used a strong acid, such as trifluoroacetic acid, TFA.[32–34,37–39] Strong acids are proposed to aid metalation of the C–H bond by increasing the electrophilicity of the Pd catalyst.[32] These biaryl CDC reactions oxidized by sodium or potassium persulfate couple simple arenes (several with directing groups), and can even be used to make phenanthridinone derivatives (*e.g.*, **1**, Scheme 6.7).[33] The Dong group has also shown that using sodium persulfate is an effective oxidant in the CDC of arenes with *O*-pheynylcarbamates.[32] Their coupling reaction achieved up to 98% yield and worked on a large library of substrates (*e.g.*, **2**, Scheme 6.7). Finally, exciting new work has come out of the labs of Yu and Seayad utilizing F^+ oxidants.[35,36] These F^+ oxidants were used along with Pd catalysts and ligand sources to couple simple arenes[35] or arylate furans[36] in yields up to 85% (*e.g.*, **3**, Scheme 6.7).

6.2.5 Stoichiometric Metal Oxidants

The most widely used conditions for biaryl CDC reactions require stoichiometric amounts of metal oxidants. Although stoichiometric metal oxidants make them the least green of all the Pd-catalyzed CDC reactions, they are still attractive due to the fact that the starting materials do not need pre-functionalization. The most frequently used metal oxidants are Ag(I) and Cu(II) salts. Both have been shown to work readily in the inter- and intramolecular coupling of simple arenes and heterocycles, and as seen in Section 6.2.2, DeBoef and Fagnou have even shown that metal oxidants influence regioselectivity.[21–27]

Silver salts have been used extensively as oxidants in many different types of biaryl CDC reactions, such as the arylation of indoles,[23,24,40–42] azoles,[43] thiophenes,[44] chromones,[45,46] xanthines,[47] uracils,[48] and pyridine-*N*-oxides.[41,42,49] They are also used in the inter- and intramolecular coupling of simple arenes.[50–55] Recently, Zhang and co-workers oxidatively coupled thiophene derivatives and fluorinated arenes, using only 2.5 mol% of Pd(OAc)$_2$ as a catalyst and AgCO$_3$ as an oxidant (Scheme 6.8).[44]

Their reaction worked on a large library of thiophene substrates, and was even shown to arylate benzofurans and indoles in the same conditions. The reaction also tolerated a vast array of functionalities including, amides, esters, aldehydes and olefins. These perfluoroarene–thiophene compounds are being used in various electronic materials including diodes and transistors,[44] making Zhang's reaction a superior way to produce these materials.

Another interesting Pd-catalyzed CDC reaction with a Ag oxidant was reported by the Sanford group in which they intermolecularly coupled arenes in high yields and high regioselectivity (Scheme 6.9).[55]

Intermolecular coupling between two non-heterocyclic arenes is less common because heterocycles are easier to regioselectively functionalize due to the intrinsic electronic differences between their C–H bonds.[55] The

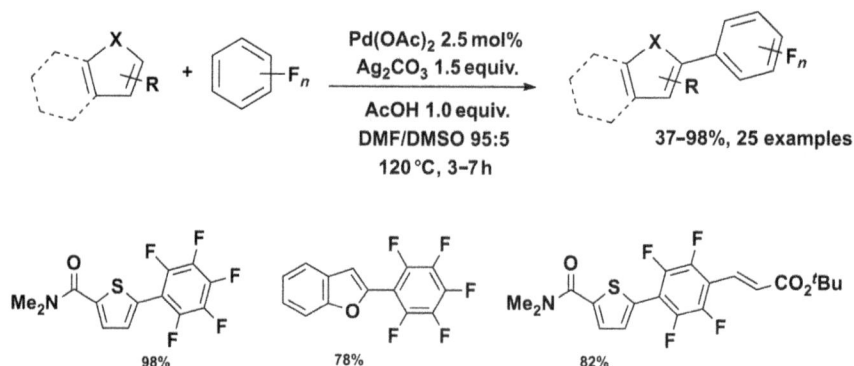

Scheme 6.8 Pd-Catalyzed CDC of thiophenes and fluorinated arenes with a Ag(i) oxidant.[44]

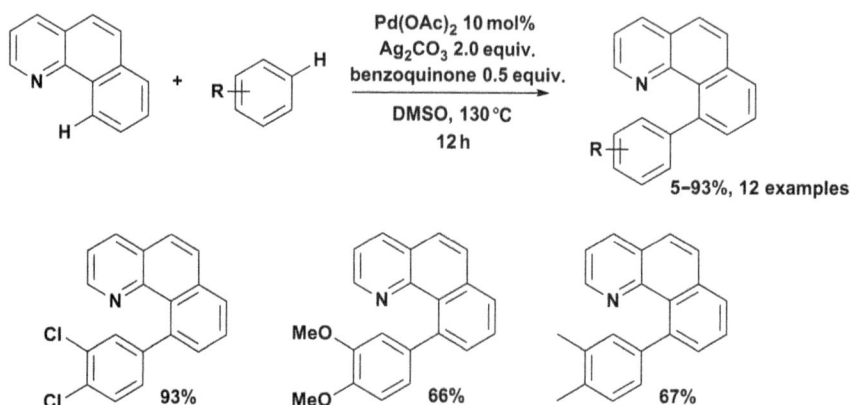

Scheme 6.9 Pd-Catalyzed CDC of benzoquinoline and arenes with a Ag(i) oxidant.[55]

Sanford group solved this problem by using benzoquinoline as one of the coupling partners. The substrate's basic nitrogen served as a directing group that co-ordinated Pd and, thus, controlled the site selectivity. They were able to achieve moderate-to-high yields, even in the presence of EWGs and EDGs on the arenes.

Copper salts are commonly used as oxidants. They are relatively inexpensive, and have been shown to work in a wide array of Pd-catalyzed CDC reactions such as the arylation of heterocycles,[22–25,27,56,57] and the intermolecular coupling of arenes.[58] However in biaryl synthesis, they are chiefly utilized in intramolecular couplings[22,59–64] and the cross-coupling of heterocycles.[64–67]

In 2011, Greaney and co-workers used a Cu(OAc)$_2$ oxidant with a Pd(OAc)$_2$ catalyst for an intramolecular CDC reaction that created medium-sized, biaryl-containing rings (Scheme 6.10).[62]

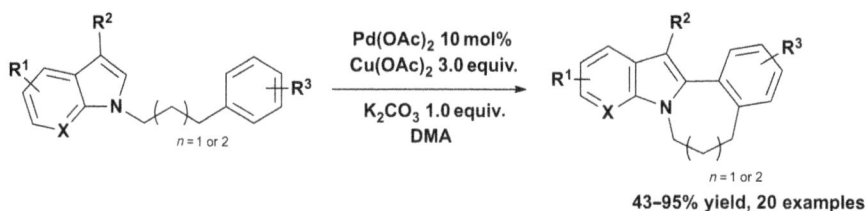

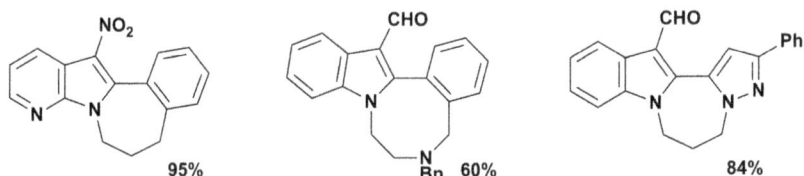

Scheme 6.10 Medium-sized ring synthesis *via* intramolecular Pd-catalyzed CDC with a Cu(II) oxidant.[62]

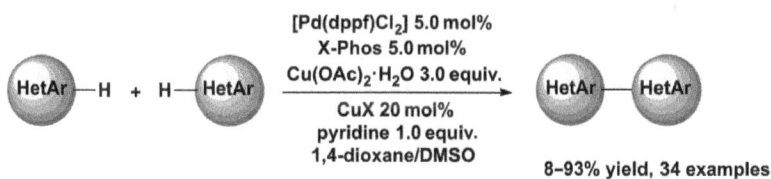

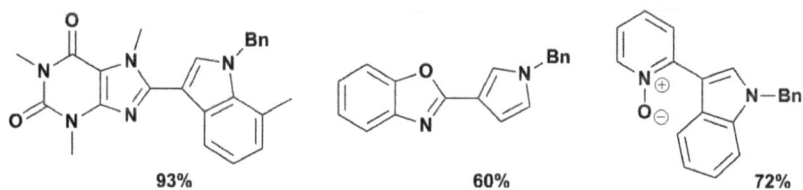

Scheme 6.11 Pd-Catalyzed cross-coupling of heterocycles with a Cu(II) oxidant.[67]

This reaction was unique in that it formed seven- and eight-membered rings, instead of the usual aromatic five-membered rings, as in the formation of carbozoles.[15,30,51,60] Medium-size rings occur naturally in many biologically active compounds, and are generally more difficult to synthesize than small rings and large macrocycles, making methods for their synthesis highly desirable.[62] Greaney's group was able to achieve the synthesis of these rings and incorporate indole into their products; indole is a common biologically active moiety. The reaction gave high yields, and worked with several other heterocycle coupling partners, like azoles. It also had an exceptional tolerance for amine and ether functionalities within the newly formed rings.

Another interesting Pd-mediated reaction with a Cu(II) oxidant comes from a study by the You group in 2011, in which they were able to cross-couple heterocycles in high yields with good regioselectivity (Scheme 6.11).[67]

The You group used a [Pd(1,1'-bis(diphenylphosphino)ferrocene (dppf))Cl₂] catalyst and a Cu(OAc)₂ · H₂O oxidant system. They also added a catalytic amount of the 2-dicyclohexylphosphino-2',4',6'-triisopropylbiphenyl (XPhos) ligand, to prevent decomposition of the indole derivatives, and a CuCl additive, that further helped control the regioselectivity. The reaction conditions were found to couple a wide array of heterocycles including, xanthines, pyridine-*N*-oxides, azoles, indoles and pyrroles in good-to-excellent yields.

6.3 Rhodium- and Ruthenium-Catalyzed CDC Systems

Catalysts based on rhodium and ruthenium are becoming increasingly popular for the synthesis of biaryls *via* CDC reactions. Palladium catalysts, although they exhibit good functional group tolerance and require mild reaction conditions, often need 10–20 mol% catalyst loading and a gross excess of one of the coupling substrates for a reaction to reach completion.[68] Rhodium and ruthenium, on the other hand, have many of the same catalytic properties as Pd,[68] but usually require less catalyst loading (1–5 mol%) and have potentially novel selectivity properties.[69] Newfound interest in these metals as catalysts for biaryl CDC has been growing since a study on Ru-catalyzed dehydrogenative coupling of 4-pyridines by Suzuki and co-workers in 2007.[70]

Rhodium-catalyzed oxidative biaryl coupling is still in its infancy, given that it first started appearing in independent studies by Glorius, Miura and You in 2012. All of these studies used either Rh(I) or Rh(III) catalysts and Cu(II) oxidants to carry out arylations of heterocycles,[68] intramolecular couplings[71] or dual heterocycle couplings.[69,72]

One example by Glorius and co-workers showed that a Rh(III) catalyst in the presence of a AgSbF₆ co-catalyst, could successfully couple two arenes with the use of a tertiary benzamide directing group (Scheme 6.12).[68]

It was suggested that the AgSbF₆ was involved in generating an electrophilic Rh(III) species as the active catalyst. This reaction was remarkable in that it took only 5 mol% catalyst loading and achieved high yields and regioselectivity. Both EDGs and EWGs were tried in different positions on the arenes resulting in moderate-to-high yields with close to a 3 : 1 *meta*/*para*-regioselectivity in most cases. It is also interesting to note that the reaction did not couple to the *ortho*-position of bromobenzene; making it possible to exclusively synthesize *meta*-substituted biaryls using 1,3-haloarenes.

Ruthenium has been used in several oxidative homocoupling reactions;[70,73,74] however, its use in oxidative cross-coupling is brand new. The You group recently published the initial findings of a novel Ru-catalyzed biaryl CDC, wherein 2-methylthiophene was successfully coupled to both benzoquinoline and quinoxaline motifs (Scheme 6.13).[72]

This reaction, although not extensively studied yet, exhibited moderate yields with high regioselectivity.

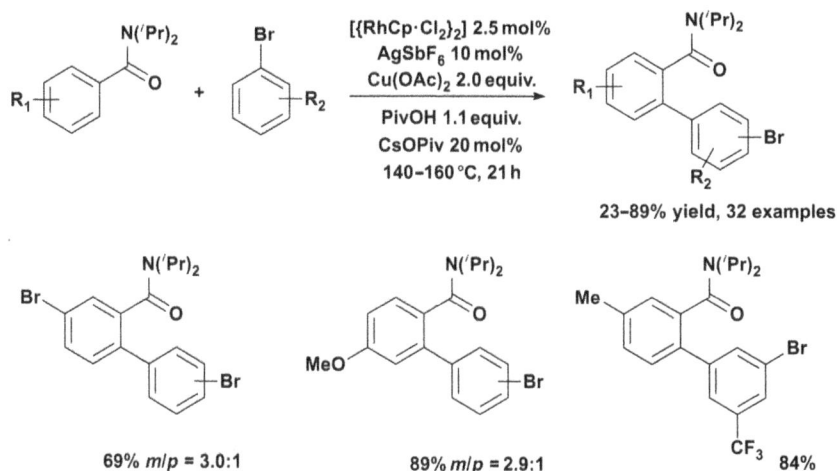

Scheme 6.12 Rh-Catalyzed biaryl CDC with a tertiary benzamide directing group.[68]

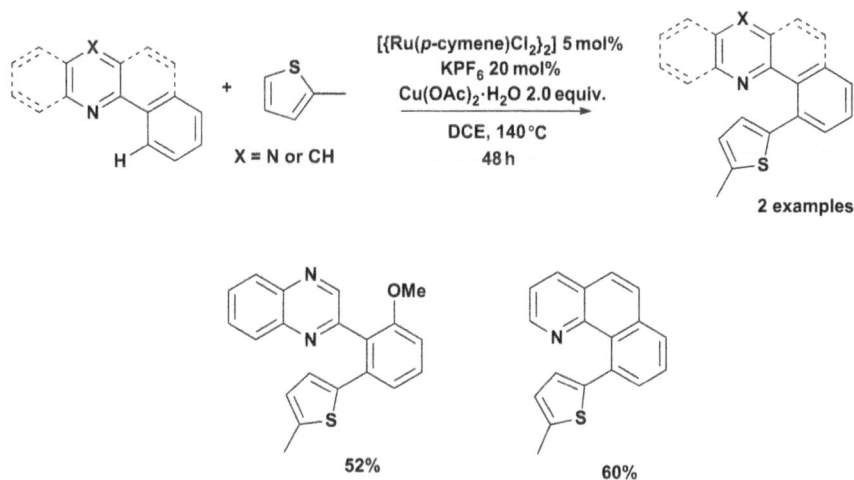

Scheme 6.13 Ru-Catalyzed CDC of methylthiophene with benzoquinolines and quinoxalines. DCE = 1,2-dichloroethane.[72]

6.4 Copper-Mediated CDC Systems

Palladium is used in the majority of biaryl CDC reactions; however copper would be an ideal economical alternative since it is abundant and relatively inexpensive.[75] It has seen historical use in oxidative naphthyl homocouplings to synthesize chiral ligands like BINOL (1,1'-bi-2-naphthol).[76–78] Although, due to the radical nature of the reaction, the heterocoupling of naphthyls has proved to be a difficult task since radical–radical recombination produces a

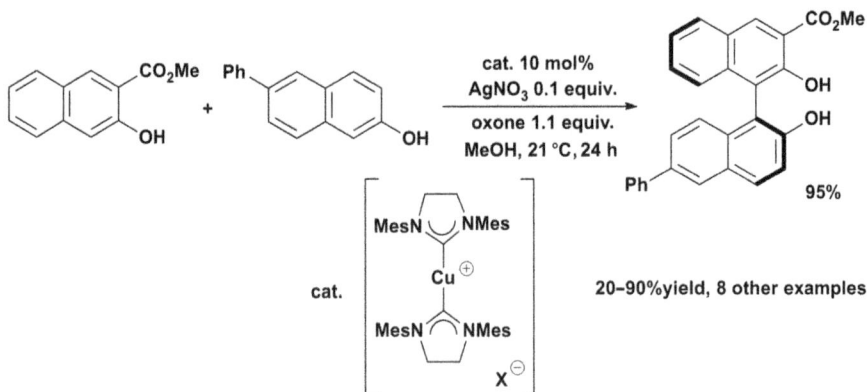

Scheme 6.14 CDC of naphthyl derivatives by an *N*-heterocyclic carbene–Cu(II) complex.[79]

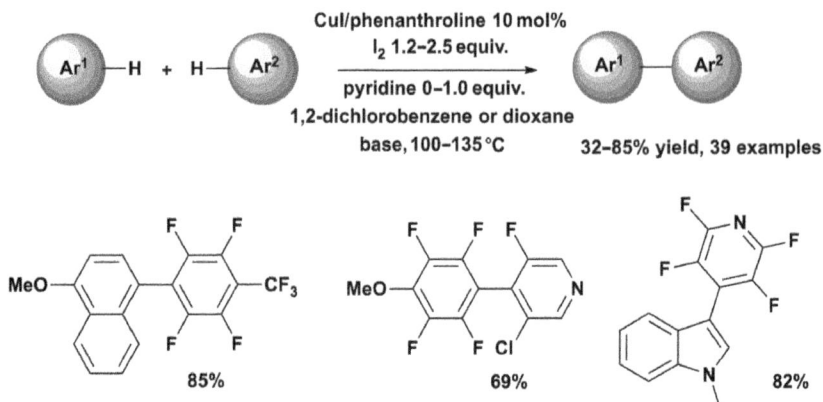

Scheme 6.15 Daugulis' conditions for iodination and Cu-directed arylation using a CuI/I$_2$ system.[85]

majority of homocoupled products.[79] Several groups have found solutions to this,[79–83] including Collins and co-workers, who used NHC (*N*-heterocyclic carbene) ligands in a bis-NHC–Cu(II) complex to catalyze the cross-coupling of naphthyl derivatives *via* a CDC (Scheme 6.14).[79]

The presence of the NHC ligands greatly effects the electronic properties of the Cu(II) making it possible to cross-couple naphthyls. The CDC was believed to proceed *via* a radical-addition mechanism.

All other Cu-mediated CDC reactions fall into two groups: Daugulis' iodination conditions,[84,85] and Miura's CDC arylation of heterocycles.[75] Daugulis' conditions include the presence of a Cu(I) salt and I$_2$ (Scheme 6.15).[85]

It is known to proceed through a one-pot iodination and then Cu-mediated direct arylation. The fact that the reaction is pre-functionalizing one of the C–H bonds perhaps makes it not a true CDC; however, the reaction requires no initial functionalization steps beforehand, giving it the outward appearance of a CDC. Daugulis and co-workers have used their CuI/I$_2$ conditions for the

32–74% yield, 19 examples

52%   69%   71%

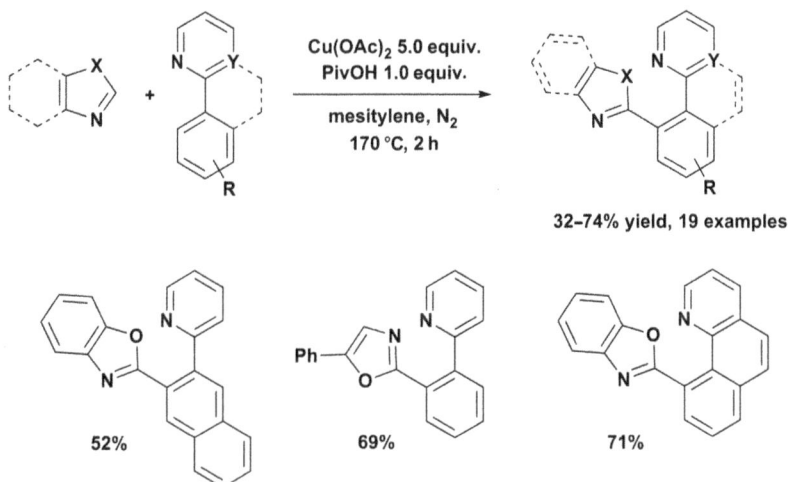

Scheme 6.16 Cu-Catalyzed CDC of pyridyl arenes and azoles.[75]

coupling of a vast array of heterocycles as well as both electron-rich and electron-poor substrates, making this a highly versatile reaction.

The final category of Cu-mediated biaryl CDC is the work by Miura and co-workers in 2011. Miura was able to cross-couple arenes with a variety of azoles using excess $Cu(OAc)_2$ (Scheme 6.16).[75]

The reaction gave moderate yields, and the use of a directing group on the arenes provided high regioselectivity. Even though this reaction used excess $Cu(OAc)_2$, it is still desirable because copper is more economical than other catalytic metals like Pd. Daugulis[86] and Mori[87] have successfully shown that catalytic Cu can be used in aryl oxidative dimerizations. Therefore, the goal of using Cu in catalytic amounts for biaryl CDC, aside from radical naphthyl or iodination coupling, seems promising.

6.5 Other CDC Systems

To date there are only a few other examples of CDC systems used for biaryl coupling. The majority of them fall under single-electron-transfer (SET) oxidants like hypervalent iodine reagents. Kita and co-workers have monopolized the field of using hypervalent iodine in biaryl CDC reactions.[88–92]

Hypervalent iodine oxidants are known to be safe,[88] mild oxidants[89] that can readily form aromatic cation radicals with electron-rich arenes through SET.[90] They usually require low temperatures and short reaction times. Unwanted dimerization can often plague these aryl CDC reactions, so the group of Kita typically uses one coupling partner like naphthalene, which is electron rich and prone to forming an aromatic cation radical with a hypervalent iodine oxidant, and a second coupling partner that is more nucleophilic so as to cause a nucleophilic attack on the radical and subsequent cross-coupling.[90] They have also shown that using hypervalent iodine reagents to couple thiophenes can form intermediate thienyl iodonium(III) salts (Scheme 6.17).[89] The formation of these intermediate salts effectively eliminates dimerization and gives excellent regioselectivity. The reaction tolerated a large library of functionalized arenes and hetero-cycles and gave moderate-to-high yields.

A few remaining CDC systems include V(v) catalytic systems,[93] and Tl(III) catalytic systems;[94,95] both of which have been used in intramolecular aryl coupling *via* free radicals, although in low yields and under harsh reaction conditions. Auto-oxidizing CDC conditions developed by Bergman are also of interest, in which one of the coupling partners, quinoline, serves as the sole oxidant as it is coupled to indole or pyrrole moieties.[96] Finally, the use of iron in biaryl CDCs is also being developed.[97–99] Katsuki and co-workers published a study in 2011 on an aerobic Fe-catalyzed cross-coupling of naphthol derivatives to make unique BINOL ligands (Scheme 6.18).[97]

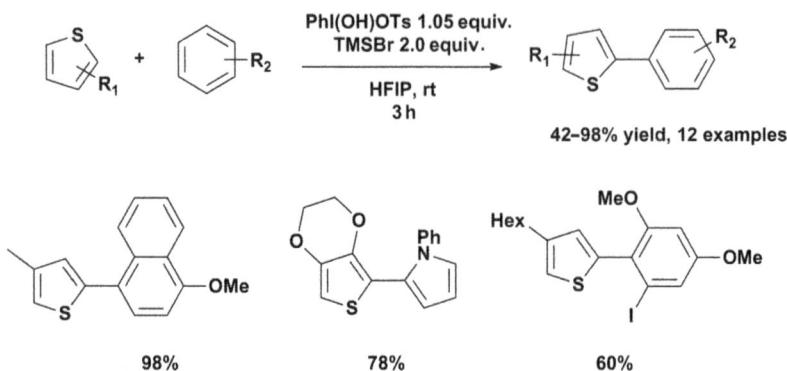

Scheme 6.17   Hypervalent iodine-mediated CDC of thiophenes with arenes and heterocycles.   TMSBr = bromotrimethylsilane   and   HFIP = hexafluoroisopropanol.[89]

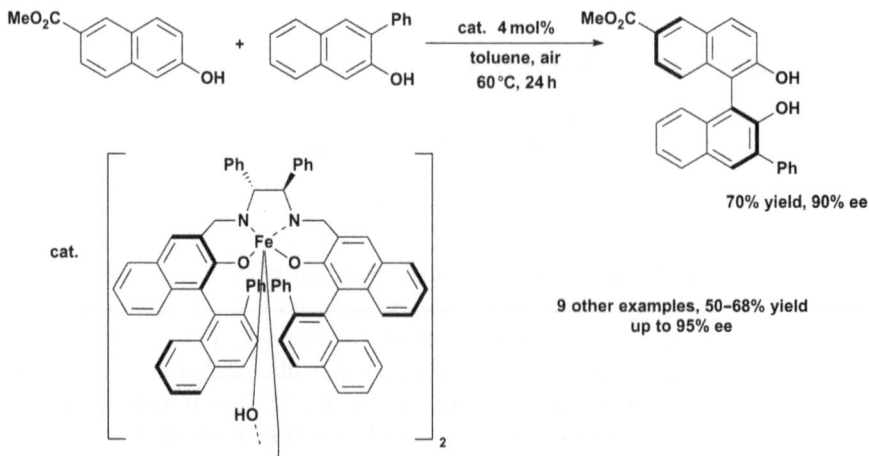

Scheme 6.18   Aerobic Fe-catalyzed CDC of naphthols.[97]

This reaction was quite green since it used an iron catalyst, which is non-toxic and inexpensive, and only a 4 mol% catalyst loading. Also, the oxidant was molecular oxygen from the air, giving off water as a by-product. The reaction was suggested to go through a radical-addition, and produced moderate yields and high enantioselectivity on a wide range of substrates. This makes it a desirable new method for synthesizing enantiomeric BINOL ligands.

6.6 Outlook

Future efforts in the CDC of arenes will perhaps include enzyme-mediated reactions, since enzymes have low toxicity, high catalytic turnovers, and are able to run in very mild conditions. In 2006,[100] d'Ischia and co-workers used peroxidase enzymes to catalyze the oxidative dimerization of indole derivatives. This technology, albeit low yielding, appears to be an excellent lead to optimize cross-dehydrogenative-coupling.

References

1. D. J. C. Constable, P. J. Dunn, J. D. Hayler, G. R. Humphrey, J. L. Leazer Jr, R. J. Linderman, K. Lorenz, J. Manley, B. Pearlman, A. Wells, A. Zaks and T. Y. Zhang, *Green Chem.*, 2007, **9**, 411.
2. S. H. Cho, J. Y. Kim, J. Kwak and S. Chang, *Chem. Soc. Rev.*, 2011, **40**, 5068.
3. T. W. Lyons and M. S. Sanford, *Chem. Rev.*, 2010, **110**, 1147.
4. R. Scholl and C. Seer, *Justus Liebigs Ann. Chem.*, 1912, **394**, 111.
5. B. T. King, J. Kroulík, J. R. Robertson, P. Rempala, C. L. Hilton, J. D. Korinek and L. M. Gortari, *J. Org. Chem.*, 2007, **72**, 2279.
6. C. S. Yeung and V. M. Dong, *Chem. Rev.*, 2011, **111**, 1215.
7. S. Stahl, *Angew. Chem., Int. Ed.*, 2004, **43**, 3400.
8. R. van Helden and G. Verberg, *Recl. Trav. Chim. Pays-Bas.*, 1965, **84**, 1263.
9. P. M. Henry, *Palladium Catalyzed Oxidation of Hydrocarbons*, ed. R. D. Ugo, Reidel Publishing Company, Dordrecht, 1980, ch. 1, p. 2.
10. Y. Tan and J. F. Hartwig, *J. Am. Chem. Soc.*, 2011, **133**, 3308.
11. C. Liu, H. Zhang, W. Shi and A. Leo, *Chem. Rev.*, 2011, **111**, 1780.
12. H. Yoshimoto and H. Itatani, *Bull. Chem. Soc. Jpn.*, 1973, **46**, 2490.
13. H. Hagelin, J. D. Oslob and B. Akermark, *Chem.–Eur. J*, 1999, **5**, 2413.
14. B. Liegault, D. Lee, M. P. Huestis, D. R. Stuart and K. Fagnou, *J. Org. Chem.*, 2008, **73**, 5022.
15. W. Toshiaki, S. Oishi, N. Fujii and H. Ohno, *J. Org. Chem.*, 2009, **74**, 4720.
16. S. K. Guchhait and S. Kandekar, *Tetrahedron Lett.*, 2012, **53**, 3919.
17. B. Liegault and K. Fagnou, *Organometallics*, 2008, **27**, 4841.
18. L. Zhou and W. Lu, *Organometallics*, 2012, **31**, 2124.
19. G. Brasche, J. Garcia-Fortanct and S. L. Buchwald, *Org. Lett.*, 2008, **10**, 2207.
20. A. Campbell, E. Meyer and S. S. Stahl, *Chem. Commun.*, 2011, **47**, 10257.

21. B. DeBoef, *U.S. Pat.*, 60/867, 641, 2006.
22. T. A. Dwight, N. R. Rue, D. Charyk, R. Josselyn and B. DeBoef, *Org. Lett.*, 2007, **9**, 3137.
23. D. R. Stuart, E. Villemure and K. Fagnou, *J. Am. Chem. Soc.*, 2007, **129**, 12072.
24. D. R. Stuart and K. Fagnou, *Science*, 2007, **316**, 1172.
25. S. Potavathri, A. S. Dumas, T. A. Dwight, G. R. Naumiec, J. M. Hammann and B. DeBoef, *Tetrahedron Lett.*, 2008, **49**, 4050.
26. S. Potavathri, K. Pereira, S. Gorelsky, A. Pike, A. LeBris and B. DeBoef, *J. Am. Chem. Soc.*, 2010, **132**, 14676.
27. K. C. Pereira, A. L. Porter, S. Potavathri, A. LeBris and B. DeBoef, *Tetrahedron*, 2013, **69**, 4429.
28. C.-Y. He, Q.-Q. Min and X. Zhang, *Organometallics*, 2012, **31**, 1335.
29. B.-J. Li, S.-L. Tian, Z. Fang and Z.-J. Shi, *Angew. Chem., Int. Ed.*, 2008, **47**, 1115.
30. H.-J. Knolker, *Chem. Lett.*, 2009, **38**, 8.
31. S. H. Lee, K. H. Lee, J. S. Lee, J. D. Jung and J. S. Shim, *J. Mol. Catal. A: Chem.*, 1997, **115**, 241.
32. X. Zhao, C. S. Yeung and V. M. Dong, *J. Am. Chem. Soc.*, 2010, **132**, 5837.
33. J. Karthikeyan and C.-H. Cheng, *Angew. Chem., Int. Ed.*, 2011, **50**, 9880.
34. V. Thirunavukkarasu and C.-H. Cheng, *Chem.–Eur. J.*, 2011, **17**, 14723.
35. X. Wang, D. Leow and J.-Q. Yu, *J. Am. Chem. Soc.*, 2011, **133**, 13864.
36. N. A. B. Juwainin, J. K. P. Ng and J. Seayad, *ACS Catal.*, 2012, **2**, 1787.
37. C. S. Yeung, X. Zhao, N. Borduas and V. M. Dong, *Chem. Sci.*, 2010, **1**, 331.
38. Y. Rong, R. Li and W. Lu, *Organometallics*, 2007, **26**, 4376.
39. R. Li, L. Jiang and W. Lu, *Organometallics*, 2006, **25**, 5973.
40. S. Potavathri, A. Kantak and B. DeBoef, *Chem. Commun.*, 2011, **47**, 4679.
41. X. Gong, G. Song, H. Zhang and X. Li, *Org. Lett.*, 2011, **13**, 1766.
42. A. D. Yamaguchi, D. Mandal, J. Yamaguchi and K. Itami, *Chem. Lett.*, 2011, **40**, 555.
43. Z. Li, L. Ma, J. Xu, L. Kong, X. Wu and H. Yao, *Chem. Commun.*, 2012, **48**, 3763.
44. C.-Y. He, S. Fan and X. Zhang, *J. Am. Chem. Soc.*, 2010, **132**, 12850.
45. K. H. Kim, H. S. Lee, S. H. Kim and J. N. Kim, *Tetrahedron Lett.*, 2012, **53**, 2761.
46. F. Chen, Z. Feng, C.-Y. He, H.-Y. Wang, Y.-L. Guo and X. Zhang, *Org Lett.*, 2012, **14**, 1176.
47. C. C. Malakar, D. Schmidt, J. Conrad and U. Beifuss, *Org. Lett.*, 2011, **13**, 1378.
48. K. H. Kim, H. S. Lee and J. N. Kim, *Tetrahedron Lett.*, 2011, **52**, 6228.
49. S. H. Cho, S. J. Hwang and S. Chang, *J. Am. Chem. Soc.*, 2008, **130**, 9254.
50. H. Li, R.-Y. Zhi, W.-J. Shi, K.-H. He and Z.-J. Shi, *Org. Lett.*, 2012, **14**, 4850.
51. S. Wang, H. Mao, Z. Ni and Y. Pan, *Tetrahedron Lett.*, 2012, **53**, 505.
52. R. Samanta and A. P. Antonchick, *Angew. Chem., Int. Ed.*, 2011, **50**, 5217.
53. H. Li, J. Liu, C.-L. Sun, B.-J. Li and Z.-J. Shi, *Org. Lett.*, 2011, **13**, 276.
54. M. Yu, Z. Liang, Y. Wang and Y. Zhang, *J. Org. Chem.*, 2011, **76**, 4987.

55. K. L. Hull and M. S. Sanford, *J. Am. Chem. Soc.*, 2007, **129**, 11904.
56. J.-B. Xia and S.-L. You, *Organometallics*, 2007, **26**, 4869.
57. G. Wu, J. Zhou, M. Zhang, P. Hu and W. Su, *Chem. Commun.*, 2012, **48**, 8964.
58. Y. Wei and W. Su, *J. Am. Chem. Soc.*, 2010, **132**, 16377.
59. L. Ackermann, R. Jeyachandran, H. Potukuchi, P. Novak and L. Buttner, *Org. Lett.*, 2010, **12**, 2056.
60. V. Sridharan, M. A. Martin and J. C. Menendez, *Eur. J. Org. Chem.*, 2009, **2009**, 4614.
61. M. Sun, H. Wu, J. Zheng and W. Bao, *Adv. Synth. Catal.*, 2012, **354**, 835.
62. D. G. Pintori and M. F. Greaney, *J. Am. Chem. Soc.*, 2011, **133**, 1209.
63. H.-J. Knolker and W. Frohner, *J. Chem. Soc., Perkin Trans. 1*, 1998, 173.
64. J. Dong, Y. Huang, X. Qin, Y. Cheng, J. Hao, D. Wan, W. Li, X. Liu and J. You, *Chem.–Eur. J*, 2012, **18**, 6158.
65. P. Xi, F. Yang, S. Qin, D. Zhao, J. Lan, G. Gao, C. Hu and J. You, *J. Am. Chem. Soc.*, 2010, **132**, 1822.
66. M. V. Varaksin, I. A. Utepova, O. N. Chupakhin and V. N. Charushin, *J. Org. Chem.*, 2012, **77**, 9087.
67. Z. Wang, K. Li, D. Zhao, J. Lan and J. You, *Angew. Chem., Int. Ed.*, 2011, **50**, 5365.
68. J. Wencel-Delord, C. Nimphius, F. W. Patureau and F. Glorius, *Angew. Chem., Int. Ed.*, 2012, **51**, 2247.
69. N. Kuhl, M. N. Hopkinson and F. Glorius, *Angew. Chem., Int. Ed.*, 2012, **51**, 8230.
70. T. Kawashima, T. Takao and H. Suzuki, *J. Am. Chem. Soc.*, 2007, **129**, 11006.
71. K. Morimoto, M. Itoh, K. Hirano, T. Satoh, Y. Shibata, K. Tanaka and M. Miura, *Angew. Chem., Int. Ed.*, 2012, **51**, 5359.
72. J. Dong, Z. Long, F. Song, N. We, Q. Guo, J. Lan and J. You, *Angew. Chem., Int. Ed.*, 2013, **52**, 580.
73. S. Oi, H. Sato, S. Sugawara and Y. Inoue, *Org. Lett.*, 2008, **10**, 1823.
74. X. Guo, G. Deng and C.-J. Li, *Adv. Synth. Catal.*, 2009, **351**, 2071.
75. M. Kitahara, N. Umeda, K. Hirano, T. Satoh and M. Miura, *J. Am. Chem. Soc.*, 2011, **133**, 2160.
76. B. Feringa and H. Wynberg, *Bioorg. Chem.*, 1978, **7**, 397.
77. J. Brussee and A. C. A. Jansen, *Tetrahedron Lett.*, 1983, **24**, 3261.
78. X. Li, B. Hewgley, C. Mulrooney, J. Yang and M. Kozlowshi, *J. Org. Chem.*, 2003, **68**, 5500.
79. A. Grandbois, M.-E. Mayer, M. Bedard, S. Collins and T. Michel, *Chem.–Eur. J*, 2009, **15**, 9655.
80. B. H. Lipshutz and Y.-J. Shin, *Tetrahedron Lett.*, 1998, **39**, 7017.
81. M. Smrcina, S. Vyskocil, B. Maca, M. Polasek, T. A. Claxton, A. P. Abbott and P. Kocovshy, *J. Org. Chem.*, 1994, **59**, 2156.
82. M. Smrcina, M. Lorenc, V. Hanus, P. Sedmera and P. Kocovsky, *J. Org. Chem.*, 1992, **57**, 1917.

83. M. Smrcina, M. Lorenc, V. Hanus and P. Kocovsky, *Synlett.*, 1991, 231.

84. H.-Q. Do and O. Daugulis, *Chem. Commun.*, 2009, **45**, 6433.

85. H.-Q. Do and O. Daugulis, *J. Am. Chem. Soc.*, 2011, **133**, 13577.

86. H.-Q. Do and O. Daugulis, *J. Am. Chem. Soc.*, 2009, **131**, 17052.

87. D. Monguchi, A. Yamamura, T. Fujiwara, T. Somete and A. Mori, *Tetrahedron Lett.*, 2010, **51**, 850.

88. Y. Kita, T. Takada, M. Gyoten, H. Tohma, M. Zenk and J. Eichhorn, *J. Org. Chem.*, 1996, **61**, 5857.

89. Y. Kita, K. Morimoto, M. Ito, C. Ogawa, A. Goto and T. Dohi, *J. Am. Chem. Soc.*, 2009, **131**, 1668.

90. T. Dohi, M. Ito, K. Morimoto, M. Iwata and Y. Kita, *Angew. Chem., Int. Ed.*, 2008, **47**, 1301.

91. Y. Kita, M. Gyoten, M. Ohtsubo, H. Tohma and T. Takada, *Chem. Commun.*, 1996, **32**, 1481.

92. T. Takada, M. Arisawa, M. Gyoten, R. Hamada, H. Tohma and Y. Kita, *J. Org. Chem.*, 1998, **63**, 7698.

93. S. M. Kupchan, A. J. Liepa, V. Kameswaran and R. F. Bryan, *J. Am. Chem. Soc.*, 1973, **95**, 6861.

94. E. C. Taylor, J. G. Andrade and A. McKillop, *Chem. Commun.*, 1977, 538.

95. E. C. Taylor, J. G. Andrade, G. J. H. Rall and A. McKillop, *J. Am. Chem. Soc.*, 1980, **21**, 6513.

96. M. Brasse, J. A. Ellman and R. G. Bergman, *Chem. Commun.*, 2011, **47**, 5019.

97. H. Egami, K. Matsumoto, T. Oguma, T. Kunisu and T. Katsuki, *J. Am. Chem. Soc.*, 2010, **132**, 1363.

98. K. Wang, M. Lu, A. Yu, X. Zhu and Q. Wang, *J. Org. Chem.*, 2009, **74**, 935.

99. K.-L. Wang, M.-Y. Lu, Q.-M. Wang and R.-Q. Huang, *Tetrahedron*, 2008, **64**, 7504.

100. A. Pezzella, L. Panzella, O. Crescenzi, A. Napolitano, S. Navaratman, R. Edge, E. J. Land, V. Barone and M. d' Ischia, *J. Am. Chem. Soc.*, 2006, **128**, 15490.

CHAPTER 7

Asymmetric Cross-Dehydrogenative-Coupling Reactions

YOSHITAKA HAMASHIMA[a] AND MIKIKO SODEOKA*[b]

[a] School of Pharmaceutical Sciences, University of Shizuoka, 52-1 Yada, Suruga-ku, Shizuoka 422-8526, Japan; [b] Synthetic Organic Chemistry Laboratory, RIKEN, 2-1 Hirosawa, Wako, Saitama 351-0198, Japan
*Email: sodeoka@riken.jp

7.1 Introduction

Cross-dehydrogenative-coupling (CDC) reactions have evolved into one of the most efficient and atom-economical methods to form carbon–carbon bonds directly from two different C–H bonds. Asymmetric versions of these reactions are highly desirable, since they would allow the concise synthesis of complex molecules from simple starting materials without any pre-functionalization of substrates prior to the coupling reaction. Indeed, the use of appropriate chiral catalysts in combination with oxidative activation of substrates with suitable oxidants, such as molecular oxygen, tert-butyl peroxide, hydrogen peroxide, tert-butyl perbenzoic acid, benzoquinone (BQ), and 2,3-dichloro-5,6-dicyanobenzoquinone (DDQ), is expected to afford optically active and highly functionalized molecules in a single step. However, oxidative C–C bond formation is generally thought to be difficult, because nucleophiles such as enolates and electron-rich arene compounds are susceptible to oxidative degradation. Additionally, the chiral catalysts

RSC Green Chemistry No. 26
From C–H to C–C Bonds: Cross-Dehydrogenative-Coupling
Edited by Chao-Jun Li

must be stable under the reaction conditions, and the use of appropriate oxidants, depending on the nature of the substrates, is essential.

Following the rapid progress in non-asymmetric CDC reactions, asymmetric versions have been gaining increasing attention.[1] To date, the number of successful examples is still limited, probably due to the problems mentioned above. Thus, further study is needed to improve the scope of the reactions. Nevertheless, high levels of asymmetric induction are already feasible in some cases. This chapter will focus on: (1) reactions of α-C–H bonds of nitrogen-containing compounds; (2) reactions of benzylic C–H bonds; (3) functionalization *via* oxidation of enamine and Breslow intermediates; (4) naphthol coupling reactions; and (5) coupling reactions of arenes with olefins (Fujiwara–Moritani reactions).

7.2 Asymmetric α-Functionalization of Amines

Since the pioneering seminal work by Li on oxidative α-functionalization of alkyl amine compounds,[2] non-asymmetric CDC reactions of alkyl amines with C-based nucleophiles have been a subject of great interest. It is desirable to make such transformations enantioselective, because optically active nitrogen-containing compounds with a stereogenic carbon centers at the α-position of the nitrogen atom are expected to be useful in the synthesis of natural and non-natural bioactive compounds. In most of the reported asymmetric CDC reactions, however, the substrates have been limited to particular compounds, such as *N*-protected tetrahydroisoquinolines (THIQs) and glycine derivatives. The iminium ions generated from them react with chiral C-based nucleophiles generated by reactions of pronucleophiles with chiral catalysts. These reactions can afford various chiral THIQs and amino acid derivatives in a highly enantioselective manner.

7.2.1 Reactions of Tetrahydroisoquinolines

C1-Substituted THIQs represent an important family of biologically active alkaloids, and have been a target for synthetic organic chemists for decades. In 2004, Li *et al.* reported a copper-catalyzed asymmetric alkynylation of *N*-aryl substituted THIQs (Scheme 7.1).[3] Among the chiral ligands and Cu salts examined, the combination of CuOTf with phenyl-substituted chiral Pybox was found to be the best catalyst for this reaction, and the substitution occurred selectively at the benzylic position in preference to the other side of the nitrogen atom. The reaction was performed with 1 equiv. of *tert*-butyl hydroperoxide (TBHP) as an oxidant at 50 °C in tetrahydrofuran (THF), and a reasonable chemical yield was obtained, despite the fact that THF tends to undergo radical oxidation. Aromatic-substituted alkynes reacted to give the desired compounds in good yield with a high enantiomeric excess (ee) of up to 74%, whereas low enantioselectivity was observed in the case of aliphatic-substituted alkynes.

As for the reaction mechanism, the actual catalytic species remain unclear, but the Cu complex is likely to play a role in activating both substrates.

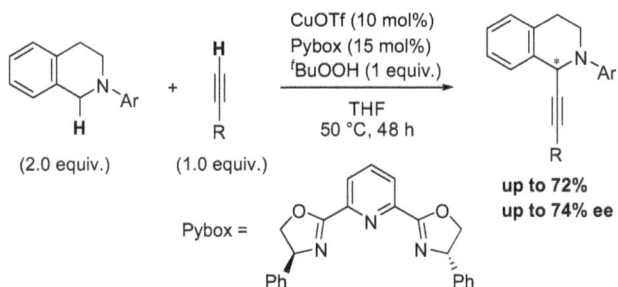

Scheme 7.1 Asymmetric Cu-catalyzed α-alkynylation of *N*-aryl THIQs.

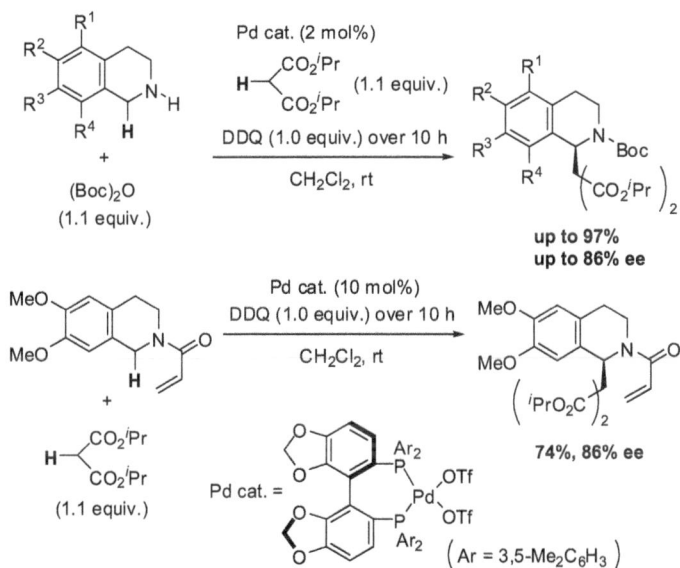

Scheme 7.2 Asymmetric Pd-catalyzed CDC reactions between *N*-Boc-protected THIQs and malonates.

Li proposed that copper acetylide is generated and undergoes nucleophilic addition to the iminium ion, which is generated by Cu-catalyzed oxidation of THIQ.

Inspired by Li's first example of the asymmetric CDC reaction of THIQs, we developed a Pd-catalyzed asymmetric oxidative Mannich-type reaction of *N-tert*-butoxycarbonyl (Boc)-protected THIQs with diisopropyl malonate using DDQ as the oxidant (Scheme 7.2).[4] THIQs could be used directly in the presence of Boc₂O, and the coupling products were obtained in good yield with up to 86% ee. In this reaction, the slow addition of DDQ was essential for high chemical yield and enantioselectivity. While *N*-phenyl-substituted substrates are normally used in this type of reaction, it is advantageous that a readily removable protecting group (Boc) and an acryloyl group were used.

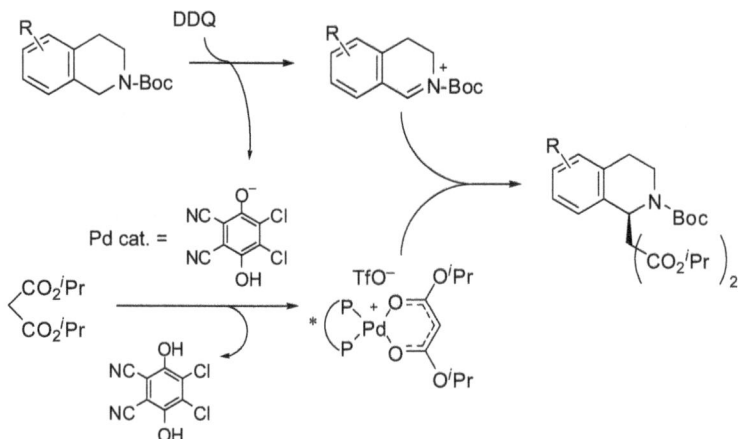

Scheme 7.3 Proposed mechanism for Pd-catalyzed CDC reactions.

We proposed a catalytic cycle for this oxidative Mannich-type reaction, as shown in Scheme 7.3. The N-Boc-THIQs undergo oxidation by DDQ to give the reactive iminium ion intermediate. The concomitantly formed phenolate anion may act as a weak base, which facilitates the formation of a chiral Pd enolate. These active species are coupled to give the desired product. We suppose that the limited substrate scope is associated with the feasibility of oxidation of N-protected THIQs, which only proceeds in the case of electron-rich substrates. It is interesting that high stereoselectivity was achieved, even though the catalyst is composed of a bisphosphine ligand that is usually sensitive to oxidative conditions.

To expand the scope of C-based nucleophiles in asymmetric CDC reactions, organocatalysis was examined. Klussmann and co-workers reported a catalytic system consisting of VO(acac)$_2$/TBHP and L-proline for cross-coupling between simple ketones and N-aryl THIQs.[5] These two catalysts are considered to play independent roles. Although the proposed mechanism involves nucleophilic addition of chiral enamines to iminium ions formed oxidatively from THIQs. The level of asymmetric induction was low ($\sim$17% ee).

In contrast, Wang *et al.* achieved excellent enantioselectivity in the CDC reactions of N-aryl THIQs with cyclic ketones using DDQ as the oxidant (Scheme 7.4).[6] It is likely that DDQ did not interfere with enamine catalysis. With phenylalanine as the best organocatalyst, N-aryl THIQs were reacted with various 6-membered cyclic ketones to afford the coupling products in good yield with high-to-excellent diastereo- and enantioselectivity. The authors hypothesized that an enamine intermediate would form an ion pair with the iminium ion as a key intermediate. N-Benzylaniline was also tested in place of THIQs, but the chemical yield was poor even after 3 days (<10% yield). The reaction pathways are assumed to involve a single-electron-transfer (SET) radical mechanism. Even though the addition of TEMPO did

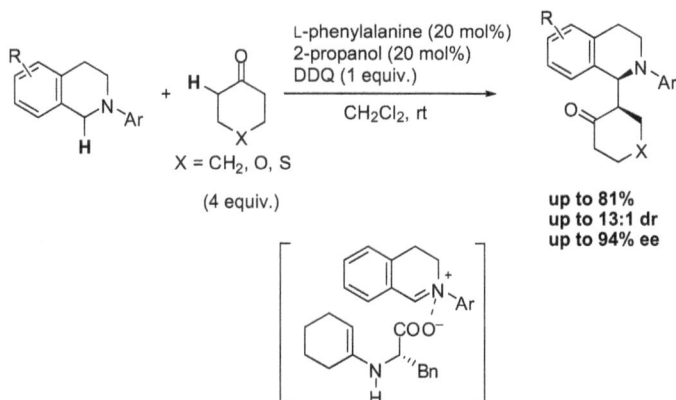

Scheme 7.4 Asymmetric CDC reaction of *N*-aryl THIQs with ketones using L-phenylalanine and DDQ.

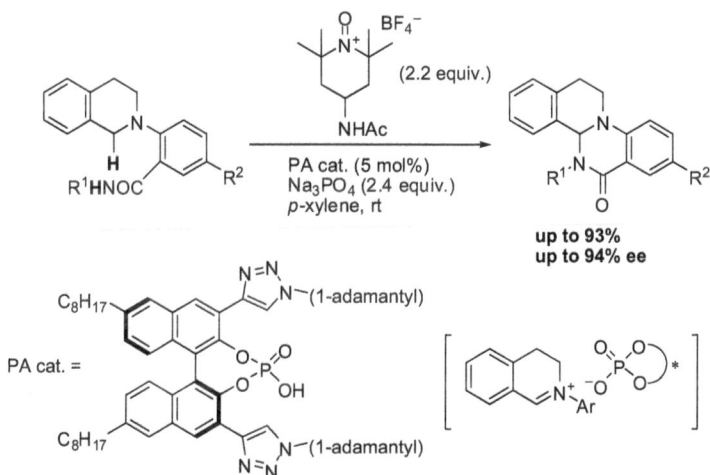

Scheme 7.5 Asymmetric intramolecular CDC reaction of THIQs under PA phase-transfer catalysis.

not induce significant change, the chemical yield was reduced considerably when 2,6-di-*tert*-butyl-4-methylphenol (BHT) was added.

Although C–N bond formation is beyond the scope of this chapter, Toste *et al.* reported a unique asymmetric reaction under chiral phosphate anion (PA) phase-transfer catalysis (Scheme 7.5).[7] It is assumed that a chiral phosphate could undergo anion exchange with an oxopiperidium salt, ensuring the formation of a tight ion-pair upon oxidation, as shown in Scheme 7.5. This oxidative amidation proceeded in a highly enantioselective manner, and the triazole unit was found to be essential for achieving high asymmetric induction. It is considered that the amide group within the substrate might have a hydrogen-bonding interaction with the triazole, resulting in additional rigidity within the chiral environment.

Chi's group developed a novel metal–organic co-operative catalytic system for asymmetric CDC reactions of *N*-aryl THIQs with simple aldehydes (Scheme 7.6).[8] In the presence of CuBr$_2$ (10 mol%) with TBHP, a diarylprolinol-based silyl ether catalyst promoted the C–C bond-forming reaction in a highly enantioselective manner. Since the solvent had a significant effect on the stereoselectivity, the mixed solvent system was carefully selected. However, it appears that intrinsic limitations on tertiary amine substrates remain a problem: for example, the reaction of an *N*-aryl pyrrolidine with propanal gave the corresponding product in low yield, albeit with 95% ee. No product was detectable in the reaction with an *N*-aryl piperidine.

Wang and co-workers reported an enantioselective metal–organic catalyzed aerobic oxidative aza-Morita–Baylis–Hilman reaction between *N*-aryl THIQs and electron-deficient olefins such as acrolein and acrylonitrile (Scheme 7.7).[9] Under co-operative catalysis with Cu(OTf)$_2$ and quinine, sp^3 C–H olefination at the benzylic position of THIQs occurred to afford the

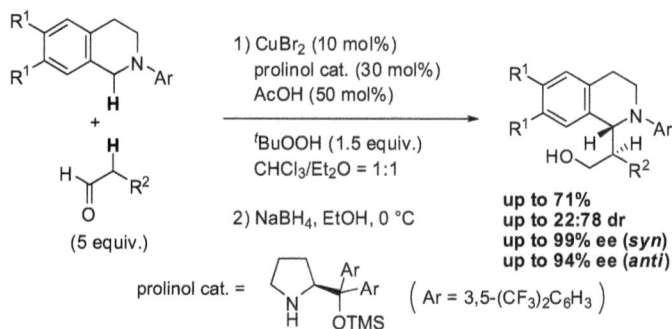

Scheme 7.6   Asymmetric CDC reaction of *N*-aryl THIQs under metal–organic co-operative catalysis.

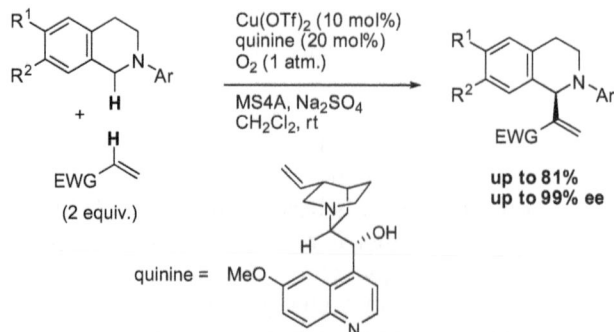

Scheme 7.7   Asymmetric CDC alkenylation of *N*-aryl THIQs under metal–organic co-operative catalysis. MS4A = 4 Å molecular sieve, EWG = electron withdrawing group.

coupling products in good yield with excellent enantioselectivity. Notably, the use of the economical and green oxidant dioxygen, with water as the only by-product, is highly appealing.

Merging visible-light photoredox catalysis and *N*-heterocyclic carbene (NHC) catalysis, Rovis' group succeeded in developing asymmetric α-acylation of tertiary amines (Scheme 7.8).[10] After careful manipulation of the reaction conditions, the desired reactivity could be realized using [Ru(bpy)$_3$]Cl$_2$ (bpy = 2,2'-bipyridyl) as the photocatalyst in the presence of *m*-dinitrobenzene (*m*DNB) as the oxidant and chiral NHC under irradiation with blue light. Aliphatic aldehydes with various functional groups were coupled with *N*-aryl THIQs in good yield with excellent enantioselectivity of up to 92%.

The proposed catalytic cycle includes the formation of a chiral Breslow intermediate from NHC and a reactive iminium ion *via* photoredox catalysis (Scheme 7.9). In this reaction, oxidation of the Breslow intermediate or NHC by [Ru(bpy)$_3$]$^{3+}$ and terminal oxidants could be problematic, because of the similar redox potentials of this catalyst scaffold and tertiary amines. To suppress such unproductive pathways, careful choice of NHC with proper steric and electronic properties and the use of a weak oxidant are important.

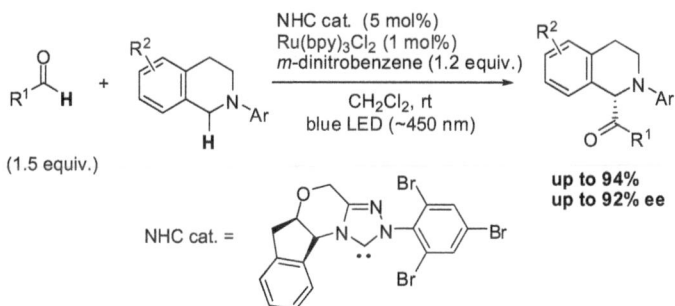

Scheme 7.8 Oxidative α-acylation of *N*-aryl THIQs under NHC and photoredox catalysis.

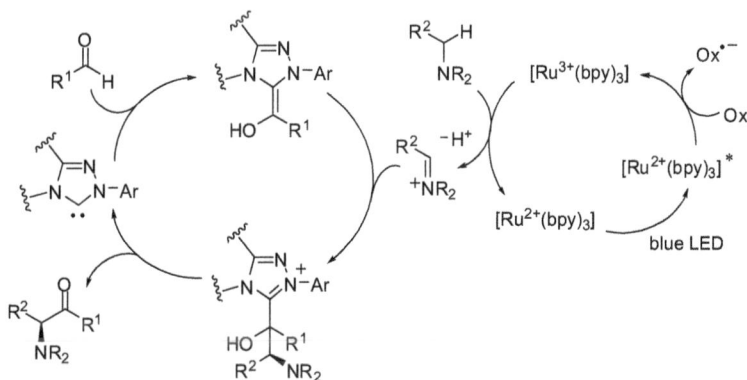

Scheme 7.9 Proposed catalytic cycle for α-acylation of THIQs.

The authors noted that reversibility of the aza-Breslow intermediate resulting from the reaction of NHC with iminium ion is also an important factor for high chemical yield.

7.2.2 Reactions of Glycine Derivatives

Since glycine is the least expensive natural amino acid, the development of a method for its direct α-C–H functionalization would provide a convenient way to synthesize various α-amino acid derivatives. In 2011, Wang *et al.* reported a DDQ-mediated CDC reaction of α-substituted β-ketoesters with *N*-aryl glycine esters in the presence of 10 mol% of Cu(OTf)$_2$ and 12 mol% of a chiral bisoxazoline (Box) ligand in THF (Scheme 7.10).[11] Although the diastereoselectivity was only modest, the desired coupling product with vicinal stereogenic centers was obtained with excellent enantioselectivity. The reaction is presumed to involve formation of the corresponding imino ester *via* SET, and the proposed catalytic cycle seems similar to that of Sodeoka's reaction (see Scheme 7.3). As a related reaction, Huang reported preliminary results on a proline ester-catalyzed oxidative coupling between acetone and *N*-aryl glycine ester in the presence of Cu(OAc)$_2$ and TBHP, but the enantioselectivity was marginal.[12]

Recently, Kanai and co-workers developed a novel catalytic system for oxidation of benzyl amines and anilines to the corresponding imines with molecular oxygen under mild conditions (Scheme 7.11).[13] Inspired by Iwabuchi's 2-azaadamantane-*N*-oxyl (AZADO) oxidation, they designed 9-azabicyclo[3.3.1]nonan-3-one *N*-oxyl (ketoABNO) as a more electron-deficient redox mediator and found that the combination with Cu(I) was effective for efficient aerobic oxidation of amines. Based on this result, they successfully applied their oxidation chemistry to CDC reactions. When the chiral Box ligand was used, an aerobic nitro-Mannich reaction between *N*-PMP glycine and 1-nitropropane proceeded with excellent diastereo- and enantioselectivity (20 : 1 diastereomeric ratio (dr) and 95% ee of the major *syn* isomer). In this case, Et$_3$N was used as a co-catalyst to accelerate the addition reaction, because tertiary amines are inert to the ketoABNO/Cu(I) system.

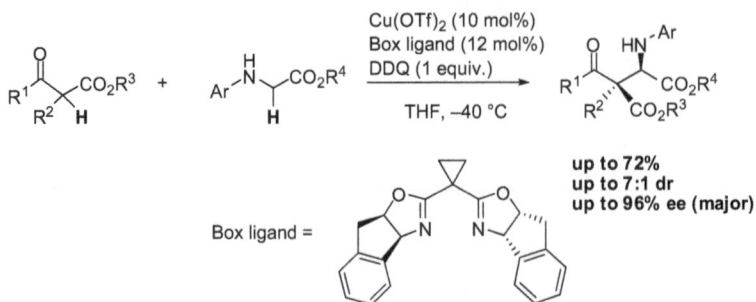

Scheme 7.10   Asymmetric CDC reaction between *N*-aryl glycine esters and β-ketoesters.

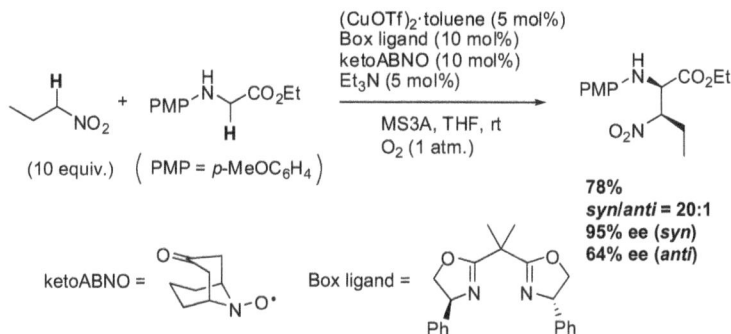

Scheme 7.11 Asymmetric CDC reaction between glycine ester and 1-nitropropane. MS3A = 3 Å molecular sieve.

7.3 Asymmetric Functionalization of Benzylic C–H Bonds

The CDC reaction of a benzylic C–H bond without an adjacent heteroatom is more challenging. Since benzylic carbocations can be generated under oxidative reaction conditions, asymmetric alkylation is considered to be possible if the carbocations are trapped with chiral enamines or chiral metal enolates.

In 2009, Cozzi *et al.* reported an organocatalytic stereoselective α-alkylation reaction of aldehydes with diarylmethanes in the presence of a MacMillan imidazolidinone catalyst (Scheme 7.12).[14] With DDQ as the terminal oxidant, alkylation products were obtained in good yield with moderate-to-good enantioselectivity. DDQ was the best oxidant, while reactions with other oxidants resulted in no reaction, or decomposition of the catalyst. It is likely that a chiral enamine formed by the reaction of an aldehyde with the catalyst reacts with the dibenzylic cation generated by DDQ oxidation of the diarylmethanes. The reaction was also applicable to diallylic, allylbenzylic and indolylarylmethane substrates.

The combination of chiral Lewis acid catalysts with oxidants is also effective for CDC reactions with diarylmethane compounds. Gong *et al.* reported a highly enantioselective oxidative coupling reaction of structurally diverse 3-arylmethylindoles with dibenzyl malonate using a chiral Cu(OTf)$_2$-Box complex and DDQ as the oxidant (Scheme 7.13).[15] Cu(OTf)$_2$ performs better than Mg(OTf)$_2$ or Zn(OTf)$_2$ in the presence of Box ligands. The proposed reaction mechanism is similar to that discussed previously (Scheme 7.3). After copper-catalyzed dehydrogenation of 3-arylmethylindole to give the vinylogous iminium cation, the resultant copper phenoxide species serves as a base to deprotonate the dibenzyl malonate. The resulting chiral enolate intermediate undergoes enantioselective conjugate addition to the vinylogous iminium cation.

Recently, Feng and co-workers reported related oxidative coupling reactions between β-ketoesters and xanthene (Scheme 7.14).[16] With TBHP as

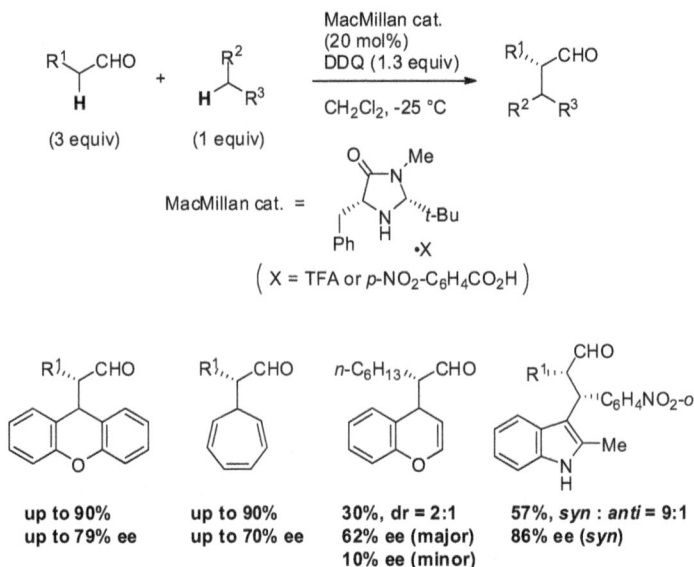

up to 90%   up to 90%   30%, dr = 2:1   57%, *syn : anti* = 9:1
up to 79% ee   up to 70% ee   62% ee (major)   86% ee (*syn*)
                                      10% ee (minor)

Scheme 7.12   Asymmetric organocatalytic CDC reaction between aldehydes and
                  sufficiently activated methylenic compounds.

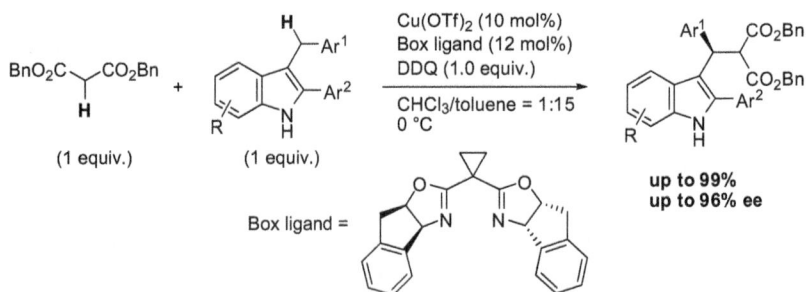

Scheme 7.13   Cu-catalyzed asymmetric alkylation of malonates using DDQ as the
                  oxidant.

the oxidant and L_1 as a ligand, indanone carboxylate underwent the coupling
reaction in a highly enantioselective manner (up to 99% ee). In the case of
tetralone carboxylates, however, the desired coupling reaction did not proceed well, although xanthene was oxidized to 9*H*-xanthen-9-one. When the
ligand L_1 was changed to L_2, the corresponding coupling product was obtained in good yield with up to 99% ee.

In the case of Feng's reaction, it is interesting that a chiral Ni/Fe bimetallic
system was essential to achieve high chemical yield and enantioselectivity
(Table 7.1). The Fe–L_1 1:1 complex (2 mol%) gave high chemical yield
without asymmetric induction. The reaction promoted by the Ni–L_1 1:1
complex (10 mol%) alone was slower, while the ee was excellent. The authors

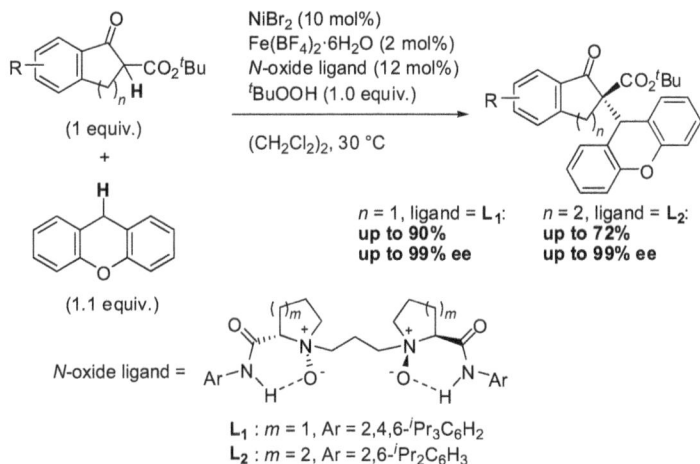

Scheme 7.14 Asymmetric CDC reaction of cyclic β-ketoesters using a bimetallic catalyst system.

Table 7.1 Control experiments.

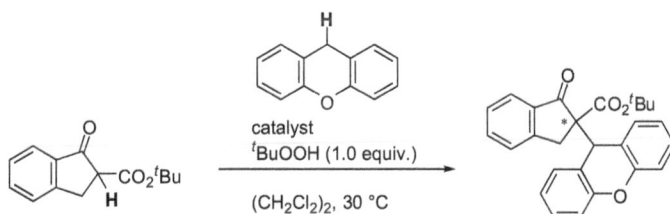

Entry	Catalyst	mol%	Yield/%	ee/%
1	$Fe(BF_4)_2$	10	47	0
2	$Fe(BF_4)_2/L_1(1:1)$	2	85	0
3	$NiBr_2/L_1(1:1)$	10	51	99
4	$NBr2/L_1(1:1) + Fe(BF_4)_2/L_1(1:1)$	10(NI) + 2(Fe)	90	99
5	$NBr_2/Fe(BF_4)_2/L_1(1:02:1)$	10(NI) + 2(Fe)	53	96

concluded that the Ni complex plays an important role in the asymmetric induction, and at the same time the Fe complex enhances the reaction rate.

7.4 Asymmetric Functionalization *via* Oxidation of Enamine and Breslow Intermediates

7.4.1 Asymmetric α-Functionalization

Organocatalysis in combination with oxidants enables intramolecular asymmetric α-arylation of carbonyl compounds. In 2009, Nicolaou and MacMillan independently reported asymmetric α-arylation of aldehydes (Scheme 7.15).[17] Using MacMillan's imidazoline catalyst and an appropriate

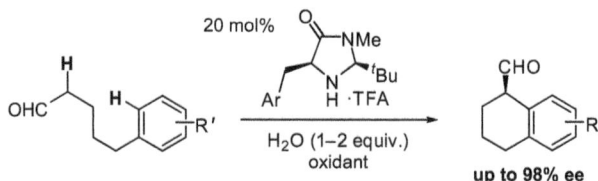

Nicolaou's conditions : CAN (2 equiv.), 1,2-dimethoxyethane (DME)

MacMillan's conditions: [Fe(phen)$_3$](PF$_6$)$_3$ (2.5 equiv.) or CAN (2 equiv.)
CH$_3$CN or acetone

Scheme 7.15 Amine-catalyzed asymmetric α-arylation of aldehydes. TFA = trifluoro-acetic acid.

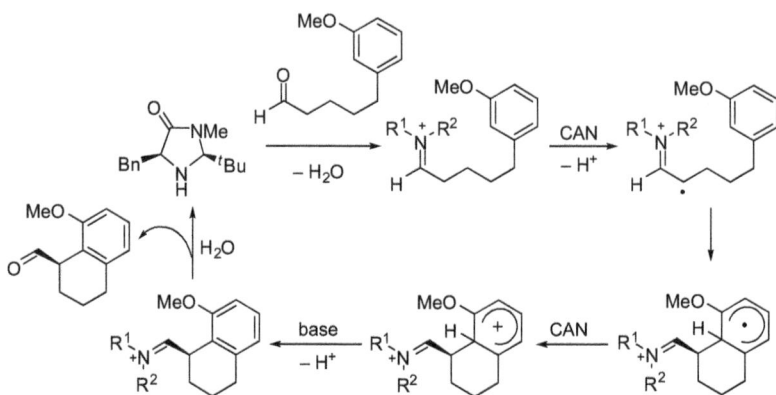

Scheme 7.16 Iron-catalyzed selective CDC reaction of benzylic C–H bonds with arenes.

oxidant (ceric ammonium nitrate (CAN) or [Fe(phen)$_3$](PF$_6$)$_3$), tetra-hydronaphthalene-1-carbaldehyde derivatives were formed with excellent enantioselectivity of up to 98% ee (Scheme 7.15). In general, electron-rich aromatic rings resulted in efficient oxidative C–C bond formation.

While a Friedel–Crafts mechanism involving ionic intermediates cannot be ruled out, MacMillan proposed an alternative mechanism that includes a radical intermediate (Scheme 7.16). An enammonium radical cation is considered to be generated from a chiral iminium ion *via* one-electron oxidation, and then undergoes a cyclization reaction. A second one-electron oxidation, followed by re-aromatization, completes the catalytic cycle. Theoretical calculations for the reaction depicted in Scheme 7.16 indicated that the radical mechanism would favor *ortho*-selective cyclization, while a Friedel–Crafts mechanism would result in *para*-selective cyclization. Since the reaction gave the *ortho*-selective cyclization product as a major product, the radical pathway is plausible.[17c]

As a related reaction, Dixon reported asymmetric arylation of active methine compounds with *in situ* generated *ortho*-quinone derivatives (Scheme 7.17).[18] When IO$_4^-$ supported on polystyrene (PS) was used as the

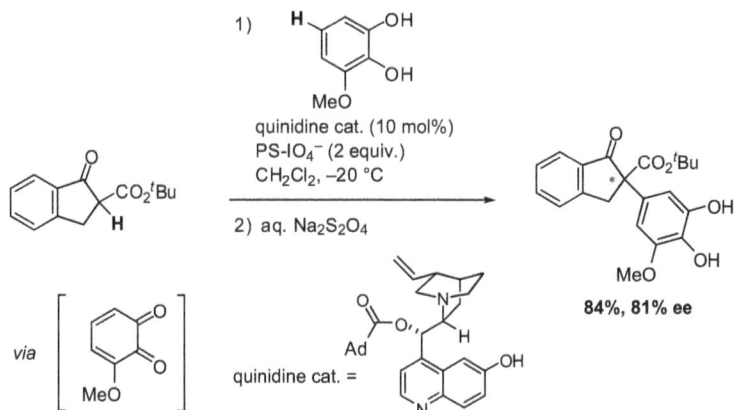

Scheme 7.17 Asymmetric oxidative α-arylation of β-ketoesters.

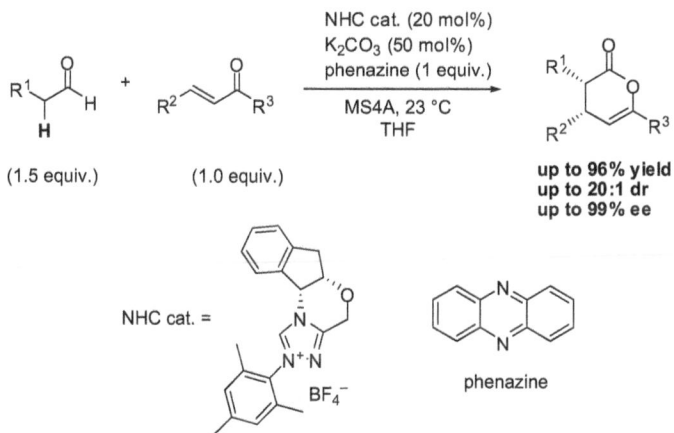

Scheme 7.18 Asymmetric oxidative α-functionalization of simple aldehydes.

stoichiometric oxidant, a quinine derivative catalyzed the coupling reaction in good yield with high enantioselectivity.

Oxidation of Breslow intermediates provides a means for unique α-functionalization of simple aldehydes. With phenazine as the oxidant and an NHC catalyst, Rovis' group reported an oxidative coupling reaction of simple aldehydes with α,β-unsaturated ketones and imines to afford the corresponding disubstituted lactones and lactams in almost optically pure form (Scheme 7.18).[19] Chi's group reported a similar reaction with a different oxidant, and the reaction mechanism is briefly discussed in the following section.[20]

7.4.2 Asymmetric β-Functionalization

Since β-functionalization of saturated carbonyl compounds is potentially a powerful synthetic method, the development of an asymmetric version is

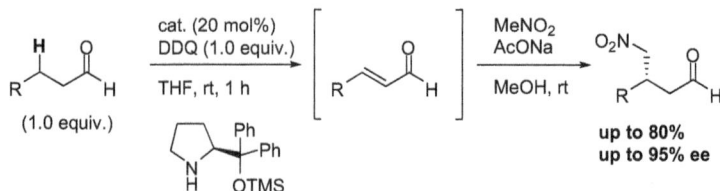

Scheme 7.19   Hayashi's asymmetric oxidative β-functionalization of saturated aldehydes.

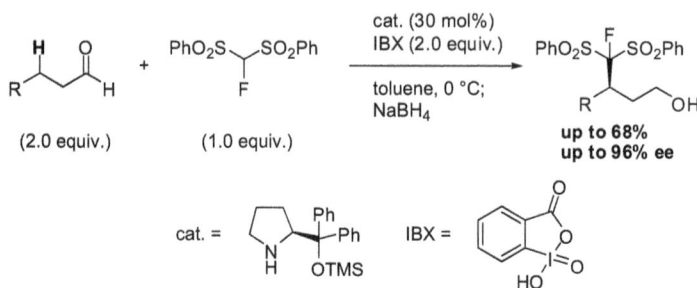

Scheme 7.20   Wang's asymmetric oxidative β-functionalization of saturated aldehydes.

highly desirable. In 2011, Hayashi and Wang independently reported similar transformations *via* oxidation of chiral enamine intermediates. Hayashi's group developed a one-pot, two-step procedure for asymmetric CDC reaction of aldehydes with nitromethane using diphenylprolinol silyl ether (Scheme 7.19).[21] In the first step, the enamine derived from the catalyst and the starting aldehyde is converted to the corresponding α,β-unsaturated iminium ion as a result of hydride abstraction by DDQ. In the second step, the released α,β-unsaturated aldehyde undergoes normal organocatalytic asymmetric conjugate addition with nitromethane, in which AcONa acts as a weak base to eliminate the resulting hydroquinone. Although this study was restricted to substrates having an aromatic or styryl group at the β-position, various coupling products were obtained in a highly enantioselective manner.

Using 2-iodoxybenzoic acid (IBX) as the oxidant, Wang *et al.* achieved one-step formal conjugate addition of active methylene compounds in moderate yield with excellent enantioselectivity (Scheme 7.20).[22] In a related study, Enders applied this β-activation of saturated aldehydes with IBX to amine-catalyzed cascade reactions for the synthesis of cyclohexene derivatives.[23] Furthermore, Xu *et al.* succeeded in replacing stoichiometric oxidants with dioxygen by combining enamine catalysis with Pd-catalyzed aerobic oxidation.[24]

Based on their earlier study on α-functionalization mentioned above, Chi and co-workers developed a novel transformation *via* β-activation of aldehydes using NHC catalysis (Scheme 7.21).[25] When the amount of oxidant was

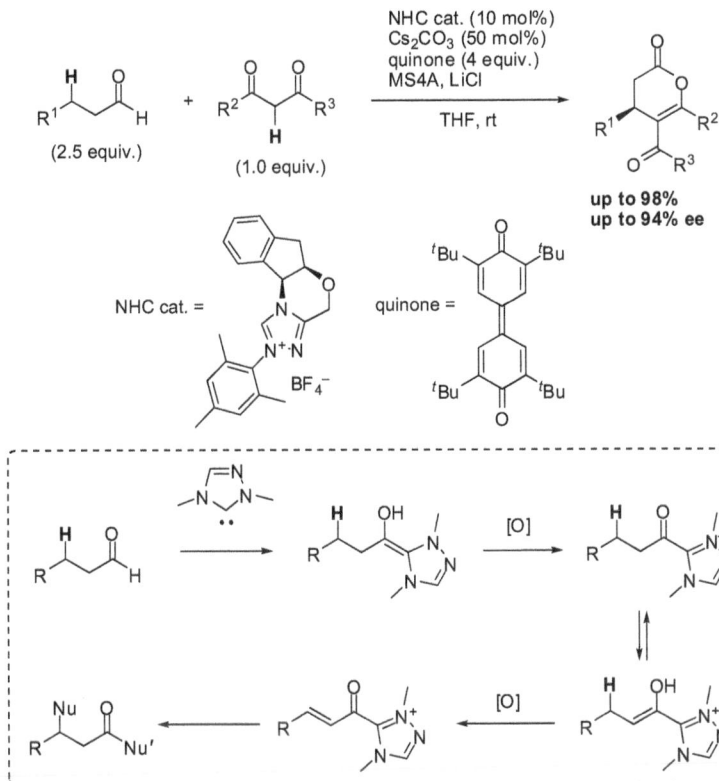

Scheme 7.21 Asymmetric functionalization of saturated aldehydes *via* oxidative NHC catalysis.

increased, an acyl triazolium salt generated by oxidation of the Breslow intermediate was further oxidized after tautomerization, affording an activated enoate equivalent. The reaction with 1,3-diketones or β-ketoesters afforded enol lactones with high enantioselectivity.

7.5 Oxidative Naphthol Coupling Reactions

Cross-coupling between two phenols (or naphthols) is useful for producing bisphenol (or binaphthol) products. Oxidative homodimerization of 2-naphthol has been well documented in the literature. However, it is difficult to achieve cross-coupling between two different phenols, and their oxidation potentials must be sufficiently different in order to oxidize one phenol selectively.

Kozlowski and co-workers found that asymmetric cross-coupling between electronically tuned 3-acyl-substituted 2-naphthols was catalyzed by a chiral Cu–diamine complex (Scheme 7.22).[26] The Cu catalyst acts as an oxidation catalyst for naphthol coupling, and the resulting reduced Cu complex is re-oxidized by O_2 as the sole oxidant.

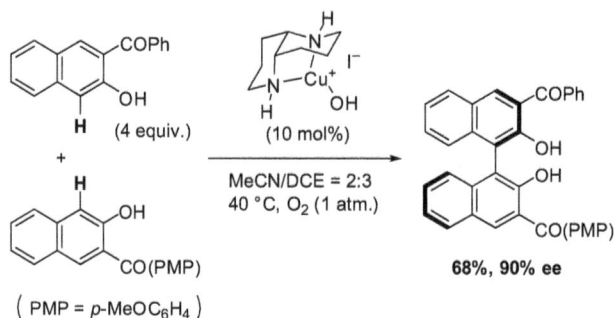

Scheme 7.22   Cu-Catalyzed asymmetric oxidative cross-coupling of 2-naphthols. DCE = 1,2-dichloroethane.

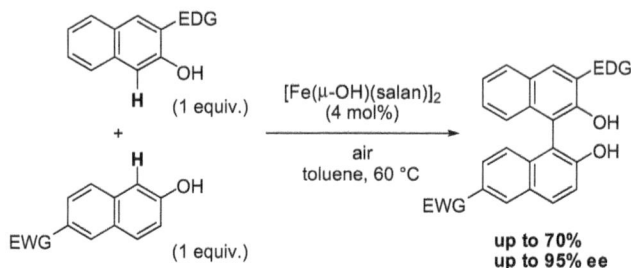

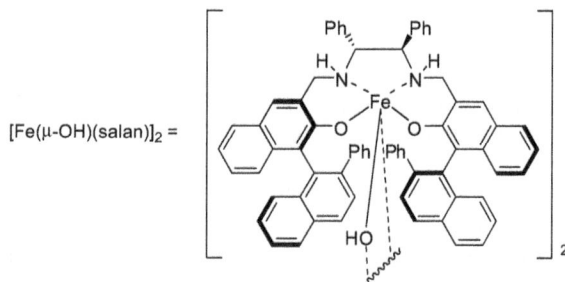

Scheme 7.23   Fe-Catalyzed asymmetric cross-coupling of 2-naphthols under aerobic conditions. EDG = electron donating group.

Recently, Katsuki and co-workers developed a more general variant of this type of transformation using a hydroxy-bridged Fe dimer complex (Scheme 7.23).[27] With a sterically demanding salan as a chiral ligand, oxidative cross-coupling of electronically differentiated 2-naphthols occurred smoothly with air as the sole oxidant. Even though the different naphthols were mixed in a ratio of 1:1, the cross-coupling reaction was favored over homocoupling reaction with excellent selectivity of up to 20:1. Based on mechanistic studies, the dimer complex is considered to dissociate to the corresponding monomer, and the reaction is proposed to proceed *via* radical-anion coupling under Fe(III/IV) catalysis. The more electron-rich substrate tends to undergo preferential one-electron oxidation by the Fe catalyst, and the naphthoxide anion generated from the more acidic 2-naphthol attacks the resulting Fe-bonded radical cation species to afford the desired cross-coupling product. To achieve high enantioselectivity, the electron-rich naphthol requires a substituent at the 3-position.

7.6 Asymmetric Oxidative Arene–Alkene Coupling (Fujiwara–Moritani) Reactions

The Fujiwara–Moritani reaction is a Pd-catalyzed oxidative coupling of an unfunctionalized arene with an alkene, in which a C–H bond of the arene is directly converted to a new C–C bond. Developing an asymmetric version of this type of reaction remains a formidable challenge in modern synthetic organic chemistry.

In 1999, Mikami's group reported the first example of an intermolecular asymmetric Fujiwara–Moritani reaction of benzene with cyclic alkenes using a Pd catalyst co-ordinating to a chiral sulfonylamide-oxazoline ligand (Scheme 7.24).[28] With PhCO$_3$^{t}Bu as the oxidant, the coupling reaction occurred with moderate enantioselectivity (up to 49% ee), although the chemical yield was low. The reaction is considered to involve the formation of a phenyl-Pd species *via* electrophilic C–H substitution by Pd(II), and Heck-type cyclization followed by re-oxidation of the resultant Pd(0) species.

Oestreich and co-workers reported an enantioselective intramolecular Fujiwara–Moritani reaction of indole and pyrrole derivatives.[29] Using a suitable oxidant (*e.g.*, PhCO$_3$^{t}Bu, benzoquinone, O$_2$), they evaluated a chiral bidentate pyridyl-oxazoline ligand (PyOx) developed by Stolz and a series of nicotine-oxazoline ligands (NicOx) with Pd(OAc)$_2$ as a catalyst precursor (Scheme 7.25). Indoles and pyrroles underwent a 5-exo-trig cyclization, but the chemical yield and enantioselectivity were generally unsatisfactory.

Based on extensive studies of C–H transformation with Pd(II), Yu *et al.* recently achieved asymmetric desymmetrization in a CDC reaction (Scheme 7.26).[30] Boc-L-Isoleucine (Boc-Ile-OH) was an effective chiral ligand for the Pd(II)-catalyzed enantioselective C–H activation of α,α-diphenylacetic acid Na salt. The resultant Pd–aryl intermediate is considered to undergo insertion into styrenes or acrylates, followed by β-hydrogen elimination to afford the coupling product with a chiral quaternary carbon center in good

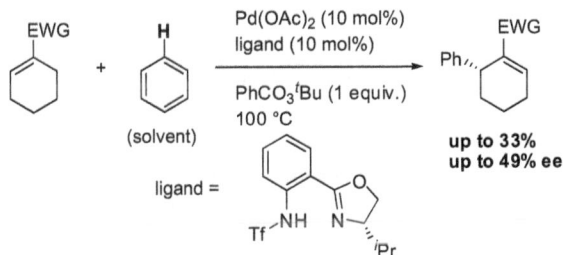

Scheme 7.24 The first example of an asymmetric Fujiwara–Moritani reaction.

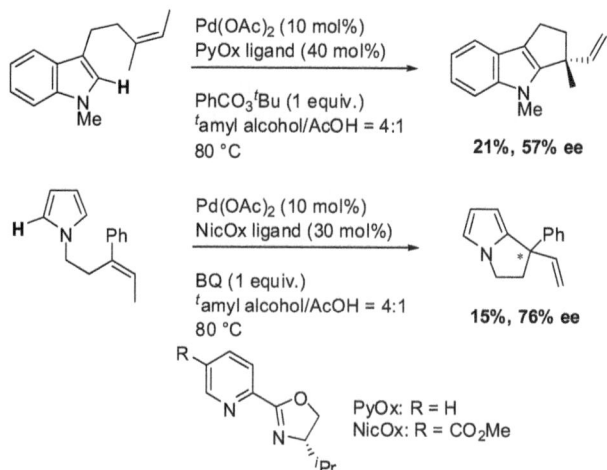

Scheme 7.25   Pd-Catalyzed asymmetric oxidative Heck-type reactions.

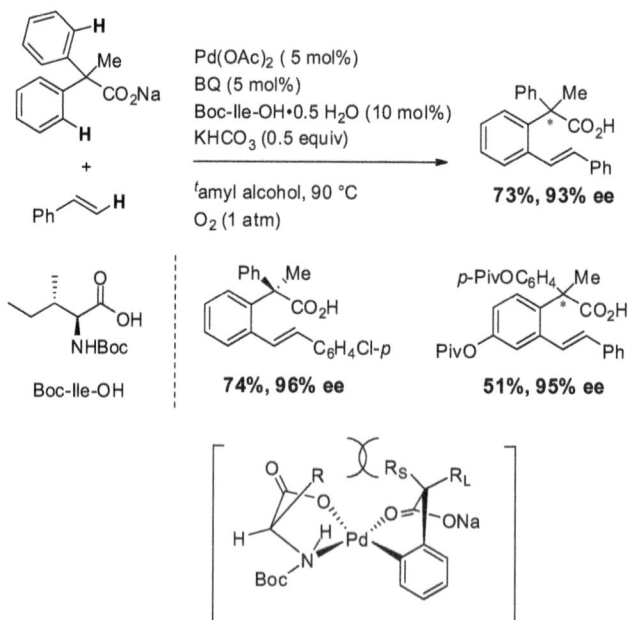

Scheme 7.26   Pd-Catalyzed asymmetric CDC reaction between diarylacetic acid Na salt and alkenes.

yield, with excellent enantioselectivity of up to 97% ee. With acrylates in place of styrenes, further cyclization *via* Michael addition occurred to give the corresponding six-membered lactones. The combined use of the carboxylate sodium salt and KHCO$_3$ was crucial for promoting the reaction efficiently. Thus, an acetic acid type starting material with an α-proton is not suitable because it undergoes racemization under the reaction conditions. It is proposed that a chiral Pd intermediate, as depicted above, would be formed with the carbonyl oxygen of the carboxylate salt co-ordinating to Pd(II).

7.7 Conclusions and Outlook

Cross-dehydrogenative-coupling (CDC) reactions have emerged as highly efficient and atom-economical methods, and the potential to achieve the concise synthesis of complex molecules from simple starting materials makes this reaction particularly attractive. Therefore, the development of catalytic asymmetric variants of CDC reactions is highly desirable. However, such reactions are difficult to achieve, because nucleophiles such as enolates and electron-rich arene compounds are susceptible to oxidative degradation, and because the chiral catalysts must be robust under the oxidative conditions used. Additionally, mild reaction conditions are expected to favor high stereoselectivity, but the process includes the energetically unfavorable cleavage of a strong C–H bond. As described in this chapter, the number of successful examples of asymmetric CDC reactions is still quite limited, and much work will be necessary to expand their scope. It seems likely that further development of asymmetric CDC reactions will require more active, but sufficiently stable, catalysts that allow C–H activation at ambient temperature, as well as the replacement of stoichiometric oxidants with molecular oxygen as the sole oxidant.

References

1. (a) L. Yang and H. Huang, *Catal. Sci. Technol.*, 2012, **2**, 1099; (b) C. S. Yeung and V. M. Dong, *Chem. Rev.*, 2011, **111**, 1215.
2. C.-J. Li, *Acc. Chem. Res.*, 2009, **42**, 335.
3. (a) Z. Li and C.-J. Li, *Org. Lett.*, 2004, **6**, 4997; (b) Z. Li, P. D. MacLeod and C.-J. Li, *Tetrahedron: Asymmetry*, 2006, **17**, 590; (c) J. Yu, Z. Li, K. Jia, Z. Jiang, M. Liu and W. Su, *Tetrahedron Lett.*, 2013, **54**, 2006.
4. (a) N. Sasamoto, C. Dubs, Y. Hamashima and M. Sodeoka, *J. Am. Chem. Soc.*, 2006, **128**, 14010; (b) C. Dubs, Y. Hamashima, N. Sasamoto, T. M. Seidel, S. Suzuki, D. Hashizume and M. Sodeoka, *J. Org. Chem.*, 2008, **73**, 5859.
5. A. Sud, D. Sureshkumar and M. Klussmann, *Chem. Commun.*, 2009, 3169.
6. G. Zhang, Y. Ma, S. Wang, W. Kong and R. Wang, *Chem. Sci.*, 2013, **4**, 2645.
7. A. J. Neel, J. P. Hehn, P. F. Tripet and F. D. Toste, *J. Am. Chem. Soc.*, 2013, **135**, 14044.
8. J. Zhang, B. Tiwari, C. Xing, X. Chen and Y. R. Chi, *Angew. Chem., Int. Ed.*, 2012, **51**, 3649.

9. G. Zhang, Y. Ma, S. Wang, Y. Zhang and R. Wang, *J. Am. Chem. Soc.*, 2012, **134**, 12334.

10. D. A. DiRocco and T. Rovis, *J. Am. Chem. Soc.*, 2012, **134**, 8094.

11. G. Zhang, Y. Zhang and R. Wang, *Angew. Chem., Int. Ed.*, 2011, **50**, 10429.

12. J. Xie and Z.-Z. Huang, *Angew. Chem., Int. Ed.*, 2010, **49**, 10181.

13. T. Sonobe, K. Oisaki and M. Kanai, *Chem. Sci.*, 2012, **3**, 3429.

14. F. Benfatti, M. G. Capdevila, L. Zoli, E. Benedetto and P. G. Cozzi, *Chem. Commun.*, 2009, 5919.

15. C. Guo, J. Song, S.-W. Luo and L.-Z. Gong, *Angew. Chem., Int. Ed.*, 2010, **49**, 5558.

16. W. Cao, X. Liu, R. Peng, P. He, L. Lin and X. Feng, *Chem. Commun.*, 2013, **49**, 3470.

17. (a) K. C. Nicolaou, R. Reingruber, D. Sarlah and S. Bräse, *J. Am. Chem. Soc.*, 2009, **131**, 2086; (b) J. C. Conrad, J. K. Kong, B. N. Laforteza and D. W. C. MacMillan, *J. Am. Chem. Soc.*, 2009, **131**, 11640; (c) J. M. Um, O. Gutierrez, F. Schoenebeck, K. N. Houk and D. W. C. MacMillan, *J. Am. Chem. Soc.*, 2010, **132**, 6001.

18. K. B. Bogle, D. J. Hirst and D. J. Dixon, *Org. Lett.*, 2007, **9**, 4901.

19. X. Zhao, K. E. Ruhl and T. Rovis, *Angew. Chem., Int. Ed.*, 2012, **51**, 12330.

20. J. Mo, R. Yang, X. Chen, B. Tiwari and Y. R. Chi, *Org. Lett.*, 2013, **15**, 50.

21. Y. Hayashi, T. Itoh and H. Ishikawa, *Angew. Chem., Int. Ed.*, 2011, **50**, 3920.

22. S.-L. Zhang, H.-X. Xie, J. Zhu, H. Li, X.-S. Zhang, J. Li and W. Wang, *Nat. Commun.*, 2011, **2**, 211.

23. X. Zeng, Q. Ni, G. Raabe and D. Enders, *Angew. Chem., Int. Ed.*, 2013, **52**, 2977.

24. Y.-L. Zhao, Y. Wang, X.-Q. Hu and P.-F. Xu, *Chem. Commun.*, 2013, **49**, 7555.

25. J. Mo, L. Shen and Y. R. Chi, *Angew. Chem., Int. Ed.*, 2013, **52**, 8588.

26. X. Xie, P. W. Phuan and M. C. Kozlowski, *J. Org. Chem.*, 2003, **68**, 5500.

27. H. Egami, K. Matsumoto, T. Oguma, T. Kunisu and T. Katsuki, *J. Am. Chem. Soc.*, 2010, **132**, 13633.

28. K. Mikami, M. Hatano and M. Terada, *Chem. Lett.*, 1999, 55.

29. (a) J. A. Schiffner, A. B. Machotta and M. Oestreich, *Synlett*, 2008, 2271; (b) J. A. Schiffner, T. H. Woste and M. Oestreich, *Eur. J. Org. Chem.*, 2010, 174.

30. B.-F. Shi, Y.-H. Zhang, J. K. Lam, D.-H. Wang and J.-Q. Yu, *J. Am. Chem. Soc.*, 2010, **132**, 460.

Cross-Dehydrogenative-Coupling Reactions without Metals

HIDETO ITO, KIRIKA UEDA AND KENICHIRO ITAMI*

Institute of Transformative Bio-Molecules (WPI-ITbM) and Graduate School of Science, Nagoya University, Chikusa, Nagoya 464-8602, Japan
*Email: itami.kenichiro@a.mbox.nagoya-u.ac.jp

8.1 Oxidants

8.1.1 Hypervalent Iodonium Salts

Hypervalent iodonium salts such as phenyliodine(III) bis(trifluoroacetate) (PIFA), phenyliodine(III) diacetate (PIDA), hydroxy(tosyl oxy)iodosylbenzene (HTIB) and pentafluorophenyliodine(III) bis(trifluoroacetate) (FPIFA) are environmentally friendly and highly efficient oxidation reagents that are often used in organic synthesis (Figure 8.1). In the past two or three decades, these hypervalent iodonium salts have been found to show reactivity for metal-free CDC reactions. Furthermore, the use of iodonium salts instead of metal reagents has a number of advantages such as low toxicity, low cost and easy handling.

8.1.1.1 sp^2–sp^3 C–C Coupling

In 1986, Tamura et al. found that intra- and intermolecular sp^2–sp^3 C–H couplings of aromatic compounds (Ar–H) and α-acylsulfides could be

RSC Green Chemistry No. 26
From C–H to C–C Bonds: Cross-Dehydrogenative-Coupling
Edited by Chao-Jun Li
© The Royal Society of Chemistry 2015
Published by the Royal Society of Chemistry, www.rsc.org

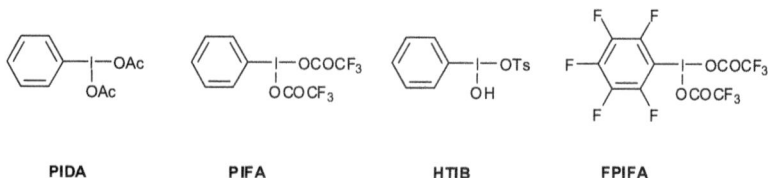

Figure 8.1   Hypervalent iodine(III) reagents.

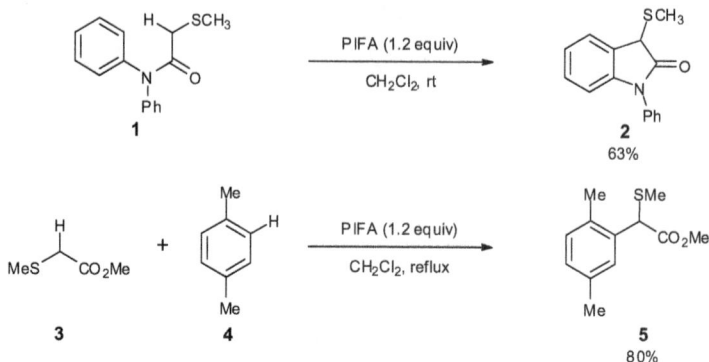

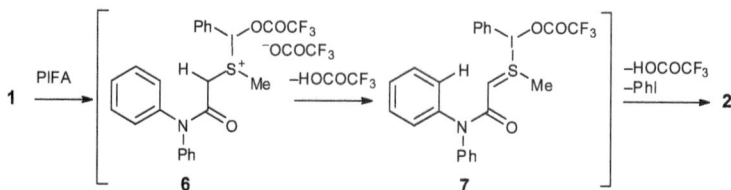

Scheme 8.1   PIFA-Mediated CDC coupling between an arene and α-acylsulfide.

promoted by employing PIFA in CH_2Cl_2.[1] These reactions seem to proceed through the Pummeer reaction intermediate **7** that can be generated from the sulfonium intermediate **6** (Scheme 8.1).

Kita *et al.* further developed PIFA-induced CDC reactions between phenyl ether derivatives and cyclic 1,3-dicarbonyl compounds as nucleophiles (Scheme 8.2).[2] The reactions with 1 equiv. of PIFA in hexafluoro-2-propyl alcohol attach nucleophiles onto the *ortho*-position of *para*-substituted phenyl ethers to afford the dehydrogenative coupling products **8** in moderate yields. UV and electron spin resonance (ESR) spectroscopic studies support a reaction mechanism involving the formation of the charge-transfer complex **9** followed by the generation of the cation radical intermediate **10**. This is the first example of the reaction of aromatic compounds with PIFA that involves the formation not of diaryliodonium(III) salt **11** but the cation radical intermediate **10** as a key intermediate.

Kita and co-workers further expanded their coupling methodology to the intramolecular version (Scheme 8.3).[3] Employing cyclic 1,3-cyclohex-anediones bearing alkyl tethers (**12**) having *para*- and/or *meta*-methoxylphenyl groups, in the presence of PIFA gives spirobenzannulated compounds (**13**)

8a
66%

8b
39%

8c
39%

8d
67%

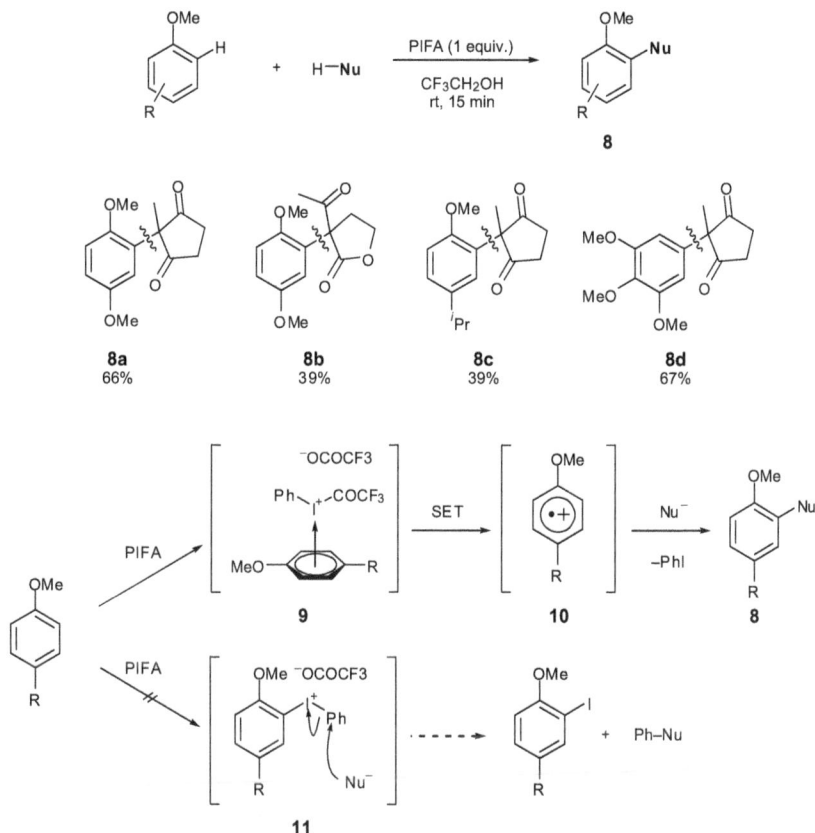

Scheme 8.2 PIFA-Induced CDC reaction of phenol ether derivatives and proposed reaction pathway.

whose structural features relate to bioactive compounds such as cannabis-piran. In contrast to the reaction of cyclic 1,3-dione, the reaction of an acyclic 1,3-dicarbonyl compound (**14**) does not afford a cyclization product, but gives the diaryl iodonium(III) compound (**15**) in good yield. Furthermore, the substrate bearing *meta*-dimethoxylphenyl groups (**16**) also give the iodonium salt **17** along with a small amount of the cyclization product **18**. They also verified the iodonium salt **17** does not convert to **18** even in the presence of base, which also supports the reaction pathway through the cation radical intermediate.

Du and Zhao demonstrated the tandem intramolecular CDC reaction of *N,N''*-diarylmalonamides (**19**) under the Kita's conditions to synthesize spiro-oxindoles (**20**) (Scheme 8.4).[4]

In 2013, Burgmann reported the CDC reaction between *N*-heterocyclic compounds (**21**) and simple alkanes (**22**) (Scheme 8.5).[5] Alkyl groups such as cycloalkyl, adamantly, 1″-mesityl, *sec*-butyl and *tert*-butyl groups can be selectively introduced at the C2-position of quinoline **21**, albeit branched alkylations also occur in the case of the reactions with branched alkanes. Furthermore, mono- and multi-cyclohexylation at electron-deficient

13a: $n = 1$ $R^1 = OMe$, $R^2 = H$, $X = CH_2$,   85%
13b: $n = 1$ $R^1 = OMe$, $R^2 = H$, $X = O$,   65%
13c: $n = 1$ $R^1 = H$, $R^2 = OMe$, $X = CH_2$,   70%
13d: $n = 2$ $R^1 = H$, $R^2 = OMe$, $X = CH_2$,   69%
13e: $n = 1$ $R^1 = OMe$, $R^2 = OMe$, $X = CH_2$,   39%

PIFA (1 equiv.)

CF₃CH₂OH
−40 °C

12   **13**

14   **15: 84%**

16   **17:65%**   +   **18: <3%**

base

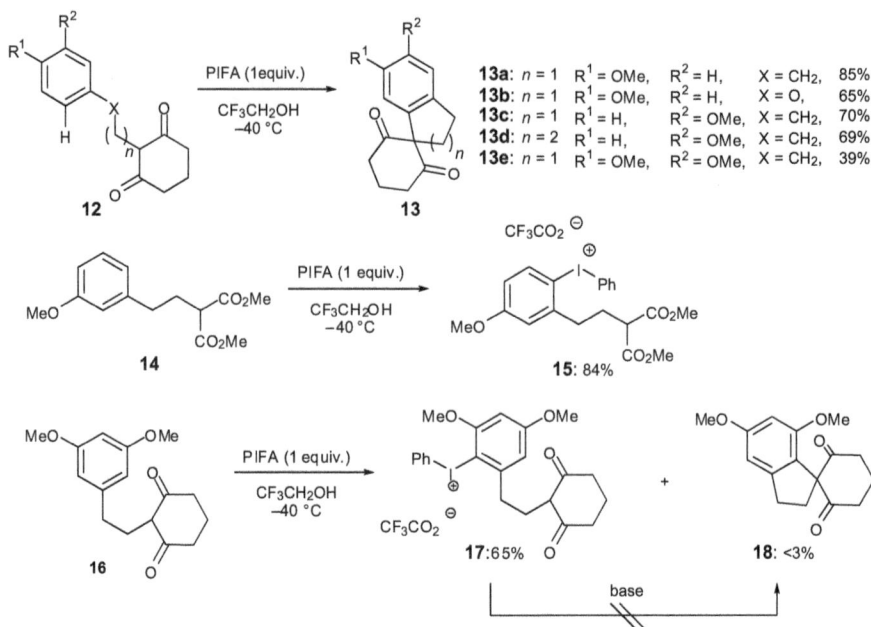

Scheme 8.3   PIFA-Induced intramolecular CDC reaction of *para*- and *meta*-substituted phenol ether derivatives.

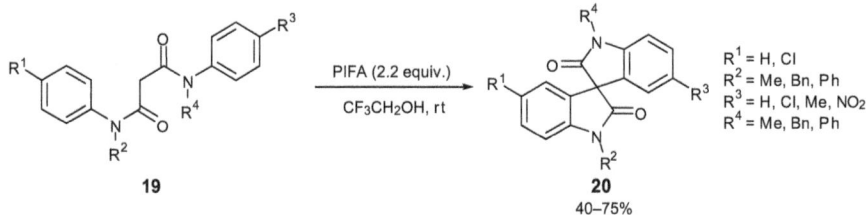

PIFA (2.2 equiv.)

CF₃CH₂OH, rt

$R^1 = H, Cl$
$R^2 = Me, Bn, Ph$
$R^3 = H, Cl, Me, NO_2$
$R^4 = Me, Bn, Ph$

19   **20**
            40–75%

Scheme 8.4   Synthesis of spiro-oxindoles by PIFA-induced tandem intramolecular CDC reaction.

positions of a variety of *N*-heteroarenes are demonstrated. Addition of both PIFA and NaN₃ is essential for the reaction to proceed, which involves the formation and addition of alkyl cation radicals.

8.1.1.2   *sp²–sp² C–C Coupling*

Hypervalent iodine reagents can also be applied to the sp²–sp² C–C coupling of electron-rich arenes such as phenyl ether derivatives. In general, heavy metal oxidizing reagents such as Ti(III) and V(V) are necessary for oxidative phenolic coupling. Kita *et al.* performed intramolecular aryl–aryl C–C coupling between two phenyl ether moieties by use of PIFA and BF₃·OEt₂ (Scheme 8.6).[6] The reactions seem to proceed through: (1) a mono-cation radical intermediate (**25**) that can induce intramolecular nucleophilic attack

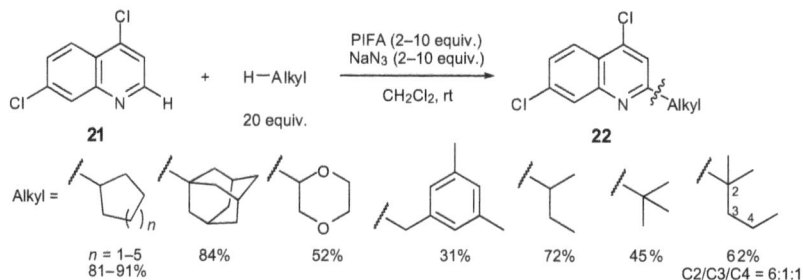

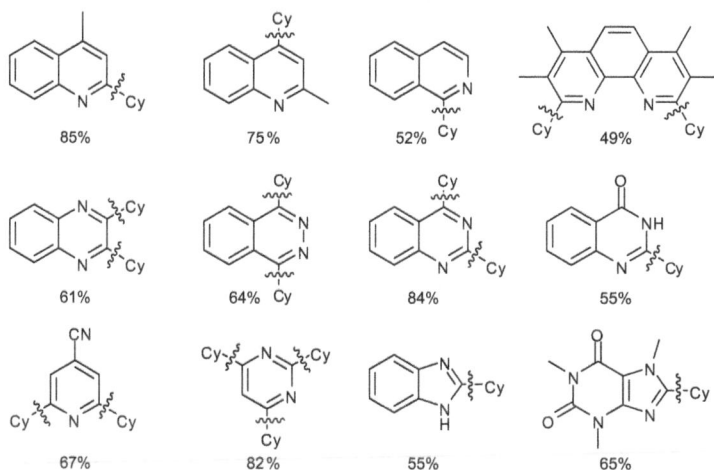

Scheme 8.5 CDC reaction between nitrogen-containing heterocycles and alkanes.

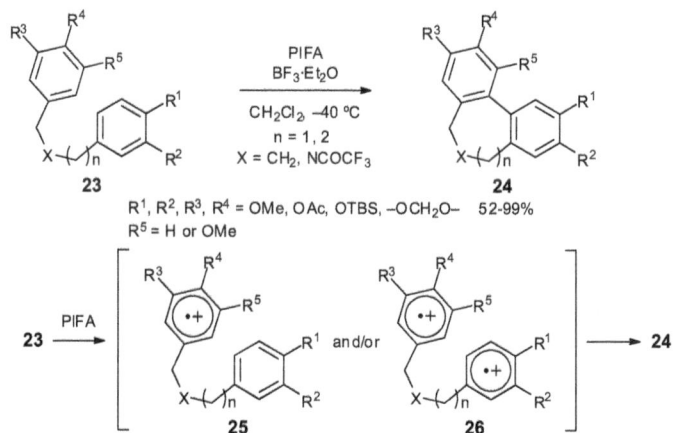

Scheme 8.6 Intramolecular oxidative coupling of phenyl ethers using PIFA.

by the other aromatic ring; or (2) a dication diradical intermediate (**26**) that can be coupled to each other. Intramolecular coupling between anisole and naphthoquinone moieties is also possible (Scheme 8.7).[7]

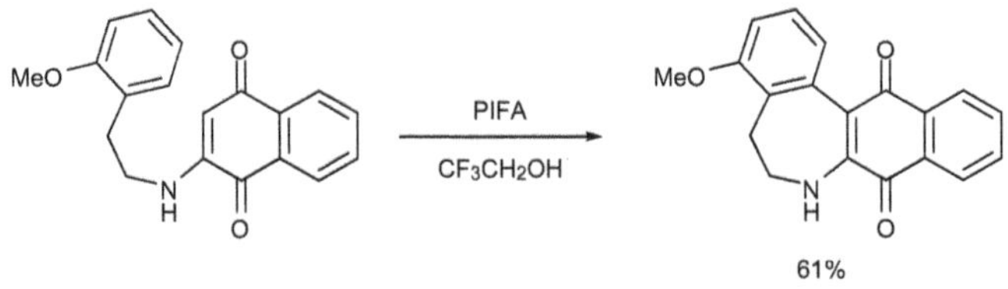

Scheme 8.7   Cyclization of an *ortho*-substituted anisole with PIFA.

Table 8.1   PIFA-Induced intramolecular oxidative coupling of norbelladine derivatives under the various reaction conditions.

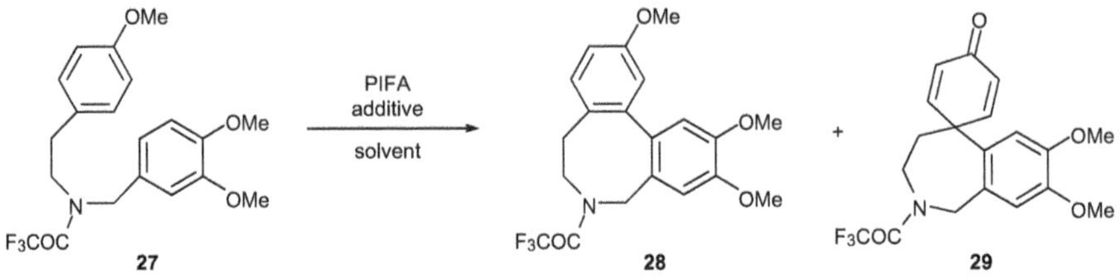

					Yield%	
Entry	Additive	Solvent	Temperature/°C	Time/h	28	29
1	BF$_3 \cdot$Et$_2$O	dry CH$_2$Cl$_2$	-40	0.16	88	0
2	BF$_3 \cdot$Et$_2$O	wet CH$_3$CN	-40	0.5	7	44
3	none	wet CH$_3$CN	-40 to rt	12	9	7
4	TfOH	wet CH$_3$CN	-40	0.5	25	42
5	TFA	wet CH$_3$CN	-40 to rt	12	14	19
6	montmorillonite K10	wet CH$_3$CN	-40 to rt	12	18	27
7	Nafion-H	wet CH$_3$CN	-40 to rt	12	16	23
8	H$_4$[SiW$_{12}$O$_{40}$]	dry CH$_3$CN	-40 to rt	0.66	0	74

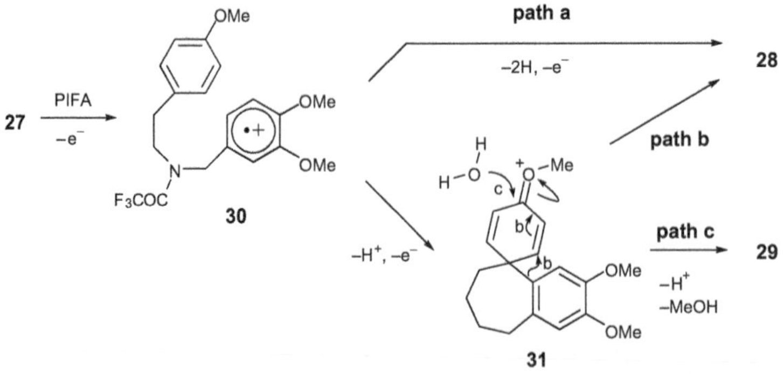

The substrate scope, the optimum reaction conditions and the presumed reaction pathway in a series of intramolecular couplings of phenyl ether derivatives have been well investigated by Kita *et al.*[8] In the reaction of norbelladine derivative **27**, which has 4-methoxyphenyl and 3,4-dimethoxy-phenyl groups, normal aryl–aryl C–C coupling products (**28**) are exclusively obtained under conditions with $BF_3 \cdot OEt_3$ and PIFA in dry CH_2Cl_2 (Table 8.1, entry 1). On the other hand, the use of wet CH_3CN as the solvent affords unprecedented spirodienone (**29**) along with **28** (entry 2). This tendency is also observed as long as wet CH_3CN is used in the reaction, regardless of the use of various Brønsted acids such as trifluoromethanesulfonic acid (TfOH), trifluoroacetic acid (TFA), montmorillonite K10 and Naflon-H (entries 3–7). When $H_4[SiW_{12}O_{40}]$ – one of the heteropolyacids – is used in the reaction in wet CH_3CN, spirodienone (**29**) forms selectively (entry 8). The plausible reaction mechanism was supported by ESR and UV–vis spectroscopic analysis (Scheme 8.8). The reactions proceed through the radical cation intermediate **30**, which is similar to one in Kita's previous report in 1994.[2] In the case of the reaction with highly methoxy-substituted starting materials such as **23**, aryl–aryl C–C coupling products such as **24** and **28** tend to form selectively through path a and/or path b *via* intermediates **30** and/or **31**. On the other hand, when one of aromatic rings is substituted with a *para*-methoxy group such as **27**, spirodienones (**29**) are dominantly produced through the intermediate **31** and path c.

Application to CDC reactions with a catalytic amount of PIFA is also possible in the presence of *meta*-chloroperbenzoic acid (*m*CPBA) as the co-oxidant (Scheme 8.9).[9] The homocoupling product **32** is synthesized from 4-bromo-1,2-dimethoxybenzene in moderate yields.

Reedijik reported that PIDA treatment in the presence of an excess of 2,6-dimethylphenol gives a biphenyl-4,4″-diol (**34**) effectively (Scheme 8.10).[10] A possible reaction mechanism can be explained by a one-electron oxidation of

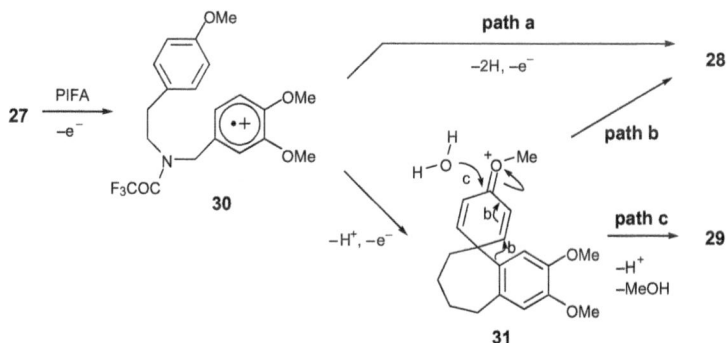

Scheme 8.8 Plausible reaction mechanisms for the reaction of **27**.

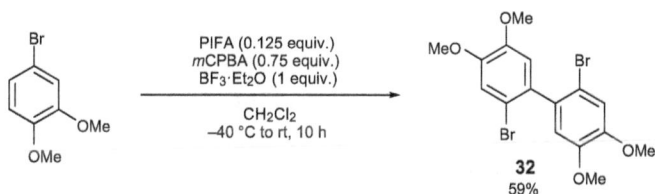

Scheme 8.9 PIFA-Catalyzed sp^2–sp^2 C–H homocoupling of 4-bromo-1,2-dimethoxybenzene.

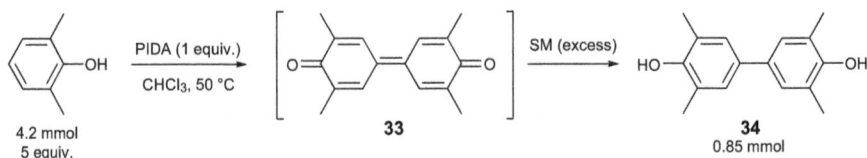

Scheme 8.10 PIDA-Promoted oxidative homocoupling of 2,6-dimethylphenol.

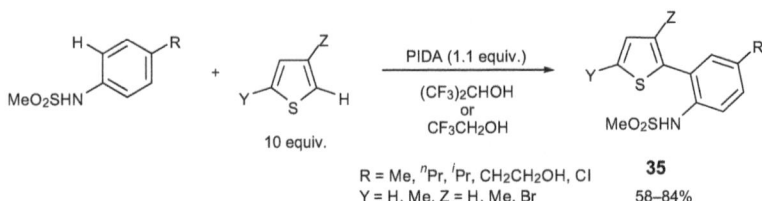

R = Me, ^{n}Pr, ^{i}Pr, CH₂CH₂OH, Cl
Y = H, Me, Z = H, Me, Br

Scheme 8.11 PIDA-Induced aromatic cross-coupling between sulfonanilides and thiophenes.

2,6-dimethylphenol by PIDA followed by homocoupling giving diphenoquinone **33** and subsequent reduction by the excess starting material (SM).

Intermolecular CDC of two different aromatic nuclei is important in organic synthesis but quite challenging due to the competitive formation of undesired homocoupling products. A practical example of intermolecular aromatic cross-coupling was reported by Canesi in 2007.[11] Treatment of sulfonanilides and thiophenes gives a variety of cross-coupling products (**35**) in good yields (Scheme 8.11).

Kita and co-workers demonstrated PIFA-promoted cross-coupling between naphthalenes and mesitylene derivatives (Scheme 8.12).[12] Based on their previous findings about reactions with hypervalent iodine(III) reagents *i.e.*; (1) the formation of a charge-transfer arene complex; (2) the generation of a cation radical intermediate through the single-electron transfer (SET) induced by the iodine(III) reagent; and (3) C–C coupling with other aromatic molecules and dehydrogenative oxidation, they chose naphthalene and electron-rich mesitylene derivatives as substrates. The conditions with PIFA and $BF_3 \cdot OEt_2$ enable highly regioselective reactions at the C2-position of naphthalene to furnish biaryls (**36**) in good yields.

Kita further developed metal-free cross-coupling between thiophenes and electron-rich arene such as 2-methoxynaphthalene, and 3,5-dimethoxytoluene using PhI(OH)OTs (HTIB) as a promoter to afford biaryls **37** (Scheme 8.13).[13] The reaction is also applicable to thiophene–pyrrole, thiophene–thiophene, and pyrrole–pyrrole cross-couplings. This reaction is expected to proceed through the formation of diaryl iodine(III) tosylate (**38**), and TMSBr may accelerate the subsequent nucleophilic attack from the other aromatic compound (H–Ar) by exchanging the –OTs group on a iodine atom with a –Br group (**39**) to give the intermediate **40**. Finally, the elimination of iodobenzene and deprotonation would furnish the coupling product **41**.

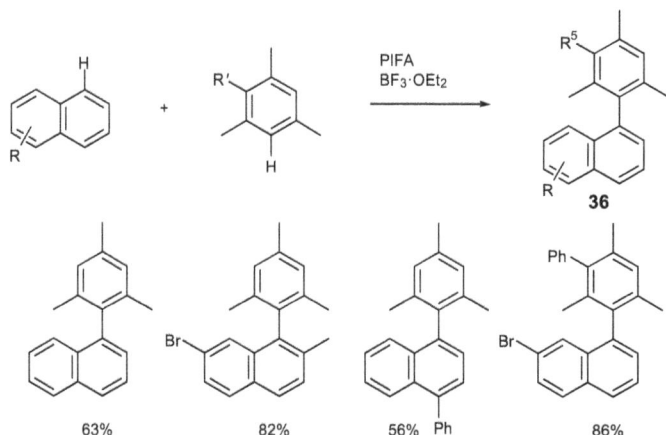

Scheme 8.12 Scope of PIFA-induced oxidative cross-coupling of different aromatics.

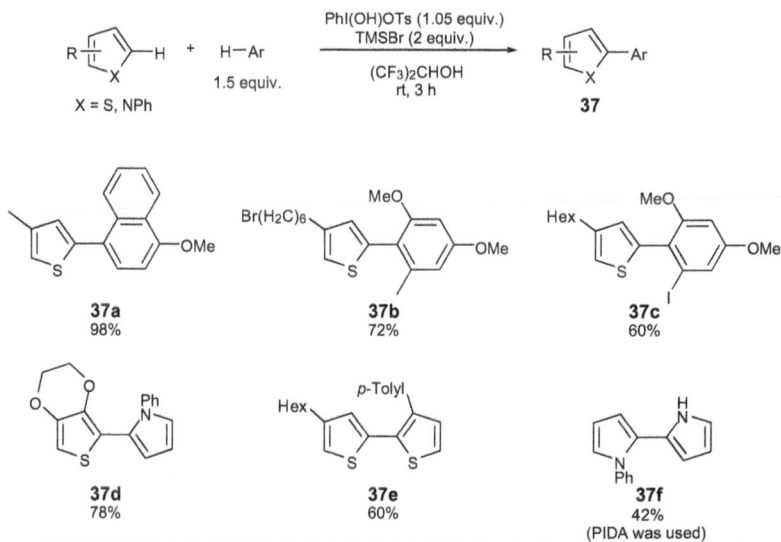

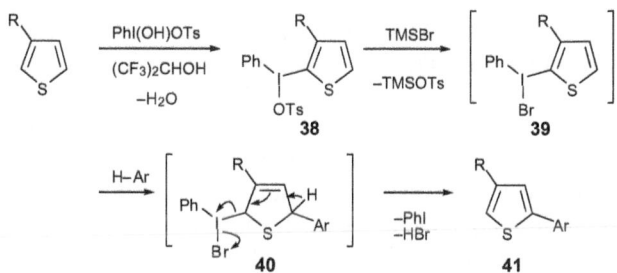

Scheme 8.13 Metal-free cross-coupling of unfunctionalized aromatics.

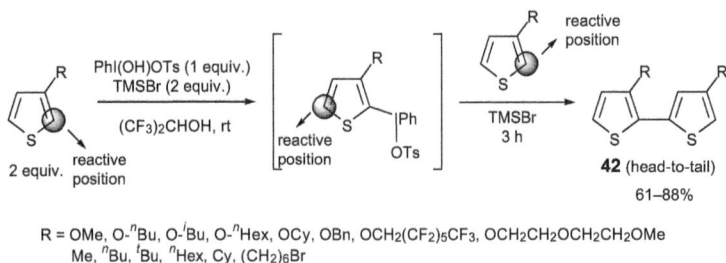

Scheme 8.14   Metal-free regioselective oxidative biaryl coupling leading to head-to-tail bithiophenes.

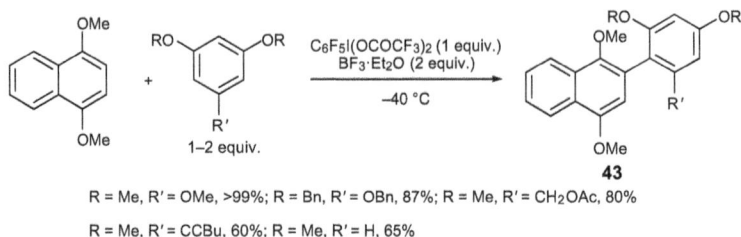

R = Me, R' = OMe, >99%; R = Bn, R' = OBn, 87%; R = Me, R' = CH₂OAc, 80%

R = Me, R' = CCBu, 60%; R = Me, R' = H, 65%

Scheme 8.15   FPIFA-Induced cross-coupling between dimethoxynaphthalene and dimethoxy/trialkoxybenzene derivatives.

Bithiophenes, especially head-to-tail dimers, are useful precursors for high quality, well-defined, regioregular, oligo- and polythiophenes and their derivatives. Kita developed a useful synthetic method for the regioselective homocoupling of thiophenes (Scheme 8.14).[14] Switching the reactive position on thiophene by employing PhI(OH)OTs, they achieved the synthesis of the head-to-tail dimer **42** with a variety of 3,3″-dialkyl- or 3,3″-dialkoxy substituents.

Cross-coupling reactions between dimethoxynaphthalene and dialkoxyl/trialkoxybenzene derivatives are also possible in the reaction with $(C_6F_5)I(OCOCF_3)_2$ (FPIFA) which is a stronger oxidant than PIFA (Scheme 8.15).[15]

A variety of substituted indoles (**45**) can be synthesized from the reaction of *N*-aryl enamines (**44**) with PIDA (Scheme 8.16).[16] The formation of the iodonium intermediate **46** followed by an intramolecular S_N2''-type cyclization gives the intermediate **47**, which tautomerizes to **45** after the formation of deprotonated **48**.

Recently, Kita demonstrated the efficient synthesis of hexahydroxytriphenylene (**49**) discotic liquid crystals, under simple and mild reaction conditions with catechol, PIDA, and methanesulfonic acid (Scheme 8.17).[17]

Hypervalent iodine(III) reagents can be applied to aryl–aryl C–H cross-couplings. Recently, Antonchick achieved CDC between various *N*-heterocycles and aldehydes (Scheme 8.18).[18] The use of PIFA and TMSN₃ is effective at promoting this reaction to give a variety of acyl *N*-heterocyclic compounds

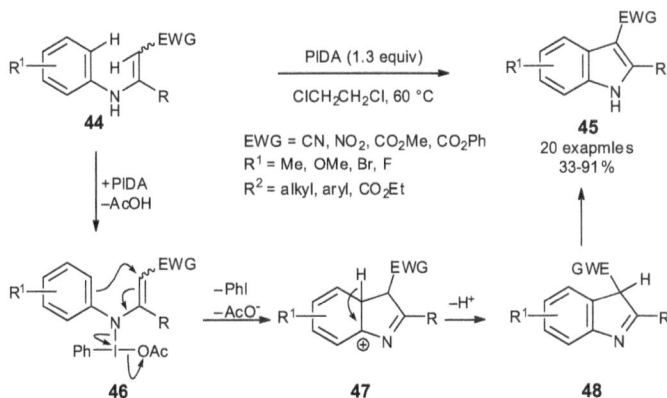

Scheme 8.16 PIDA-Mediated synthesis of indoles from *N*-aryl enamines. EWG = electron-withdrawing group.

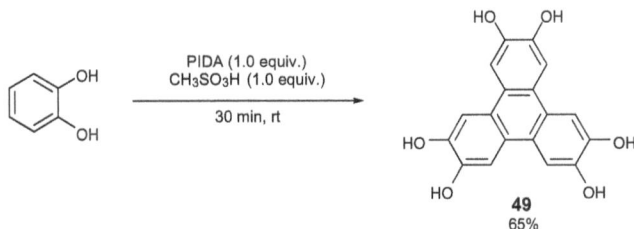

Scheme 8.17 PIDA-Induced facile oxidative trimerization to hexahydroxy-triphenylene.

50 with broad scope. The utility of this method was highlighted by the synthesis of several natural products and related compounds such as papaveraldine, pulcheotine, thalimicrinone and 1-acetyl-β-carboline. The key to this reaction is the formation of a nucleophilic acyl radical inter-mediate in the presence of PIFA and TMSN$_3$.

8.1.1.3 *sp^3–sp^3 C–H Coupling*

Coupling between sp^3 C–H bonds is an attractive but challenging transfor-mation in organic synthesis. Oxidative cyclopropanation involving formal sp^3–sp^3 C–H coupling with 1,3-dicarbonyl compounds **51** and PIDA was achieved by Fan *et al.* (Scheme 8.19).[19] The reaction seems to proceed *via* the iodonium intermediate **53** produced by the reaction of enol tautomer **51-enol** with acetyl hypoiodite (AcOI) which could be generated by the reaction of PIDA with Bu$_4$NI. The successive intramolecular nucleophilic attack of an azinate moiety forms nitrocyclopropane (**54**) which finally give the product **52**. They also performed a one-pot asymmetric Michael addition of 1,3-dicarbonyl compounds to nitro-olefins catalyzed with bifunctionalized thiourea (**55**) and the subsequent oxidative cyclopropanation with high diastereo- and enantioselectivities.

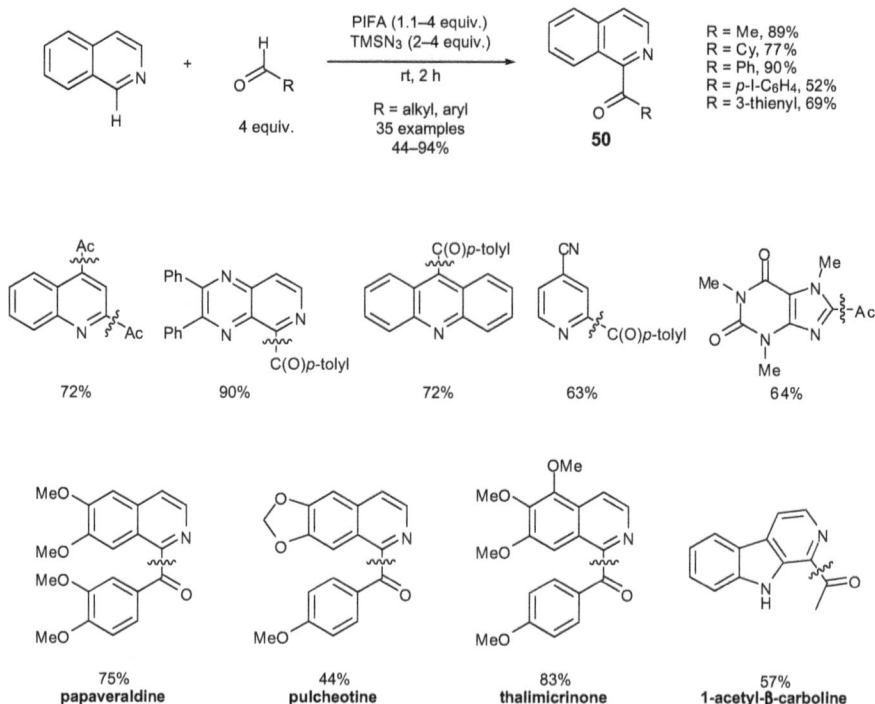

R = Me, 89%
R = Cy, 77%
R = Ph, 90%
R = p-I-C₆H₄, 52%
R = 3-thienyl, 69%

PIFA (1.1–4 equiv.)
TMSN₃ (2–4 equiv.)

rt, 2 h

R = alkyl, aryl
35 examples
44–94%

50

4 equiv.

72%   90%   72%   63%   64%

75%   44%   83%   57%
papaveraldine   pulcheotine   thalimicrinone   1-acetyl-β-carboline

Scheme 8.18 PIFA- and TMSN₃-induced CDC reaction between *N*-heterocycles and aldehydes.

The same type of reactions can be carried out by using tricarbonyl compounds (**56**) with iodosobenzene (PhIO), giving triacyl cyclopropanes (**57**) in good yields (Scheme 8.20).[20]

Liang performed the PIDA-induced sp³ C–H bond functionalization of tetrahydroisoquinolines (THIQs) (**58**) with nitroalkanes or malonates as the nucleophile (Scheme 8.21).[21] The reactions involve the oxidation of **58** to an iminium intermediate (**60**) and nucleophilic attack to furnish the coupling products **59** and **60**.

8.1.2 Peroxides

Bringmann and Lin used *tert*-butylperoxide (DTBP) in the oxidative dimerization of carbazole (**62**) to synthesize **63**, which is further oxidized with pyridinium cholorochromate (PCC) to give bismurrayaquinone A (Scheme 8.22).[22]

An electron-rich arene (**64**) is also reactive toward oxidative dimerization by DTBP to give an analog of mastigophorenes A and B (**65**) in good yield (Scheme 8.23).[23]

Li and Itami demonstrated a DTBP-promoted CDC reaction between pyridine-*N*-oxide and cycloalkanes in the absence of any transition metals (Scheme 8.24).[24] Unactivated sp³-hybridized C–H bonds on cycloalkanes, norbornane, and 1,4-dioxane can also undergo this reaction, giving the

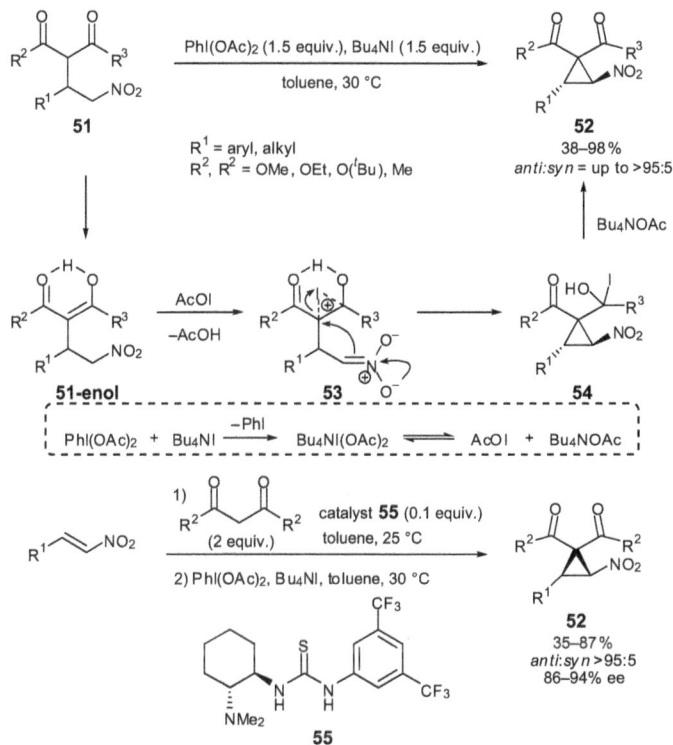

Scheme 8.19 PhI(OAc)$_2$/Bu$_4$NI-Induced oxidative cyclopropanation of Michael adducts.

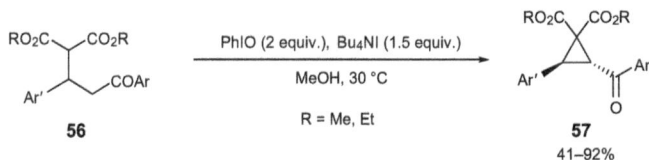

Scheme 8.20 Oxidative cyclization of Michael adducts of malonates with chalcones.

2,6-dicycloalkylpyridine-*N*-oxide (**66**) in good yields, although the 2,4,6-dicycloalkylpyridine-*N*-oxide (**67**) also forms to some extent. In some cases of using substituted substrates, selective mono-alkylations and di-alkylations are possible.

A metal-free CDC reaction between azoles and alcohols/ethers was reported by Wang and co-workers (Scheme 8.25).[25] The cross-couplings occur at the C2-position of the azoles and at the α-positions of alcohols or ethers in the presence of 4 equiv. of *tert*-butyl hydroperoxide (TBHP) to afford a variety of C2-alkylated benzoxazoles (**68**) in good yields. A possible reaction mechanism involves the free-radical process shown in Scheme 8.25. The authors

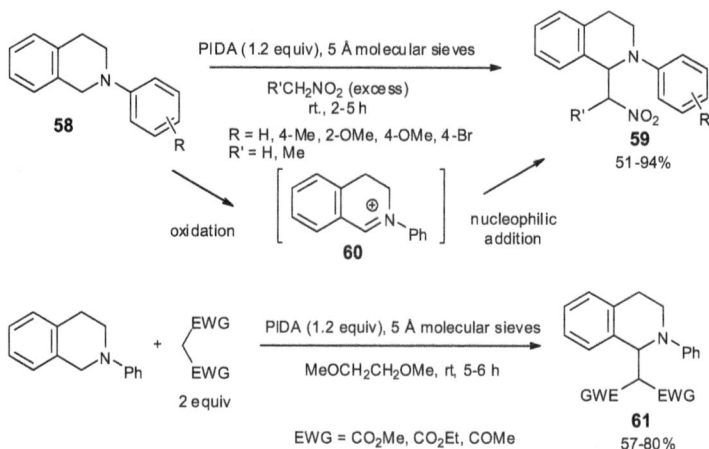

Scheme 8.21 PIDA-Mediated cross-coupling of THIQs with nitroalkanes and malonates.

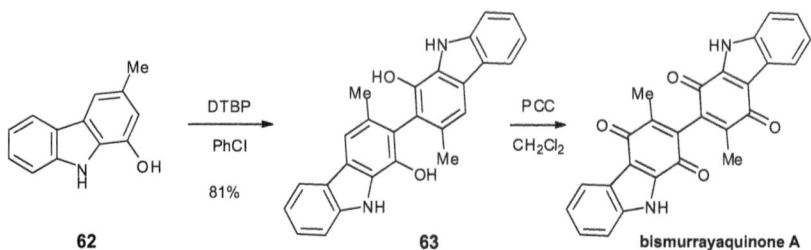

Scheme 8.22 Synthesis of bismurrayaquinone A using DTBP-induced oxidative dimerization of carbazole.

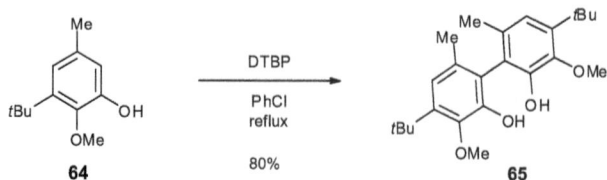

Scheme 8.23 DTBP-Induced oxidative coupling of a mono-phenolic derivative.

verified that the addition of 2,2,6,6-tetramethylpiperidine-1-oxyl (TEMPO) or ascorbic acid, which are radical scavengers, suppressed the reaction.

Qu and Guo applied TBHP-promoted CDC to the reaction of cycloalkanes and various heteroaromatics such as purines (nucleosides), benzothiazoles, and benzoxazoles (Scheme 8.26).[26] The reaction shows broad substrate scope, which is represented by the reaction with purine **69**. Cycloalkylated products such as **70** were obtained in good yields.

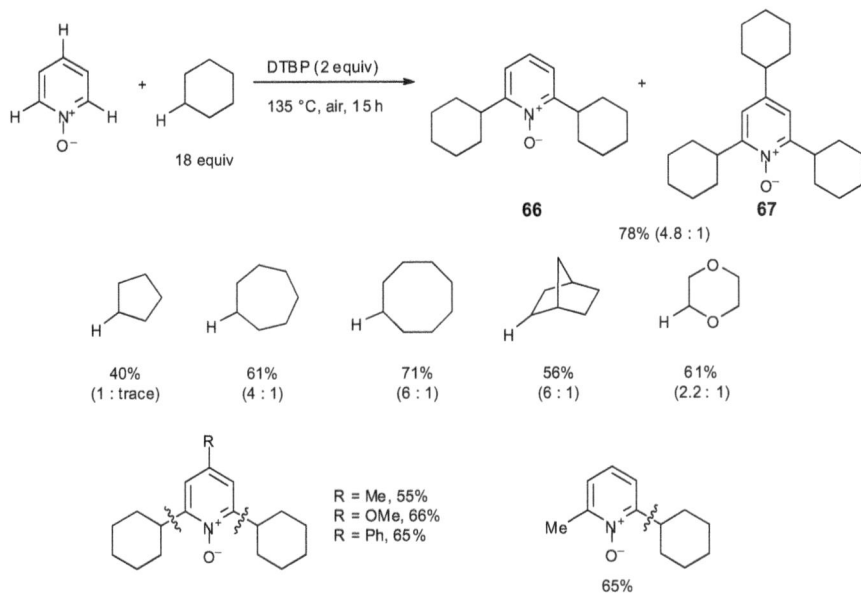

Scheme 8.24 Transition metal free cross-coupling of pyridine-*N*-oxide with cycloalkanes.

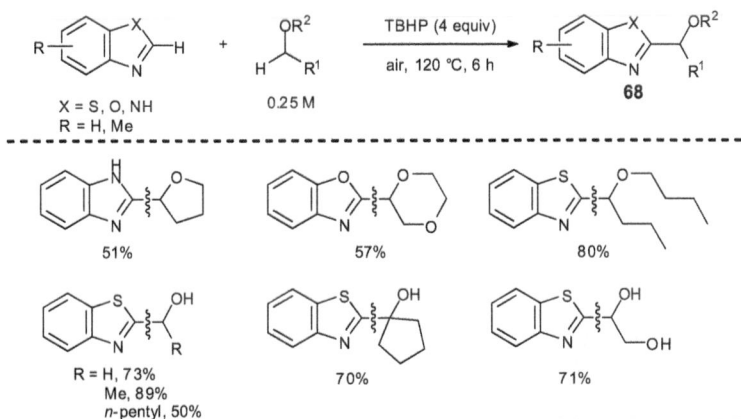

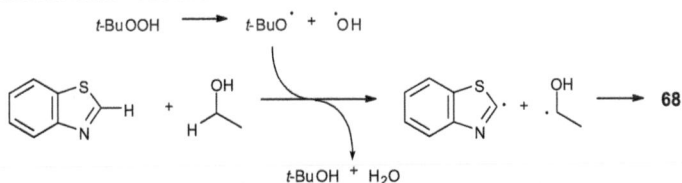

Scheme 8.25 CDC reaction between azoles and alcohols/ethers in the presence of TBHP.

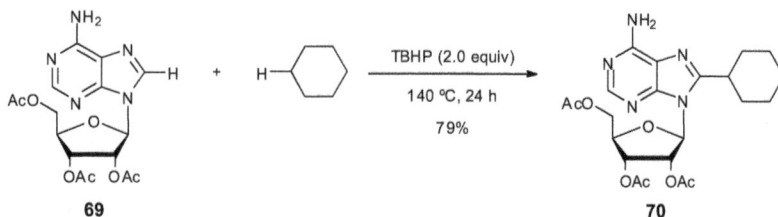

Scheme 8.26 CDC of purine derivative **69** with cyclohexane.

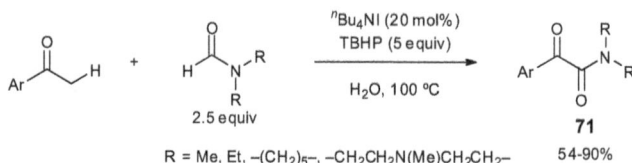

R = Me, Et, –(CH$_2$)$_5$–, –CH$_2$CH$_2$N(Me)CH$_2$CH$_2$–   54–90%

Scheme 8.27 ^{n}Bu$_4$NI-Catalyzed synthesis of α-ketoamides from aryl methyl ketones and *N,N*-dialkylformamide using TBHP as the oxidant.

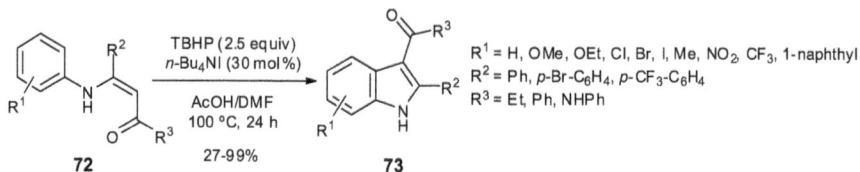

R^1 = H, OMe, OEt, Cl, Br, I, Me, NO$_2$, CF$_3$, 1-naphthyl
R^2 = Ph, *p*-Br-C$_6$H$_4$, *p*-CF$_3$-C$_6$H$_4$
R^3 = Et, Ph, NHPh

Scheme 8.28 Iodide ion catalyzed CDC reaction for the synthesis of indole derivatives.

Mai and Qu demonstrated formal C–H coupling between aryl methyl ketones and *N,N*-dialkylformamide using TBHP and a catalytic amount of ^{n}Bu$_4$NI to form α-ketoamides (**71**) (Scheme 8.27).[27] The plausible reaction mechanism is expected to be the radical oxidative coupling.

Nagano and Li reported an intramolecular C–C bond-forming CDC reaction for the synthesis of indole derivatives (Scheme 8.28).[28] Cyclization of aniline derivatives (**72**) effectively proceeds in the presence of a catalytic amount of tetrabutylammonium iodide and 2.5 equiv. of TBHP to afford a variety of 3-acyl-2-aryl indoles (**73**). The catalytic formation of iodine (I$_2$) by TBHP seems to be essential for the reaction to occur.

8.1.3   Quinones

3-Indolylquinone substructures can be found in a number of biologically significant natural products such as asterriquinone. Pirrung developed an efficient synthesis of 3-indolylquinones (**74**) from various indoles and 2,5-dichlorobenzoquinone in the presence of DDQ (Scheme 8.29).[29] The subsequent hydrolysis with NaOH gives 2,5-dihydroxyl-3-indolylquinones (**75**) which are the substructure of asterriquinone.

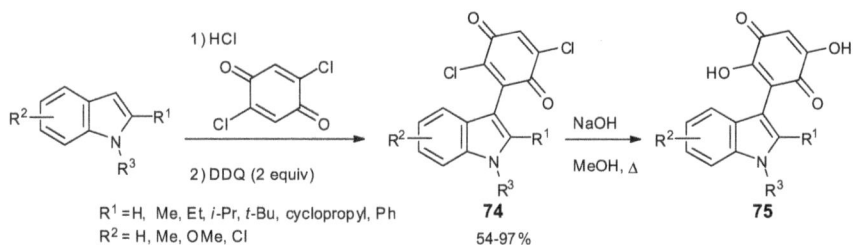

Scheme 8.29 Acid-catalyzed condensation of indole with 2,5-dichlorobenzoqui-none and subsequent oxidation with DDQ.

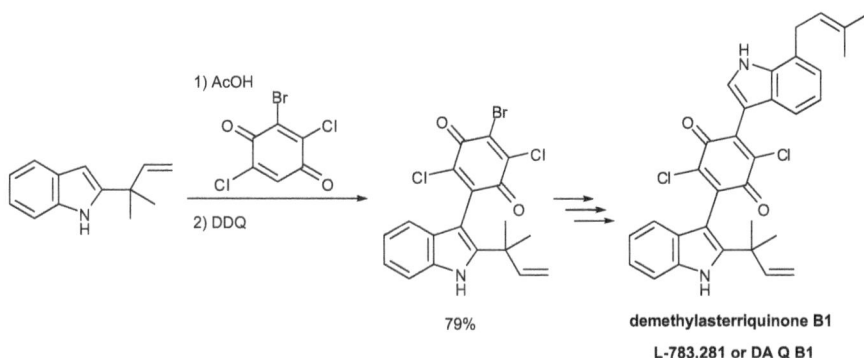

Scheme 8.30 Total synthesis of demethylasterriquinone B1.

They also applied this acid-catalyzed condensation and oxidation strategy to the total synthesis of demethylasterriquinone B1, which inhibits transmembrane signaling by the epidermal growth factor (EPG) oncogene product (Scheme 8.30).[30] Willis also synthesized demethylasterriquinone A1 in the same manner.[30]

Li achieved metal-free CDC reaction between simple ketones and benzyl ethers by the simple addition of DDQ (Scheme 8.31).[31] The oxidation of benzyl ether to a benzoxy cation by DDQ followed by the enolization of the ketone and its nucleophilic attack is proposed as the reaction mechanism to afford the β-alkoxy-β-arylketone (**76**).

The Pummerer- and Knoevenagel-type reactions of benzyl sulfide with 1,3-dicarbonyl compounds in the presence of o-chloranil were achieved by Z. Li et al. (Scheme 8.32).[32]

The combination of N-phenyl THIQs and nitromethane is also possible, to give the corresponding C–C coupling products (**77**) (Scheme 8.33).[33]

Todd revealed that the three-component iminium ion complex **79** is a reactive intermediate in the oxidative C–C coupling reaction between iso-quinolines and DDQ, and the yields of product **78** from the 'indirect' re-action are much higher than those from the 'direct' reaction (Scheme 8.34).[34]

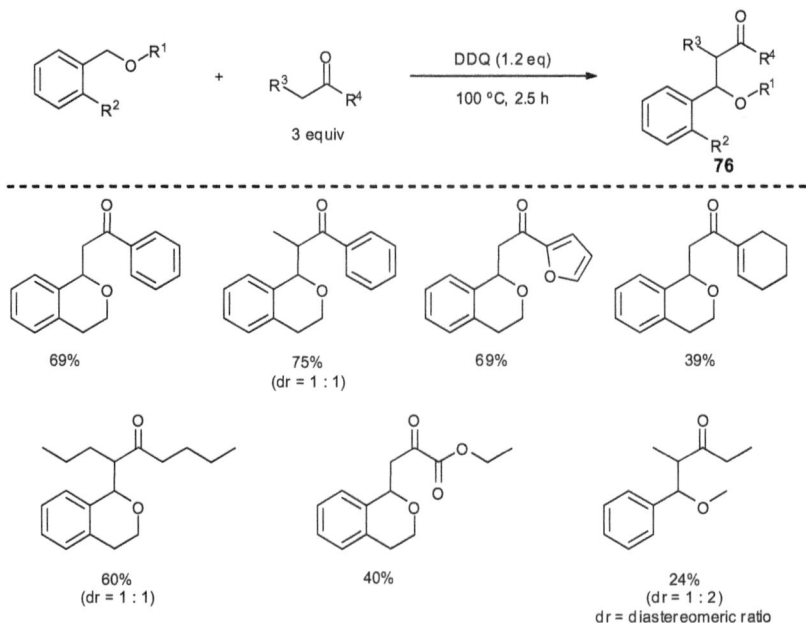

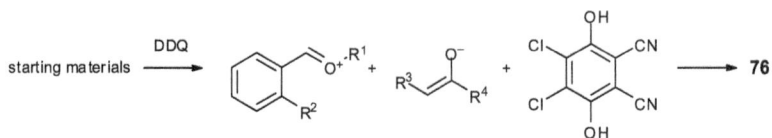

Scheme 8.31   DDQ-Mediated CDC reaction between benzyl ethers and ketones.

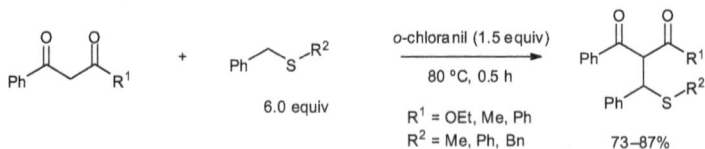

Scheme 8.32   *o*-Chloranil-mediated Pummerer- and Knoevenagel-type reactions.

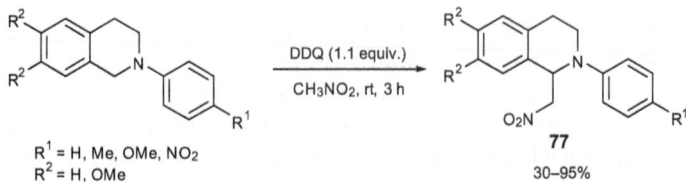

Scheme 8.33   Oxidative coupling of *N*-phenyl THIQs and nitromethane with DDQ.

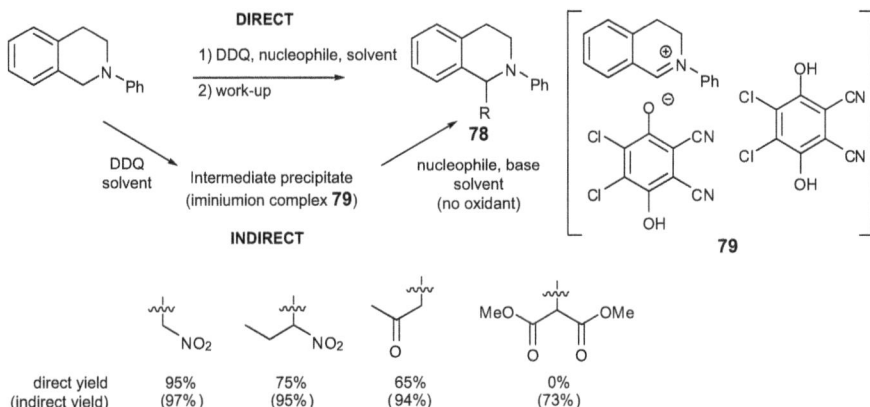

Scheme 8.34 DDQ-Mediated oxidative addition reactions between isoquinolines, or the intermediate iminium ion complex (**79**), and a range of nucleophiles.

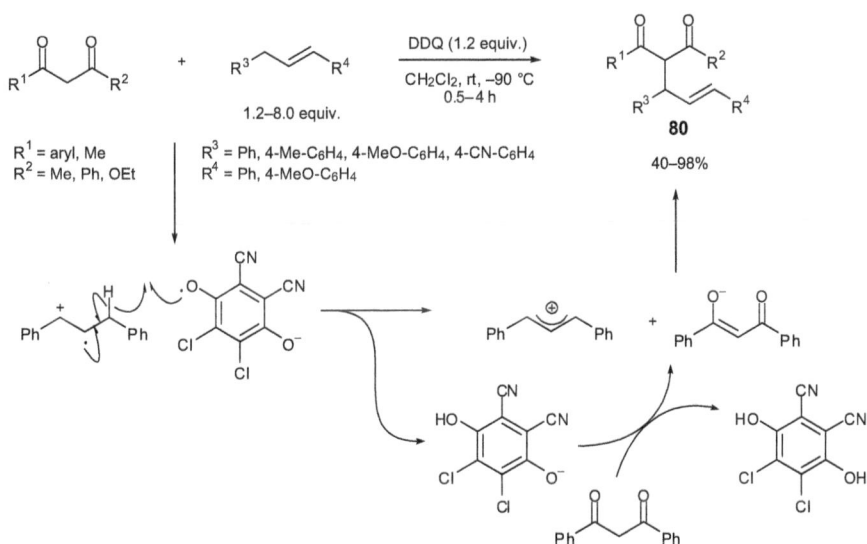

Scheme 8.35 Allylation of 1,3-dicarbonyl compounds by DDQ.

Allylation of 1,3-dicarbonyl compounds proceeds effectively to afford 2-ally-1,3-diketone **80** when using diarylpropene and DDQ (Scheme 8.35).[35] The oxidation of diarylpropene by DDQ followed by the nucleophilic attack of enolate is proposed as the possible reaction mechanism.

Similar to the reaction with diarylpropene, the use of diarylpropynes enables propargylation of 1,3-dicarbonyl compounds to give **81** (Scheme 8.36).[36]

Oxidative C–C bond formations arc also possible at the benzylic position in the case of reaction with 1,3-diaryl-1,3-diketones, and allylated and benzylated compounds **82** are obtained in moderate yields (Scheme 8.37).[37]

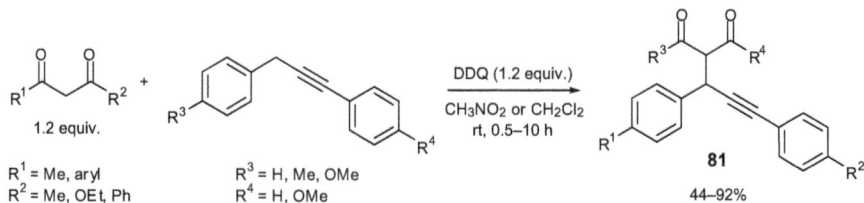

Scheme 8.36   Propargylation of 1,3-dicarbonyl compounds by DDQ.

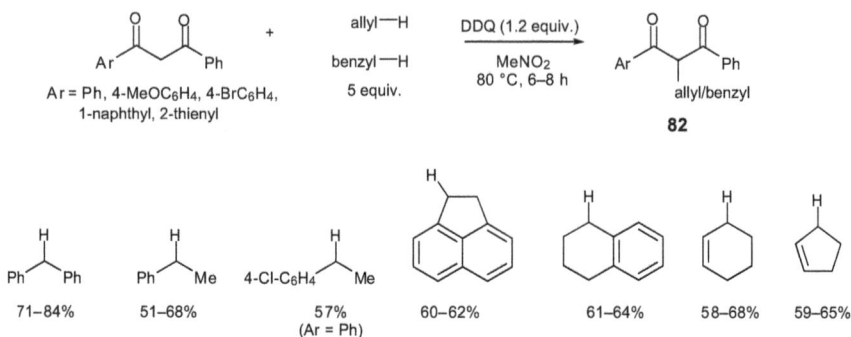

Scheme 8.37   Oxidative coupling between benzylic or allylic C–H bond and various 1,3-diaryl-1,3-diketones.

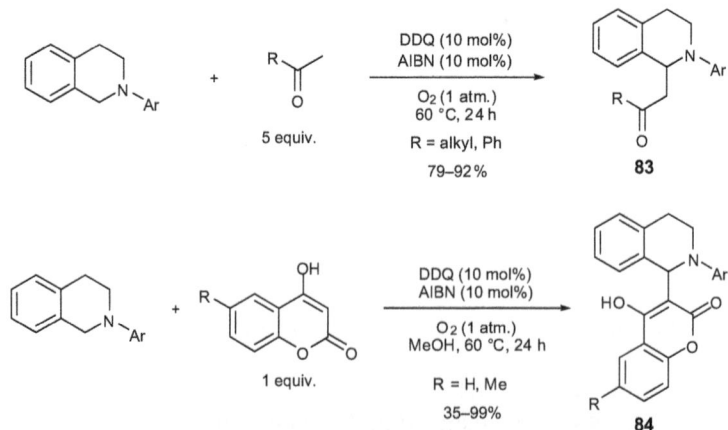

Scheme 8.38   DDQ-Catalyzed Mannich reaction of *N*-aryl THIQs.

Ketones and 4-hydroxycoumarins are applicable as nucleophiles in the CDC reaction of *N*-aryl THIQs (Scheme 8.38).[38] In addition, DDQ- and azobisisobutyronitrile (AIBN)-catalyzed coupling can be achieved under aerobic conditions to give the corresponding coupling products **83** and **84**.

Scheme 8.39 Synthesis of ring-fused THIQs.

Kim demonstrated a DDQ-mediated intramolecular CDC reaction in the synthesis of ring-fused THIQs (**85**) bearing a formyl group (Scheme 8.39).[39]

The Scholl reaction is one of the oldest and most useful C–C coupling reactions, and is often used in the synthesis of polycyclic aromatic hydrocarbons (PAHs). In general, a large amount of metal oxidant (such as $FeCl_3$) is employed in this reaction, and chlorinated by-products are produced in many cases. Rathore found that DDQ and protic acid effectively promote the Scholl reaction of various 1,2-diarylbenzenes (**86**) to give triphenylenes (**87**) in excellent yields (Scheme 8.40).[40] The advantages of this reaction over typical metal oxidant promoted Scholl reactions are: (1) an excess of DDQ is not required; (2) no chlorinated by-products are produced; and (3) recovery and re-use of reduced hydroquinone **DDQ-H$_2$** is possible. The reaction of other diaryls tethered with methylene and propylene, and the intermolecular reaction also proceeds effectively. Furthermore, the treatment of hexaarylbenzene (**88**) with DDQ quantitatively affords collonene (**89**).

The possible reaction pathways in the DDQ-mediated Scholl reaction were also well investigated by Rathore, and the authors propose that the reaction likely proceeds through a radical cation intermediate (Path A) (Scheme 8.41).[41] They also referred to the existence of Path B which involves the formation of an arenium ion intermediate.

Recently, a remarkable example of a DDQ-induced Scholl reaction was demonstrated by Itami and Scott (Scheme 8.42).[42] Employing pentabiphenylcorannulene (**90**) with DDQ/triflic acid gives a glossy warped nanographene (**91**) which is an entirely new class of nanocarbon, and the largest PAH other than fullerenes and their derivatives.

8.1.4   Other Oxidants

In 1980, Szántay found that tetraethylammonium[di(acyloxy)iodate(I)] derivatives effectively promote the CDC reaction of **92** to give **93** (Scheme 8.43).[43]

Barluenga reported that the intramolecular formal Friedel–Crafts acylation of arenes with aldehydes can be promoted by IPyBF$_4$ and HBF$_4$

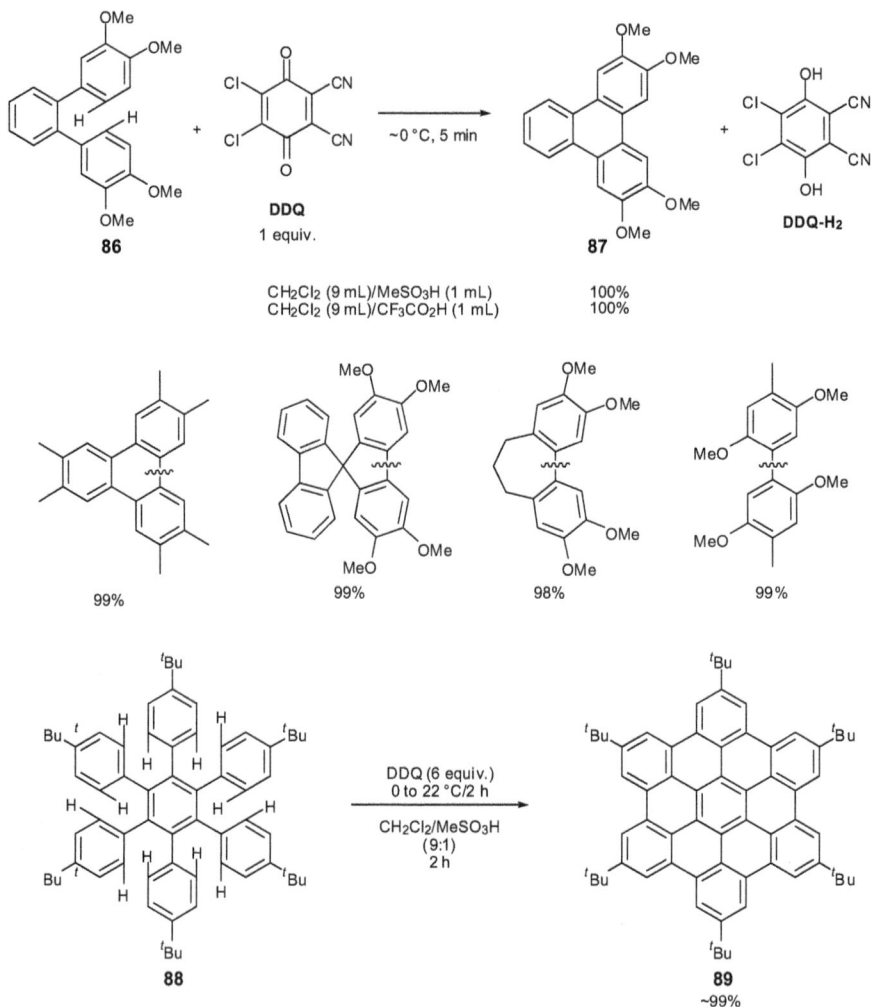

Scheme 8.40 DDQ- and protic acid promoted Scholl reactions of diaryl compounds.

(Scheme 8.44).[44] This is a straightforward method for synthesizing benzo-cyclic ketones (**94**) from aldehydes and arenes.

The CDC reaction between indoles and quinones or benzoquinone is possible with a catalytic amount of I_2 under ultrasonic irradiation, giving (3-indolyl)quinones in 30–90% yields (Scheme 8.45).[45]

The group of Wu reported that the CDC reaction between aryl methyl ketones and electron-rich arenes, such as aniline derivatives, can be pro-moted by I_2 and CF_3SO_3H to afford α,α-diarylketones **96** (Scheme 8.46).[46]

The combination of I_2 and ^tBuOK is effective for producing oxindole de-rivatives (**97**) (Scheme 8.47).[47] A radical process by I_2 is expected as one of the

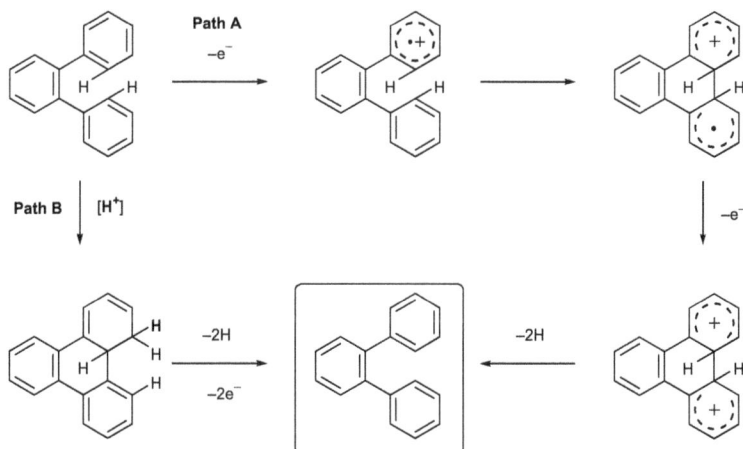

Scheme 8.41 Possible reaction pathway of the DDQ-mediated Scholl reaction.

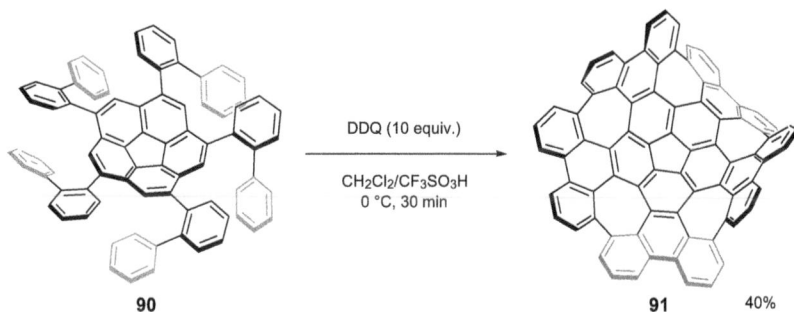

Scheme 8.42 DDQ-mediated synthesis of warped nanographene.

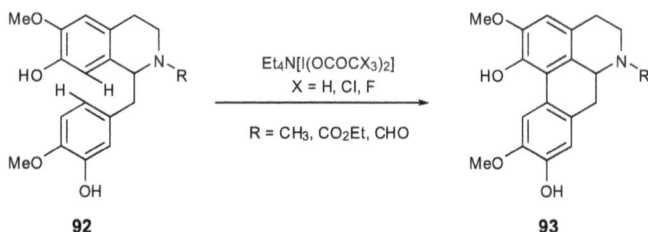

Scheme 8.43 Intramolecular CDC reaction between two phenol moieties.

possible reaction mechanisms. The reaction of pyrrolidonecarboxamide (**98**) gives spiro-fused oxindole (**99**), resembling the cores of coerulescine and horsfiline.

One example of iodine-promoted CDC coupling, using 5-bromoindole and acetophenone was reported by Chen (Scheme 8.48).[48]

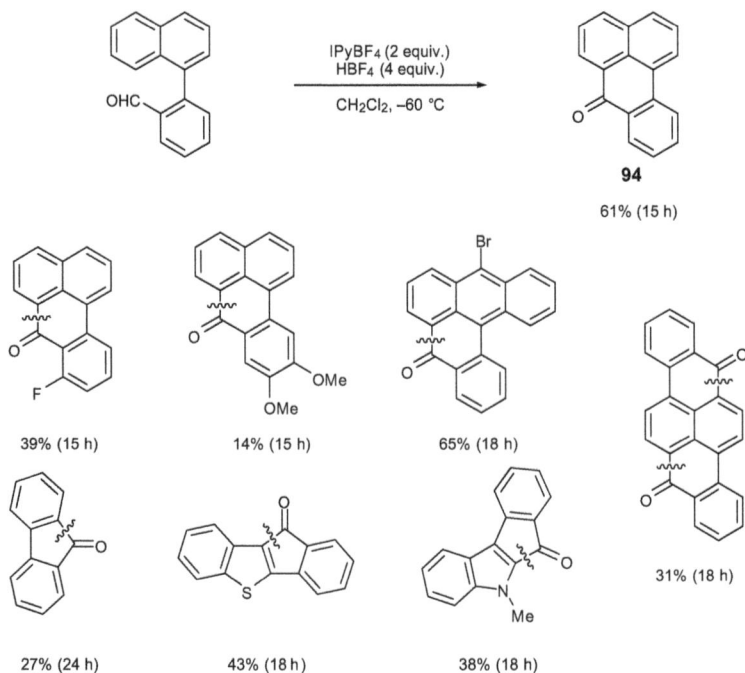

Scheme 8.44   Polycyclic aromatic ketones obtained by the formal intramolecular Friedel–Crafts acylation of arenes with aldehydes.

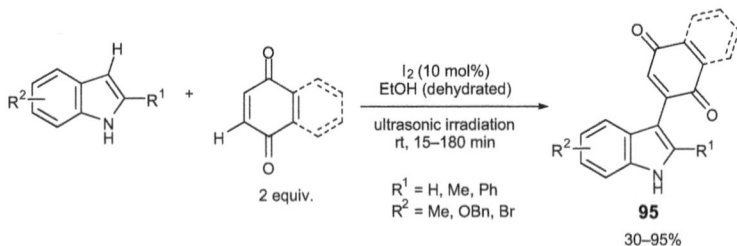

Scheme 8.45   Iodine-catalyzed CDC reaction between indoles and quinones.

Using a catalytic amount of I_2 in the presence of H_2O_2, CDC reaction of THIQs was achieved by Itoh (Scheme 8.49).[49] Nitroalkane, ketone, and dimethyl malonate can be introduced to produce **100** in good yields. The generation of HOI from the oxidation of I_2 or HI by H_2O_2 is expected to play an important role in the formation of the iminium ion intermediate **101**, which reacts with the nucleophiles.

A similar I_2-catalyzed CDC reaction was also reported by the group of Prabhu at almost the same time (Scheme 8.50).[50] They conducted the reactions under aerobic conditions or in the presence of TBHP as the

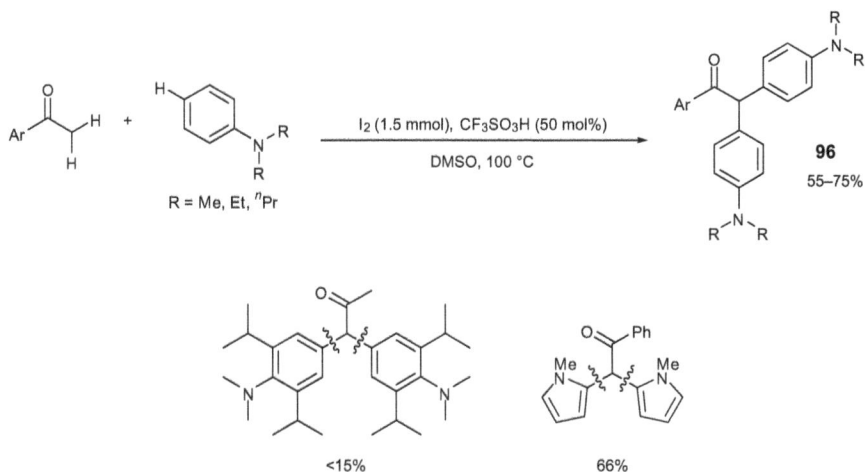

Scheme 8.46 Scope of the CDC reaction between aryl methyl ketones and electron-rich arenes.

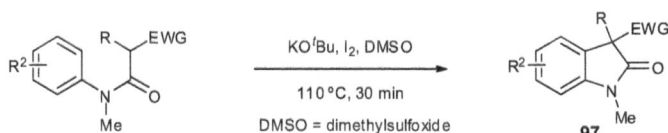

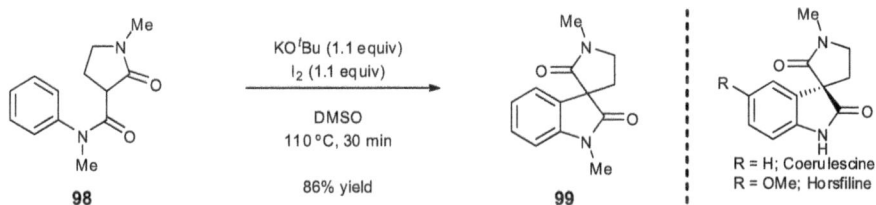

Scheme 8.47 Iodine-promoted intramolecular CDC reaction involving sp^2–sp^3 C–C bond formation.

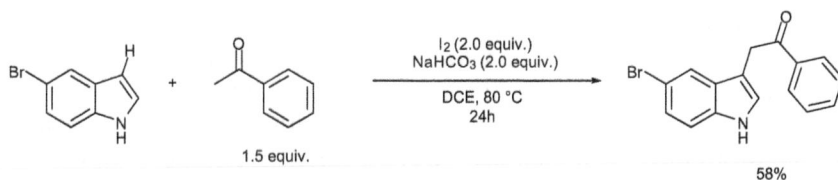

Scheme 8.48 Iodine-promoted CDC reaction between 5-bromoindole and acetophenone.

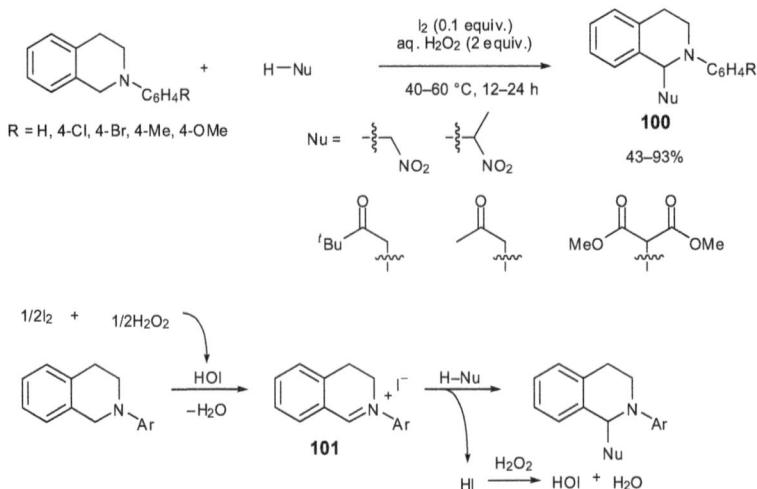

Scheme 8.49   Iodine-catalyzed CDC reaction of THIQs with various nucleophiles.

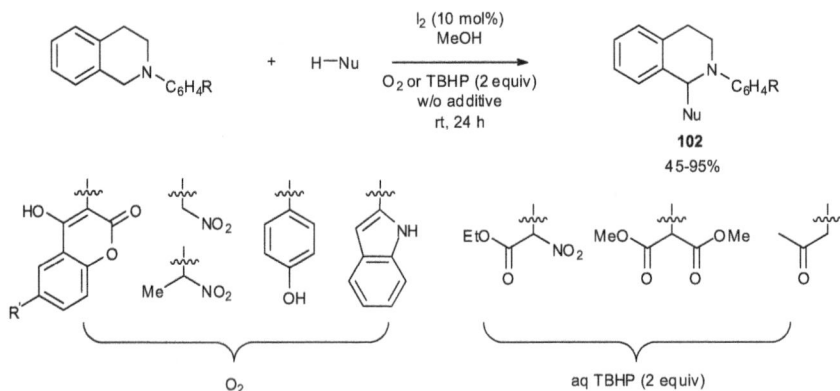

Scheme 8.50   Iodine-catalyzed CDC reaction of THIQs with various nucleophiles under aerobic conditions.

oxidant, and demonstrated the availability of a broad scope of nucleophiles such as 4-hydroxycoumarine, nitroalkane, phenol, indole and carbonyl compounds to give a variety of functionalized THIQs **102**.

Catalytic amounts of TEMPO are effective for the oxidative C–C coupling between 9,10-dihydroacridines with nitroalkanes, ketones, dimethylmalonate and malononitrile to afford coupling products **103** (Scheme 8.51).[51]

The simple aerobic CDC reaction between xanthene and various ketones or 1,3-dicarbonyl compounds was demonstrated by Klussmann (Scheme 8.52).[52] A catalytic amount of methanesulfonic acid is essential for this reaction. Coupling of acridanes and THIQs with ketones is also possible under these conditions.

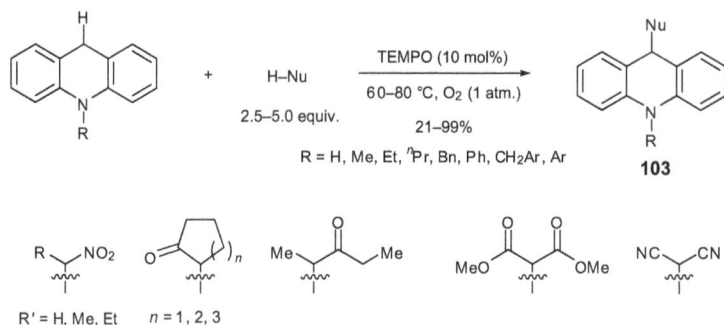

Scheme 8.51 TEMPO-Catalyzed aerobic oxidative C–C coupling of 9,10-dihydroacridines with carbon nucleophiles.

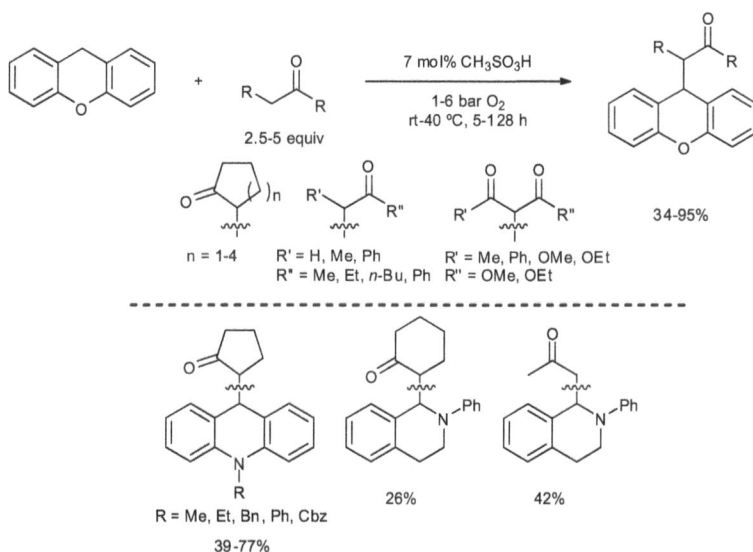

Scheme 8.52 Acid-catalyzed aerobic CDC reaction between xanthene and carbonyl compounds.

Selectfluor® works as the oxidant in the Mannich-type reaction of a piperidine derivative (**104**) (Scheme 8.53).[53] The piperidine derivative can be oxidized by Selectfluor® to form an iminium ion intermediate that can be attacked by nucleophiles such as furan or *N*-methylpyrrole to form the formal CDC products **105**.

The metal- and solvent-free CDC coupling of quinolines **106** with indoles and pyrroles proceeds smoothly in the presence of stoichiometric amounts of HCl (Scheme 8.54).[54] The reaction mechanism involves the nucleophilic attack of indole on the quinolinium ion (**108**) followed by the oxidation of the coupling product by **108** to give coupling products **107**.

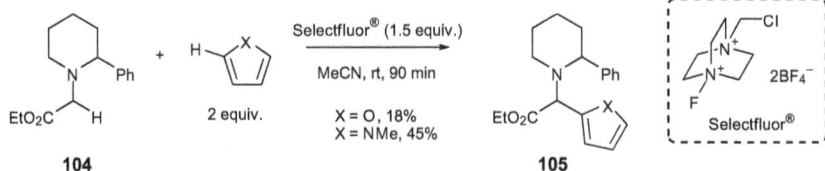

Scheme 8.53 Selectfluor®-promoted Mannich-type reaction of a piperidine derivative.

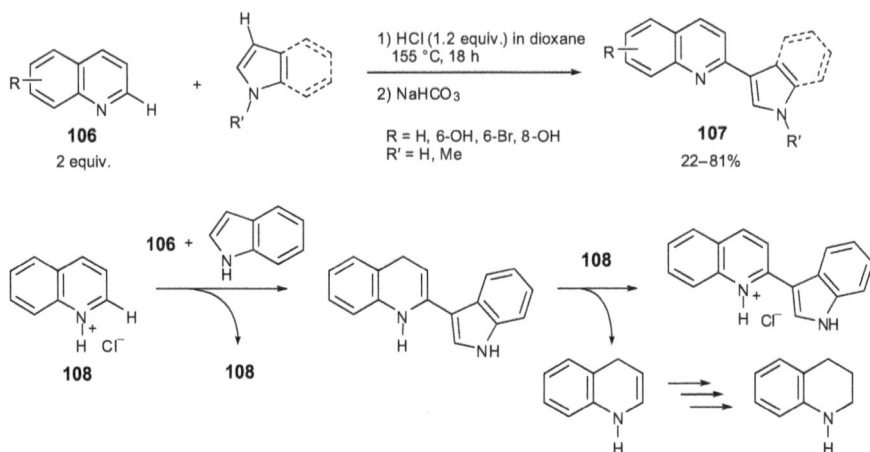

Scheme 8.54 Metal- and solvent-free, auto-oxidative coupling of quinolines with indoles and pyrroles.

8.2 Electrochemical Reactions

In 1999, the "cation pool" method debuted for the first time in a CDC reaction (Scheme 8.55).[55] Yoshida carried out the electrolysis of carbamate (109) at low temperature, and the generated cation pool of carbocations (or iminium ions) (110) was further reacted with various nucleophiles, affording pipelidine derivatives 111 in moderate yields. This is the conventional and direct method for oxidative C–C bond formation.

Yoshida's cation pool method is also applicable to the coupling of aromatics (Scheme 8.56).[56] The radical cation pool (112), which is generated by the anodic oxidation of naphthalene and pyrene, has been added to a variety of aromatics such as substituted benzenes and indoles to provide the CDC products (113) in good yields.

Lessard and Li applied the cation pool method to the nitro-Mannich reaction of N-phenyl THIQ with nitromethane which is well known to proceed in the presence of various oxidants (Scheme 8.57).[57] The reaction is effectively promoted in the ionic liquid [BMIm][BF$_4$] to afford 114 in good yield.

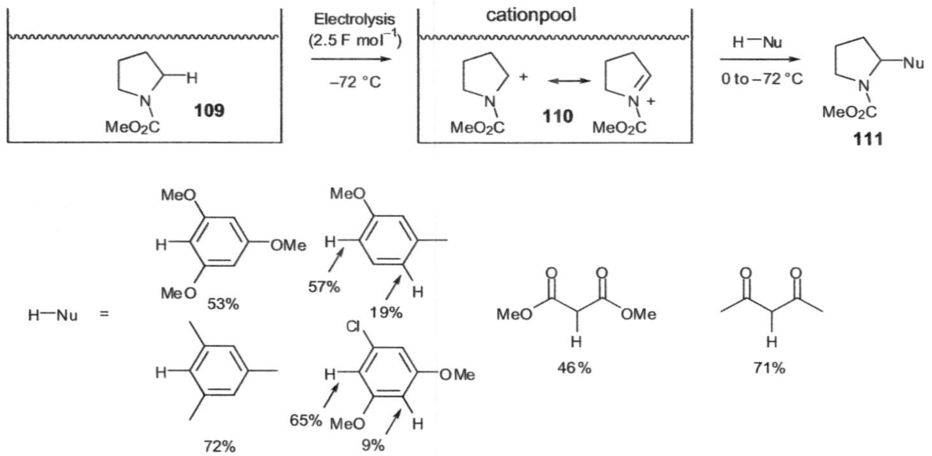

Scheme 8.55 Direct carbon–carbon bond formation of carbamate with aromatic compounds and 1,3-dicarbonyl compounds using the cation pool method.

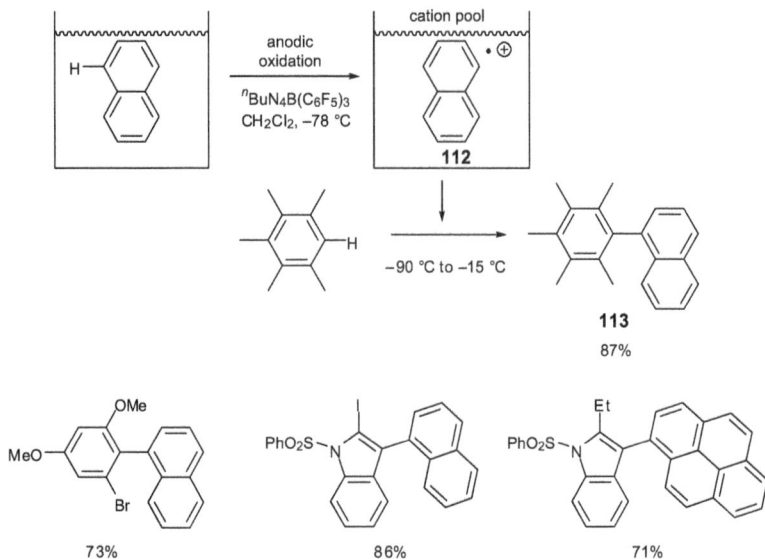

Scheme 8.56 CDC reactions of aromatic compounds by the cation pool method.

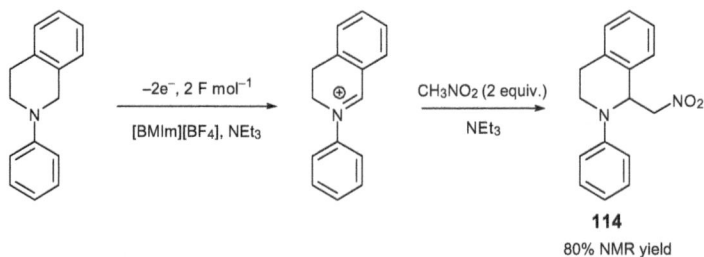

Scheme 8.57 Nitro-Mannich reaction by the cation pool method in an ionic liquid.

Electroreductive coupling was conducted using 1-benzyl-2-hydrofullerene (**115**) by Gao (Scheme 8.58).[58] Mechanistic investigations revealed that the two-electron reduction, the homolytic cleavage of a hydrogen radical on C$_{60}$–H, and the two-electron abstraction give two chiral radicals (**116**), which undergo homo- and heterocouplings to give three type of fullerene dimers (**117**).

Waldvogel's protocol is related to the anodic direct C–C coupling between phenol derivatives and arenes (Scheme 8.59).[59] The use of a boron-doped diamond (BDD) as an electrode in fluorinated media resulted in the formation of non-symmetrical biaryls (**118**) in good yields.

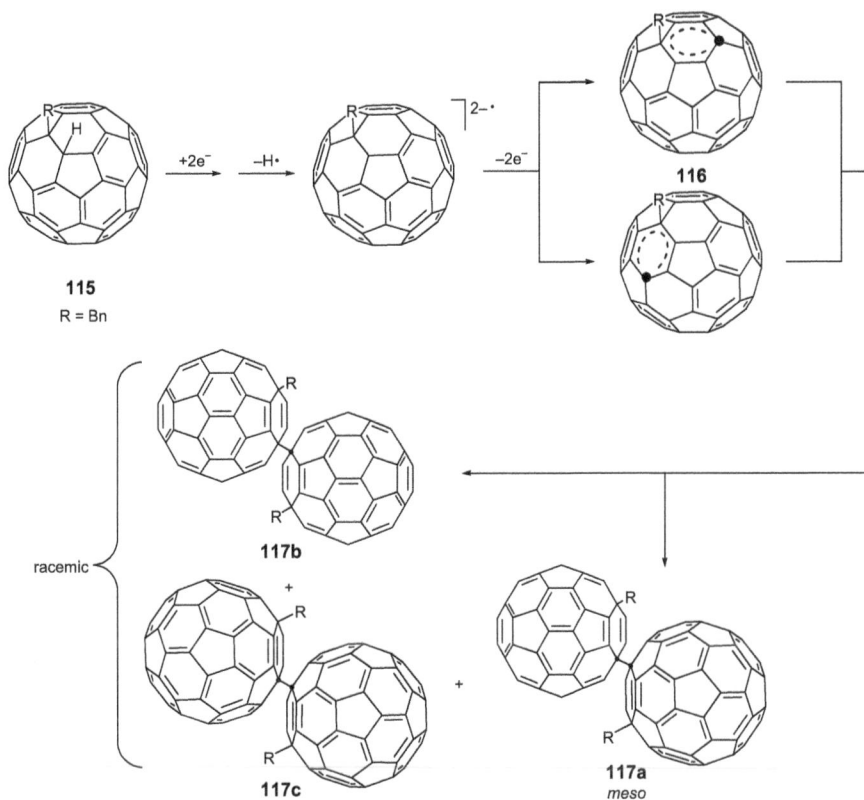

Scheme 8.58 Proposed mechanism for the formation of singly bonded RC_{60}–$C_{60}R$ dimers electroreductive activation of 1,2-$(PhCH_2)HC_{60}$.

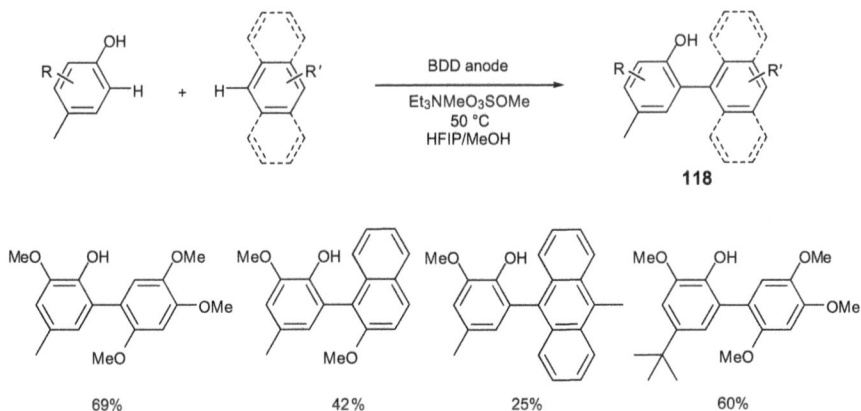

Scheme 8.59 Anodic cross-coupling reactions of various phenols with different arenes in the presence of HFIP/MeOH.

8.3   Light Irradiation

Light irradiation is a simple yet powerful tool for CDC reactions. This section describes some representative examples of intra- and intermolecular C–H/C–H couplings without metal oxidants under light irradiation conditions.

In the 1960s and 1970s, the synthesis of PAHs by photocyclization was frequently investigated.[60] For example, benzoperylene (**119**) can be obtained from several precursors under light irradiation with/without iodine (Scheme 8.60).

In 1970, Sato reported the synthesis of dibenzopyrene (**122**) from **120** or **121** by the irradiation of a low-pressure Hg lamp in the presence of a stoichiometric amount of I$_2$ (Scheme 8.61).[61] The electrocyclization and oxidation of the cyclized product **123** by iodine are expected as the proposed reaction mechanism to give the CDC product **124**.

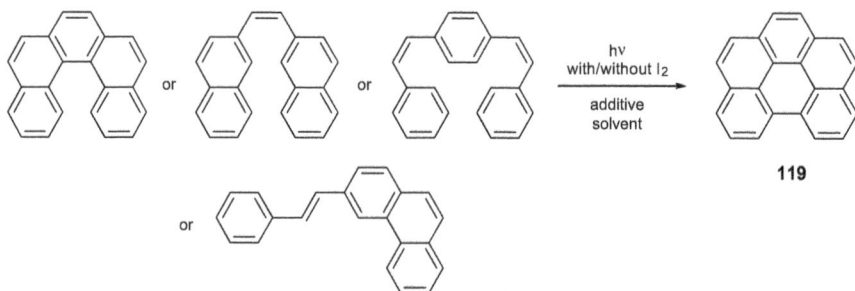

Scheme 8.60   Photochemical synthesis of benzoperylene from various precursors.

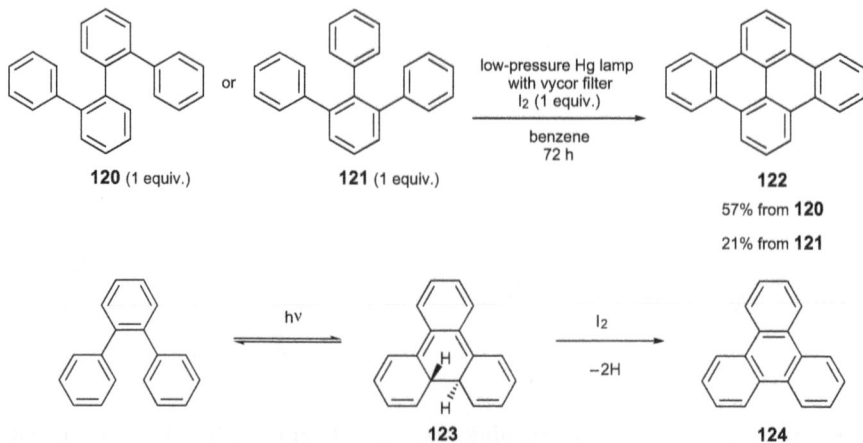

Scheme 8.61   Photochemical CDC reactions for the synthesis of PAHs.

Photochemical CDC reactions were first carried out for the synthesis of [7]circulene by Yamamoto, Kai, and Kasai in 1988 (Scheme 8.62).[62] Cyclophadiene (125) was irradiated with a high-pressure Hg lamp in the presence of a small amount of iodine to furnish the dehydrohexahelicene 126 in 42% yield, which was successfully converted to [7]circulene.

The irradiation of 3,4-diaryl-5,6-dihydropyridinones (127) gives the cyclization products 128 in moderate yields (Scheme 8.63).[63a] These compounds can be transformed to (±)-tyrophorine and (±)-cryptopleurine, which have biological and pharmacological activities. The same type of reactions are reported using diarylethenes 129 to give phenanthrene derivatives 130 in moderate yields.[63b]

The Mannich-type reaction of THIQs (131) with various nucleophiles was achieved under the irradiation of visible light (Scheme 8.64).[64] In these reactions, eosin Y or eosin-Y bis(tetrabutylammonium salt) (TBA-eosin Y) Y were used as photocatalysts to give the coupling products 132.

125   126   [7]circulene
42%

high-pressure Hg lump
I_2
cyclohexane, 2 h

Scheme 8.62 Photochemical CDC reaction for the synthesis of [7]circulene.

high-pressure Hg lamp
I_2
dioxane or dioxane
rt., 20 h
R = H, OMe
n = 1, 2

127   128
48–55%

tyrophorine: n = 1, R = H
cryptopleurine: n = 2, R = OMe

hv
I_2, air
benzene
R = H, OMe
EWG = CO_2Me, CN

steps → tyrophorine, cryptopleurine

129   130
65–78%

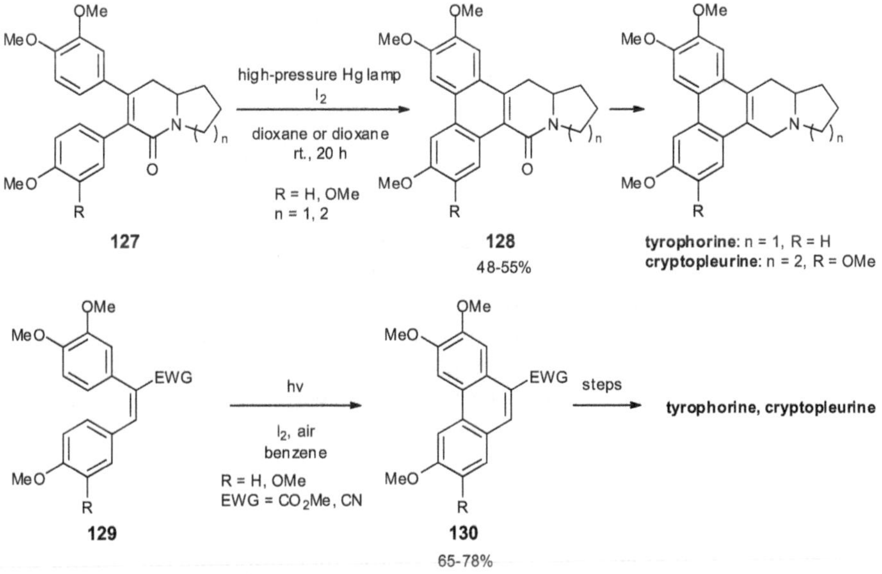

Scheme 8.63 Synthesis of (±)-tyrophorine and (±)-cryptopleurine by photochemical CDC reaction.

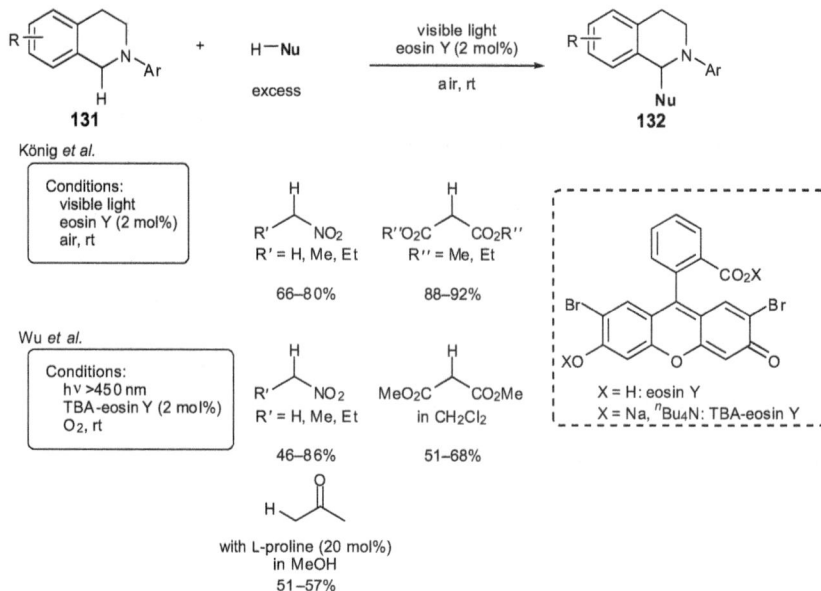

Scheme 8.64   Photo-induced Mannich-type reaction.

8.4   Organocatalysts

Organocatalysts are especially useful for asymmetric C–C bond formations such as asymmetric aldol reaction, Michael addition, and so on. Organocatalysts are also useful in CDC reactions with various oxidants.

Cozzi employed stereoselective CDC reactions promoted by DDQ and the MacMillan catalyst **133** (Scheme 8.65).[65] Various compounds having a benzylic and/or allylic carbon can be converted into the corresponding C–C coupling products **134–137** in low-to-moderate yields with moderate-to-high enantioselectivities.

The same types of CDC reactions with a MacMillan catalysts **138** and **140** have been reported using anodic oxidation[66] (Scheme 8.66) or aerobic oxidation[67] (Scheme 8.67) to afford xanthenes, thioxanthenes and dihydroacridines bearing various aldehydes (**139** and **141**). The enantiomeric excesses (ee) of the coupling products in these reactions are not so high.

In the CDC reaction between naphthoquinones and oxindoles, the high yields and enantioselectivities of **143** are achieved by employing a thiourea organocatalyst (**142**) under air (Scheme 8.68).[68]

8.5   Enzymes

Horseradish peroxidase, which is one of the metalloporphyrin enzymes, is known to catalyze various organic and inorganic reactions. As shown in Figure 8.2, this enzyme shows activity toward the oligomerization/polymerization of 5,6-dihydroxyindole to afford various CDC reaction products which

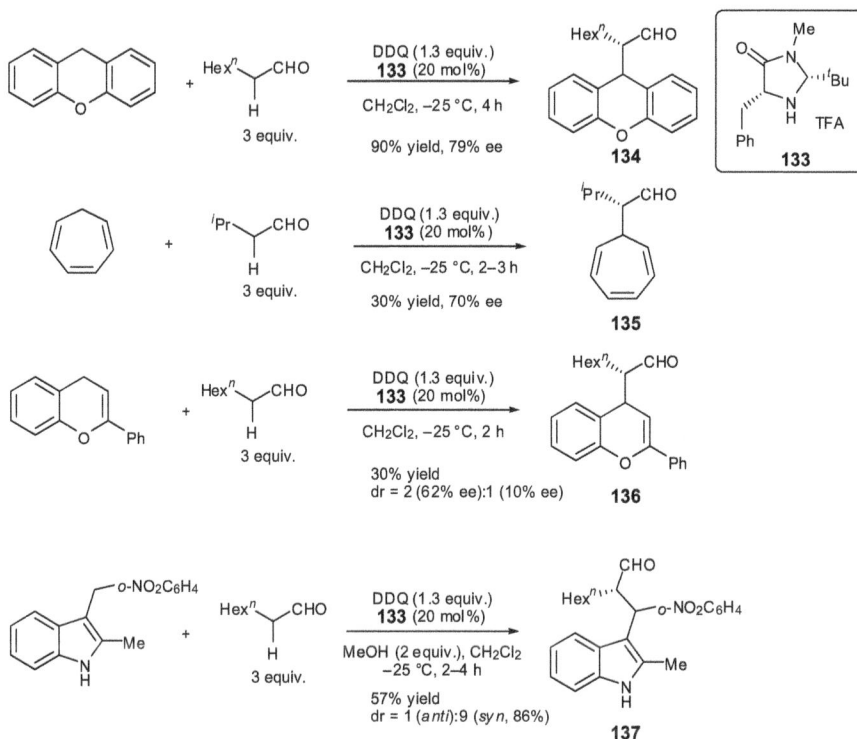

Scheme 8.65 Organocatalytic CDC reactions with the MacMillan catalyst.

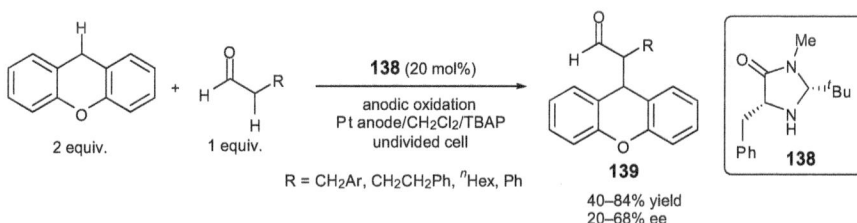

Scheme 8.66 CDC reactions between xanthene and aldehyde with MacMillan catalysts and anodic oxidation.

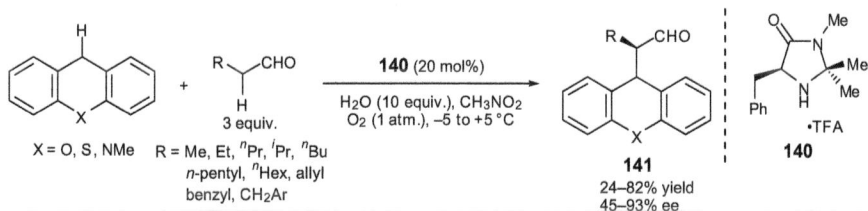

Scheme 8.67 MacMillan catalyst- and aerobic oxidation-promoted CDC reactions of xanthene, thioxanthene and 10-methyl-9,10-dihydroacridine with various aldehydes.

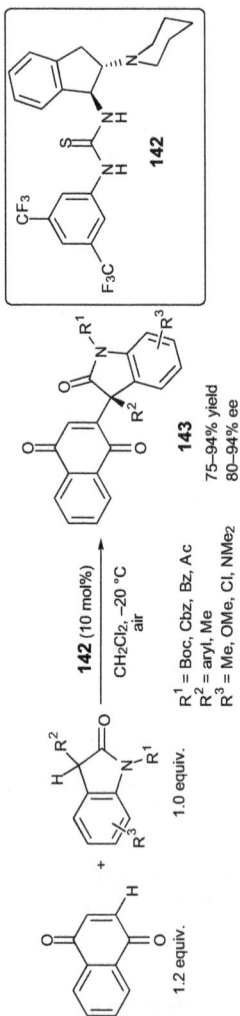

Scheme 8.68 Catalytic enantioselective CDC reaction between oxindoles and naphthoquinones.

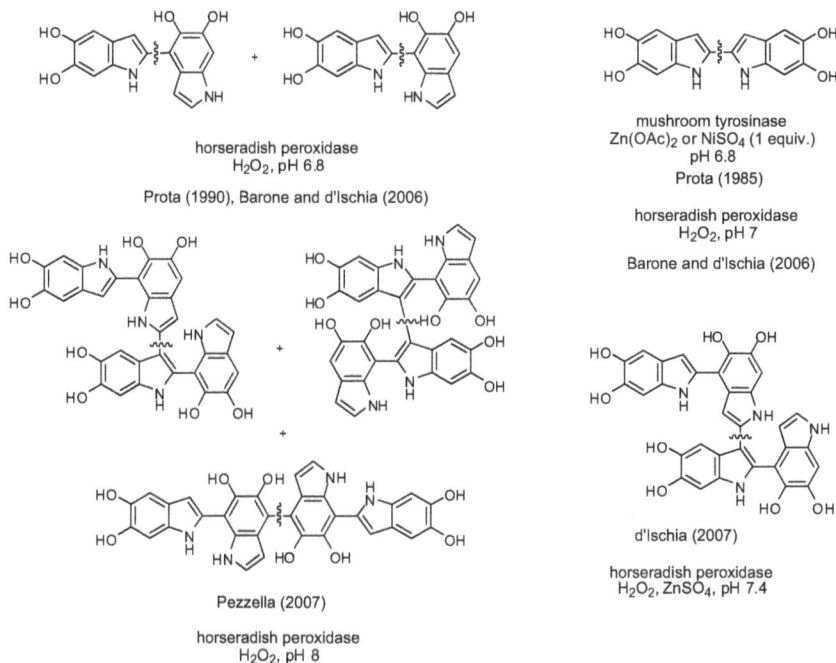

Figure 8.2 Products in the enzymatic oxidative coupling of 5,6-dihydroxyindole.

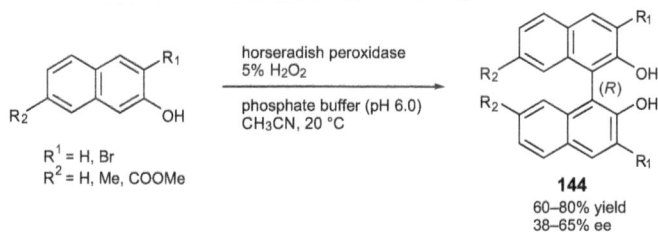

Scheme 8.69 Enzymatic enantioselective CDC reaction of naphthols by horse-radish peroxidase.

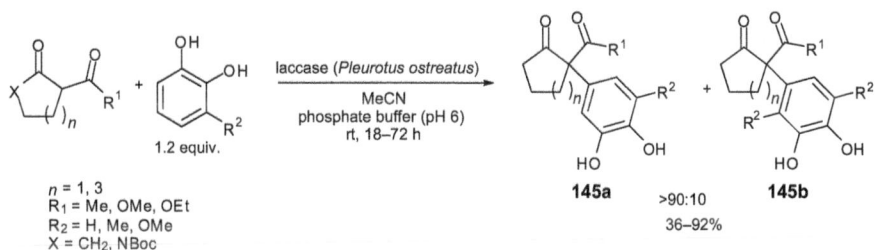

Scheme 8.70 Laccase-catalyzed arylation of cyclic β-dicarbonyl compounds with catechols.

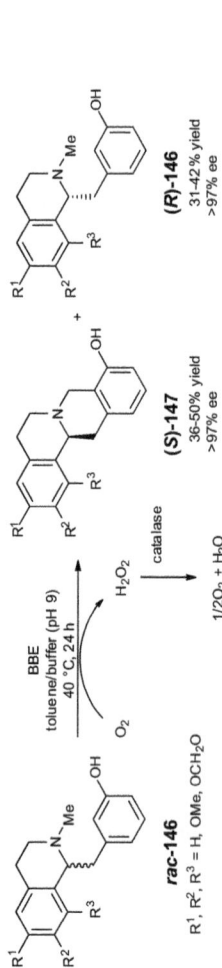

Scheme 8.71 Preparative oxidative C–C coupling employing BBE.

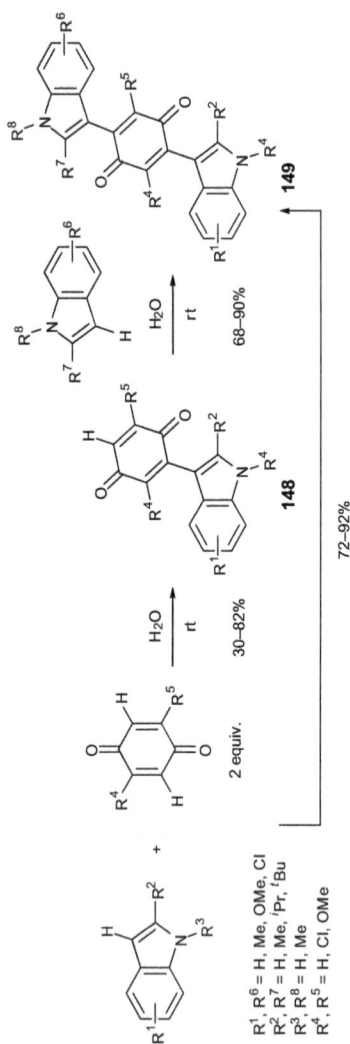

Scheme 8.72 Water-promoted CDC reaction between indoles and benzoquinones: synthesis of mono- and bis(indolyl)-1,4-benzoquinones.

are believed to model the later stages of the buildup of eumelanins, the key functional components of the human pigmentary system (Figure 8.2).[69]

Horseradish peroxidase is also an attractive enzyme in terms of its applicability toward the enantioselective homo-coupling of naphthols (Scheme 8.69).[70] In the presence of peroxidase and 5% H_2O_2 in phosphate buffer (pH 6) at 20 °C, CDC proceeds effectively to give the chiral binaphthols (**144**) with moderate enantioselectivities.

Laccase-catalyzed CDC reaction between cyclic β-dicarbonyl compounds and catechols was developed by Pietruszka, achieving the environmentally friendly synthesis of the arylated compounds **145** (Scheme 8.70).[71]

The berberine bridge enzyme (BBE) was employed for the first preparative oxidative biocatalytic C–C coupling (Scheme 8.71).[72] Racemic *N*-methyl THIQs (*rac*-**146**) was enantioselectively converted by BBE in toluene/buffer co-solvent under aerobic conditions. The sp³–sp² C–H coupling products (*S*)-**147**, berbine derivatives, and unreacted (*R*)-**146** were obtained with high ee, accomplishing highly efficient kinetic resolution.

8.6 Others

Li demonstrated the water-promoted CDC reaction between indoles and benzoquinones without any catalyst, organic solvent, or additives (Scheme 8.72).[73] This water-promoted process enables a highly efficient and direct transformation to mono- and bis(indolyl)benzoquinones (**148** and **149**).

The tetrahydrocarbazole **150** reacts with DMSO and trichloroacetic anhydride to enable the CDC reaction with *N*-methylindole to afford the coupling product **151** in excellent yield (Scheme 8.73).[74]

The NaOH-catalyzed aerobic direct dimerization of functionalized hydrofullerenes (**152**) was reported by Jin and Yamamoto (Scheme 8.74).[75] The reaction is assumed to proceed through deprotonation and one-electron oxidation of **152**, giving the isomeric fullerene dimers **153** in high yields.

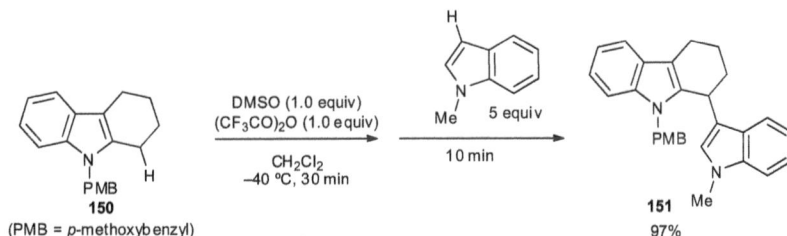

Scheme 8.73 Introduction of indole using DMSO and $(CF_3CO)_2O$ to a tetrahydro-carbazole derivative.

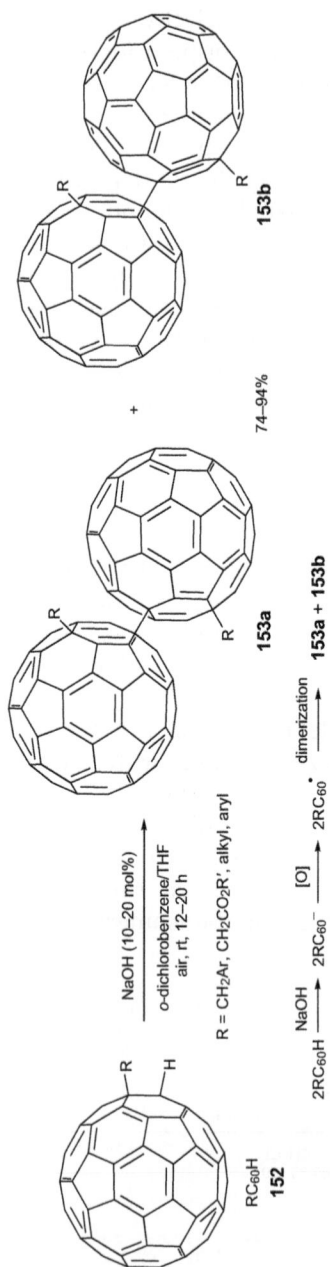

Scheme 8.74 NaOH-Catalyzed aerobic dimerization of mono-substituted hydrofullerenes.

Further Reading

V. Resch, J. H Schrittwieser, E. Siirola, W. Kroutil and *Curr. Opin. Chem. Biol.*, 2011, **22**, 793.

G. Broggini and G. Zecchi, *Synthesis*, 1999, **6**, 905.

R. J. Sundberga, M.-H. Théreta and L. Wright, *Org. Prep. Proced. Int.*, 1994, **26**, 386.

D. Walker and J. D. Hiebert, *Chem. Rev.*, 1967, **67**, 153.

C. S. Yeung and V. M. Dong, *Chem. Rev.*, 2011, **111**, 1215. M. Klussmann and D. Sureshkumar, *Synthesis*, 2011, **3**, 353.

C.-J. Li, *Acc. Chem. Res.*, 2009, **42**, 335.

T. Dohi, M. Ito, N. Yamaoka, K. Morimoto, H. Fujioka and Y. Kita, *Tetrahedron*, 2009, **65**, 10797.

C. J. Scheuermann, *Chem. Asian J.*, 2010, **5**, 436.

K. M. Jones, M. Klussmann, *Synlett*, 2012, **23**, 159.

E. A. Merritt and B. Olofsson, *Synthesis*, 2011, **4**, 517.

Chanda and V. V. Fokin, *Chem. Rev.*, 2009, **109**, 725.

M.-O. Simon and C.-J. Li, *Chem. Soc. Rev.*, 2012, **41**, 1415.

Y. Kita, T. Doi and K. Morimoto, *Gekkankagaku*, 2011, **66**, 12.

References

1. Y. Tamura, T. Yakura, Y. Shirouchi and J. Haruta, *Chem. Pharm. Bull.*, 1986, **34**, 1061.

2. Y. Kita, H. Tohma, K. Hatanaka, T. Takada, S. Fujita, S. Mitoh, H. Sakurai and S. Oka, *J. Am. Chem. Soc.*, 1994, **116**, 3684.

3. M. Arisawa, N. G. Ramesh, M. Nakajima, H. Tohma and Y. Kita, *J. Org. Chem.*, 2001, **66**, 59.

4. J. Wang, Y. Yuan, R. Xiong, D. Zhang-Negrerie, Y. Du and K. Zhao, *Org. Lett.*, 2012, **14**, 2210.

5. A. P. Antonchick and L. Burgmann, *Angew. Chem., Int. Ed.*, 2013, **52**, 3267.

6. (a) Y. Kita, M. Gyoten, M. Ohtsubo, H. Tohma and T. Takada, *Chem. Commun.*, 1996, 1481; (b) Y. Kita, T. Takada, M. Gyoten, H. Tohma, M. H. Zenk and J. Eichhorn, *J. Org. Chem.*, 1996, **61**, 5857.

7. Y. Kita, T. Takada, M. Ibaraki, M. Gyoten, S. Mihara, S. Fujita and H. Tohma, *J. Org. Chem.*, 1996, **61**, 223.

8. H. Hamamoto, G. Anilkumar, H. Tohma and Y. Kita, *Chem.–Eur. J.*, 2002, **8**, 5377.

9. T. Dohi, A. Maruyama, M. Yoshimura, K. Morimoto, H. Tohma and Y. Kita, *Angew. Chem., Int. Ed.*, 2005, **44**, 6193.

10. C. Boldron, G. Aromi, G. Challa, P. Gameza and J. Reedijk, *Chem. Commun.*, 2005, 5808.

11. A. Jean, J. Cantat, D. Bérard, D. Bouchu and S. Canesi, *Org. Lett.*, 2007, **9**, 2553.

12. T. Dohi, M. Ito, K. Morimoto, M. Iwata and Y. Kita, *Angew. Chem., Int. Ed.*, 2008, **47**, 1301.

13. Y. Kita, K. Morimoto, M. Ito, C. Ogawa, A. Goto and T. Dohi, *J. Am. Chem. Soc.*, 2009, **131**, 1668.

14. K. Morimoto, N. Yamaoka, C. Ogawa, T. Nakae, H. Fujioka, T. Dohi and Y. Kita, *Org. Lett.*, 2010, **12**, 3804.

15. T. Dohi, M. Ito, I. Itani, N. Yamaoka, K. Morimoto, H. Fujioka and Y. Kita, *Org. Lett.*, 2011, **13**, 6208.

16. W. Yu, Y. Du and K. Zhao, *Org. Lett.*, 2009, **11**, 2417.

17. K. Morimoto, T. Dohi and Y. Kita, *Eur. J. Org. Chem.*, 2013, 1659.

18. K. Matcha and A. P. Antonchick, *Angew. Chem., Int. Ed.*, 2013, **52**, 2082.

19. R. Fan, Y. Ye, W. Li and L. Wang, *Adv. Synth. Catal.*, 2008, **350**, 2488.

20. Y. Ye, C. Zheng and R. Fan, *Org. Lett.*, 2009, **11**, 3156.

21. X.-Z. Shu, X.-F. Xia, Y.-F. Yang, K.-G. Ji, X.-Y. Liu and Y.-M. Liang, *J. Org. Chem.*, 2009, **74**, 7464.

22. (a) G. Bringmann, A. Ledermann, M. Stahl and K.-P. Gulden, *Tetrahedron*, 1995, **51**, 9353; (b) G. Lin and A. Zhang, *Tetrahedron Lett.*, 1999, **40**, 341.

23. G. Bringmann, T. Pabst and S. Busemann, *Tetrahedron*, 1998, **54**, 1425.

24. G. Deng, K. Ueda, S. Yanagisawa, K. Itami and C.-J. Li, *Chem.–Eur. J.*, 2009, **15**, 333.

25. T. He, L. Yu, L. Zhang, L. Wang and M. Wang, *Org. Lett.*, 2011, **13**, 5016.

26. R. Xia, H.-Y. Niu, G.-R. Qu and H.-M. Guo, *Org. Lett.*, 2012, **14**, 5546.

27. W.-P. Mai, H.-H. Wang, Z.-C. Li, J.-W. Yuan, Y.-M. Xiao, L.-R. Yang, P. Mao and L.-B. Qu, *Chem. Commun.*, 2012, **48**, 10117.

28. Z. Jia, T. Nagano, X. Li and A. S. C. Chan, *Eur. J. Org. Chem.*, 2013, 58.

29. (a) M. C. Pirrung, K. Park and Z. Li, *Org. Lett.*, 2001, **3**, 365; (b) M. C. Pirrung, L. Deng, Z. Li and K. Park, *J. Org. Chem.*, 2002, **67**, 8374.

30. (a) M. C. Pirrung, Z. Li, K. Park and J. Zhu, *J. Org. Chem.*, 2002, **67**, 7919; (b) M. C. Pirrung, Y. Liu, L. Deng, D. K. Halstead, Z. Li, J. F. May, M. Wedel, D. A. Austin and N. J. G. Webster, *J. Am. Chem. Soc.*, 2005, **127**, 4609; (c) A. J. Fletcher, M. N. Baxb and M. C. Willis, *Chem. Commun.*, 2007, 4764.

31. Y. Zhang and C.-J. Li, *J. Am. Chem. Soc.*, 2006, **128**, 4242.

32. Z. Li, H. Li, X. Guo, L. Cao, R. Yu, H. Li and S. Pan, *Org. Lett.*, 2008, **10**, 803.

33. A. S.-K. Tsang and M. H. Todd, *Tetrahedron Lett.*, 2009, **50**, 1199.

34. A. S.-K. Tsang, P. Jensen, J. M. Hook, A. S. K. Hashmi and M. H. Todd, *Pure Appl. Chem.*, 2011, **83**, 655.

35. D. Chenga and W. Bao, *Adv. Synth. Catal.*, 2008, **350**, 1263.

36. D. Cheng and W. Bao, *J. Org. Chem.*, 2008, **73**, 6881.

37. D. Ramesh, U. Ramulu, S. Rajaram, P. Prabhakar and Y. Venkateswarlu, *Tetrahedron Lett.*, 2010, **51**, 4898.

38. K. Alagiri, P. Devadig and K. R. Prabhu, *Chem.–Eur. J.*, 2012, **18**, 5160.

39. S. H. Gwon and S.-G. Kim, *Tetrahedron: Asymmetry*, 2012, **23**, 1251.

40. L. Zhai, R. Shukla and R. Rathore, *Org. Lett.*, 2009, **11**, 3474.
41. L. Zhai, R. Shukla, S. H. Wadumethrige and R. Rathore, *J. Org. Chem.*, 2010, **75**, 4748.
42. K. Kawasumi, Q. Zhang, Y. Segawa, L. T. Scott and K. Itami, *Nat. Chem.*, 2013, **5**, 739.
43. C. Szántay, G. Blaskó, M. Bárczai-Beke, P. Péchy and G. Dörnyei, *Tetrahedron Lett.*, 1980, **21**, 3509.
44. J. Barluenga, M. Trincado, E. Rubio and J. M. González, *Angew. Chem., Int. Ed.*, 2006, **45**, 3140.
45. B. Liua, S.-J. Jia, X.-M. Sua and S.-Y. Wang, *Synth. Commun.*, 2008, **38**, 1279.
46. Y.-P. Zhu, F.-C. Jia, M.-C. Liu, L.-M. Wu, Q. Cai, Y. Gao and A.-X. Wu, *Org. Lett.*, 2012, **14**, 5378.
47. S. Ghosh, S. De, B. N. Kakde, S. Bhunia, A. Adhikary and A. Bisai, *Org. Lett.*, 2012, **14**, 5864.
48. W. Chen, R. Yan, D. Tan, S. Guo, X. Meng and B. Chen, *Tetrahedron*, 2012, **68**, 7956.
49. T. Nobuta, N. Tada, A. Fujiya, A. Kariya, T. Miura and A. Itoh, *Org. Lett.*, 2013, **15**, 574.
50. J. Dhineshkumar, M. Lamani, K. Alagiri and K. R. Prabhu, *Org. Lett.*, 2013, **15**, 1092.
51. B. Zhang, Y. Cuia and N. Jiao, *Chem. Commun.*, 2012, **48**, 4498.
52. (a) A. Pinter, A. Sud, D. Sureshkumar and M. Klussmann, *Angew. Chem., Int. Ed.*, 2010, **49**, 5004; (b) A. Pinter and M. Klussmann, *Adv. Synth. Catal.*, 2012, **354**, 701.
53. M. H. Daniels and J. Hubbs, *Tetrahedron Lett.*, 2011, **52**, 3543.
54. M. Brasse, J. A. Ellman and R. G. Bergman, *Chem. Commun.*, 2011, **47**, 5019.
55. J.-I. Yoshida, S. Suga, S. Suzuki, N. Kinomura, A. Yamamoto and K. Fujiwara, *J. Am. Chem. Soc.*, 1999, **121**, 9546.
56. T. Morofuji, A. Shimizu and J.-I. Yoshida, *Angew. Chem., Int. Ed.*, 2012, **51**, 7259–7262.
57. O. Baslé, N. Borduas, P. Dubois, J. Chapuzet, T.-H. Chan, J. Lessard and C.-J. Li, *Chem.–Eur. J.*, 2010, **16**, 8162.
58. W.-W. Yang, Z.-J. Li and X. Gao, *J. Org. Chem.*, 2011, **76**, 6067.
59. A. Kirste, B. Elsler, G. Schnakenburg and S. R. Waldvogel, *J. Am. Chem. Soc.*, 2012, **134**, 3571.
60. (a) E. V. Blackburn, C. E. Loader and C. J. Timmons, *J. Chem. Soc. Chem. Commun.*, 1970, 163; (b) M. Scholz, M. Mühlstädt and F. Dietz, *Tetrahedron Lett.*, 1967, **8**, 665; (c) W. J. Carruthers, *J. Chem. Soc. Chem. Commun.*, 1967, 1525; (d) L. Liu and T. J. Katz, *Tetrahedron Lett.*, 1991, **32**, 6831; (e) W. H. Laarhoven, T. J. H. M. Cuppen and R. J. F. Nivard, *Tetrahedron*, 1970, **26**, 1069; (f) F. Dietz and M. Scholz, *Tetrahedron*, 1968, **24**, 6845; (g) F. B. Mallory and C. W. Mallory, *J. Org. Chem.*, 1983, **48**, 526.
61. (a) T. Sato, S. Shimada and K. Hata, *J. Chem. Soc. Chem. Commun.*, 1970, 766; (b) T. Sato, S. Shimada and K. Hata, *Bull. Chem. Soc. Jpn.*, 1971, **44**, 2484.

62. K. Yamamoto, T. Harada, Y. Okamoto, H. Chikamatsu, M. Nakazaki, Y. Kai, T. Nakao, M. Tanaka, S. Harada and N. Kasai, *J. Am. Chem. Soc.*, 1988, **110**, 3578.

63. (a) H. Iida and C. Kibayashi, *Tetrahedron Lett.*, 1981, **22**, 1913; (b) H. Iida, Y. Watanabe, M. Tanaka and C. Kibayashi, *J. Org. Chem.*, 1984, **49**, 2412; (c) S. Yamashita, N. Kurono, H. Senboku, M. Tokuda and K. Orito, *Eur. J. Org. Chem.*, 2009, 1173.

64. (a) D. P. Hari and B. König, *Org. Lett.*, 2011, **13**, 3852; (b) Q. Liu, Y.-N. Li, H.-H. Zhang, B. Chen, C.-H. Tung and L.-Z. Wu, *Chem.-Eur. J.*, 2012, **18**, 620.

65. F. Benfatti, M. G. Capdevila, L. Zoli, E. Benedetto and P. G. Cozzi, *Chem. Commun.*, 2009, 5919.

66. X.-H. Ho, S.-I. Mho, H. Kang and H.-Y. Jang, *Eur. J. Org. Chem.*, 2010, 4436.

67. B. Zhang, S.-K. Xiang, L.-H. Zhang, Y. Cui and N. Jiao, *Org. Lett.*, 2011, **13**, 5212.

68. W.-Y. Siau, W. Li, F. Xue, Q. Ren, M. Wu, S. Sun, H. Guo, X. Jiang and J. Wang, *Chem.-Eur. J.*, 2012, **18**, 9491.

69. (a) A. Napolitano, M. G. Corradini and G. Prota, *Tetrahedron Lett.*, 1985, **26**, 2805; (b) M. d'Ischia, A. Napolitano, K. Tsiakas and G. Prota, *Tetrahedron*, 1990, **46**, 5789; (c) A. Pezzella, L. Panzella, O. Crescenzi, A. Napolitano, S. Navaratman, R. Edge, E. J. Land, V. Barone and M. d'Ischia, *J. Am. Chem. Soc.*, 2006, **128**, 15490; (d) L. Panzella, A. Pezzella, A. Napolitano and M. d'Ischia, *Org. Lett.*, 2007, **9**, 1411; (e) A. Pezzella, L. Panzella, A. Natangelo, M. Arzillo, A. Napolitano and M. d'Ischia, *J. Org. Chem.*, 2007, **72**, 9225.

70. M. Sridhar, S. K. Vadivel and U. T. Bhalerao, *Tetrahedron Lett.*, 1997, **38**, 5659.

71. J. Pietruszka and C. Wang, *Green Chem.*, 2012, **14**, 2402.

72. J. H. Schrittwieser, V. Resch, J. H. Sattler, W.-D. Lienhart, K. Durchschein, A. Winkler, K. Gruber, P. Macheroux and W. Kroutil, *Angew. Chem., Int. Ed.*, 2011, **50**, 1068.

73. H.-B. Zhang, L. Liu, Y.-J. Chen, D. Wang and C.-J. Li, *Eur. J. Org. Chem.*, 2006, 869.

74. M. Tayu, K. Higuchi, M. Inaba and T. Kawasaki, *Org. Biomol. Chem.*, 2013, **11**, 496.

75. S. Lu, T. Jin, M. Bao and Y. Yamamoto, *Org. Lett.*, 2012, **14**, 3466.

Cross-Dehydrogenative-Coupling Reactions with Molecular Oxygen as the Terminal Oxidant

O. BASLÉ

Ecole Nationale Superieure de Chimie de Rennes, UMR CNRS 6226,
11 allée de Beaulieu, 35708 Rennes, France
Email: olivier.basle@ensc-rennes.fr

9.1   Introduction

Cross-dehydrogenative-coupling (CDC) between two different C–H bonds represents one of the most attractive strategies for constructing carbon–carbon bonds.[1] Indeed, the need for green and sustainable chemistry has given rise to much interest in the development of innovative technologies that can decrease overall waste levels and offer more efficient processes.[2] With this in mind, oxidative CDC reactions avoid both starting material pre-functionalization and the subsequent production of harmful metal salts that arises from traditional cross-coupling reactions.[3] Moreover, a stoichiometric amount of oxidant is necessary for the reaction to proceed, and when taking atom economy into account, the use of molecular oxygen as the terminal oxidant constitutes the ideal choice with only the generation of water as a by-product [eqn (9.1)].[4]

$$\text{C–H} + \text{H–C} \xrightarrow{\text{cat. M, O}_2} \text{C–C} + \text{H}_2\text{O} \tag{9.1}$$

RSC Green Chemistry No. 26
From C–H to C–C Bonds: Cross-Dehydrogenative-Coupling
Edited by Chao-Jun Li
© The Royal Society of Chemistry 2015
Published by the Royal Society of Chemistry, www.rsc.org

This chapter focuses on CDC reactions that utilize oxygen as the terminal oxidant including methodologies which necessitate the presence of a sub-stoichiometric amount of another oxidizing agent that can potentially act as hydrogen acceptor, but also play the role of ligand for the metal catalyst. Some of the examples described in this chapter constitute useful alternatives to well-known cross-coupling reactions, such as the Mizoroki–Heck, Suzuki–Miyaura, or Sonogashira reactions. The CDC strategy offers opportunities for the development of novel transformations with expected future broad applications.

9.2 Aerobic CDC Reactions Between sp^2 C–H Bonds

9.2.1 Alkenylation of Aromatic Compounds

As expected, the use of molecular oxygen as a terminal oxidant was considered during the early stages of CDC. Following the seminal report by Fujiwara and Moritani,[5] Shue described in 1971 a palladium-catalyzed dehydrogenative olefination of benzene in the presence of oxygen as the only oxidant.[6] Despite the low yield, interesting catalytic activities, up to turnover number (TON) $= 11$, were obtained in benzene as the solvent for the formation of the desired stilbene product (Scheme 9.1). The synthetic potential of this transformation, compared to the famous Mizoroki–Heck reaction, has attracted significant interest in the past 30 years,[7] but it is only during the last decade that major breakthroughs (using O_2) have been realized.

The direct preparation of styrene from the oxidative arylation of ethylene constitutes a highly desirable reaction, notably from an industrial point of view. In 2000, Matsumoto and Yoshida reported a rhodium-catalyzed oxidative coupling of benzene and ethylene under oxygen.[8] Rhodium(I) complexes such as Rh(acac)(CO)$_2$, Rh(acac)(C$_2$H$_4$)$_2$, [Rh(cod)Cl]$_2$, and Wilkinson's catalyst [RhCl(PPh$_3$)$_3$] afforded similar catalytic activities in the presence of acetylacetone (acacH) and oxygen pressure, while a decreased rate was observed with Rh(III) pre-catalysts and no reaction in the absence of acacH (Scheme 9.2). Additionally, the C–H bond activation was demonstrated to be the rate-limiting step.[9]

A year later, Milstein and co-workers described a ruthenium-catalyzed oxidative coupling of arenes with olefins.[10] The reaction in which O_2 can be directly used as the oxidant required CO pressure (6.1 atm.) and proceeded with good activities (up to TON $= 88$). Both, electron-donating substituents on the arene, and electron-deficient alkenes (*i.e.*, methyl acrylate) appeared to be the most efficient coupling partners. Low activities were obtained for

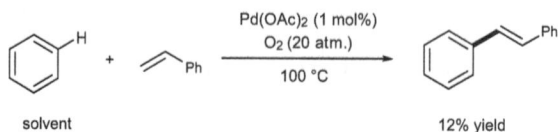

Scheme 9.1

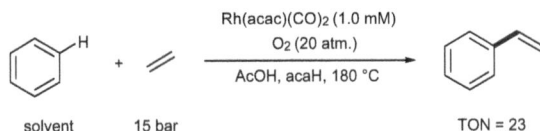

Scheme 9.2

Scheme 9.3

the oxidative coupling of benzene and ethylene (3% yield). It is important to note that the reaction also proceeded in the absence of oxygen with the olefin itself serving as the H_2 acceptor (Scheme 9.3).

9.2.1.1 Palladium-Based Catalytic Systems

Among the three transition metal presented above, palladium has been a major part of the research effort, and palladium-catalyzed reactions have witnessed tremendous improvements over recent years.[11] Following reports from Nagashima[12] and Fujiwara[13] that used peroxide as the hydrogen acceptor; in 2003 two independent research groups described efficient $Pd(OAc)_2$-based catalytic systems for the olefination of aromatic compounds with oxygen as the terminal oxidant. On the one hand, Jacobs and co-workers demonstrated an accelerating effect when benzoic acid was present as an additive, suggesting that the reaction may occur through electrophilic attack on the aromatic compound.[14] On the other hand, Ishii and co-workers found that the oxidative coupling of benzenes with electron-deficient alkenes was facilitated by the use of a catalytic amount of molybdovanado-phosphoric acid (HPMoV) under one oxygen atmosphere.[15]

In 2006, Gaunt and co-workers described a regioselective alkenylation of pyrroles under mild aerobic oxidative conditions.[16] A catalytic amount of palladium acetate was able to selectively functionalize *tert*-butoxycarbamate (BOC)-protected pyrrole at the C2-position while a C3-selectivity was achieved with triisopropylsilyl (TIPS)-protected pyrroles (Scheme 9.4). The aerobic conditions worked well for reactive alkenes such as acrylate derivatives, nevertheless the use of peroxide (^{t}BuOOBz) appeared necessary in order to achieve good yields with more challenging substrates (*i.e.*, methyl vinyl sulfone). Interestingly, intermolecular alkenylation of pyrrole can also be effected with complete selectivity and good yields.

Intramolecular annulation of heterocycles has been used as the key step in several total syntheses.[17] Nevertheless, only rare examples refer to the use of

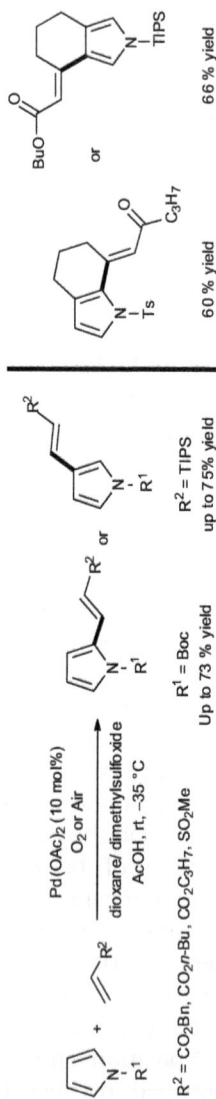

Scheme 9.4

a catalytic amount of palladium metal under aerobic conditions for this highly atom-economical transformation. In 2003, Stoltz and co-workers described an efficient aerobic oxidative annulation of indoles under mild conditions using ethyl nicotinate as the ligand (Scheme 9.5).[18] The successive palladation/olefin insertion/β-H elimination mechanism proposed in this communication was in agreement with those generally postulated in related oxidative Heck-type reactions.[7]

Later, Yu and co-workers reported regioselective alkenylations of electron-deficient arenes.[19] The palladium-catalyzed methodology in the presence of the 2,6-dialkylpyridine ligand afforded good activity and good-to-excellent *meta*-selectivity for the desired cross-dehydrogenative olefination products (Scheme 9.6). The observed reactivity was potentially attributed to the special ligand design affording stable active species in solution.

In 2011, the same group reported unprecedented C3-selective functionalization of pyridine derivatives.[20] The oxidative coupling catalyzed by palladium acetate/1,10-phenanthroline/Ag_2CO_3 under air afforded the desired olefination product in up to good yields (Scheme 9.7). Moreover, a significant kinetic isotope effect was observed, which is consistent with metal-mediated C–H bond cleavage.

Based on the chelation-assisted C–H bond activation strategy,[21] Yu's group developed a highly efficient palladium-catalyzed *ortho*-alkenylation of phenylacetic acid derivatives using molecular oxygen as the terminal oxidant.[22] The versatility and synthetic utility of this reaction was demonstrated with a broad substrate scope, including commercially available drugs. Under the standard conditions, optically pure naproxen was efficiently alkenylated, although with eroded enantiomeric excess (ee). On the other hand, the use of Li_2CO_3 prevented racemization but also lowered the yield (Scheme 9.8). In the same report, the authors demonstrated the possibility of tuning the steric properties of the metal catalyst through coordination of protected amino acid ligands to afford enhanced selectivity. To their delight, they observed that certain amino acids could improve the reactivity allowing for the alkenylation of challenging substrates such as 3-phenylpropionic acid and electron-deficient phenylacetic acids (Scheme 9.9). Moreover, addition of the appropriate amino acid ligand also efficiently afforded 1,2,3-trisubstituted products *via* either simultaneous or sequential di-alkenylation processes.[23] The discovery of amino acid accelerated C–H bond alkenylation made fast reactions (20 min) possible, while decreasing the palladium catalyst loading to as low as 0.2 mol%. The observed reactivity was explained by a ligand-promoted acceleration of the C–H bond cleavage *via* a proton abstraction mechanism.[24] This strategy was recently applied in a convergent total synthesis of (+)-lithospermic acid.[25]

Asymmetric C–H bond activation with chiral amino acid ligands, to desymmetrize an α,α-diphenylacetic acid substrate has also been evaluated (Scheme 9.10). Yu and co-workers obtained good yields and up to excellent enantioselectivities in the cross-coupling with styrene derivatives using Boc-Ile-OH as a chiral ligand.[26]

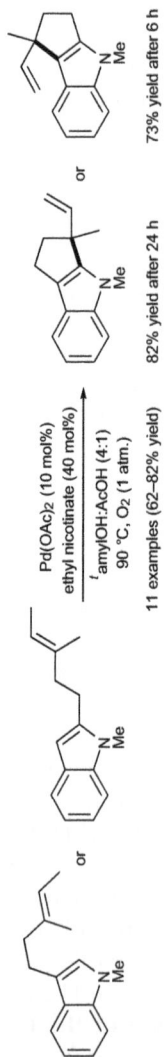

Scheme 9.5

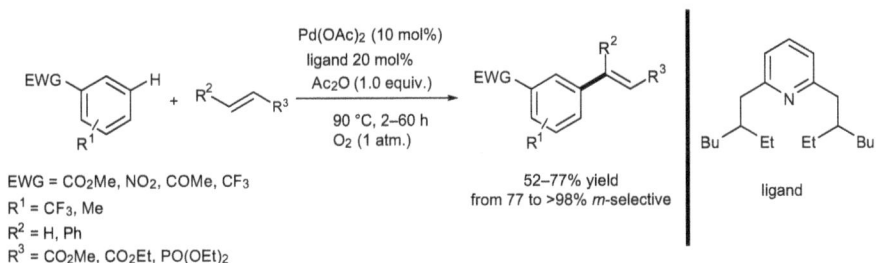

EWG = CO₂Me, NO₂, COMe, CF₃
R¹ = CF₃, Me
R² = H, Ph
R³ = CO₂Me, CO₂Et, PO(OEt)₂

52–77% yield
from 77 to >98% *m*-selective

ligand

Scheme 9.6

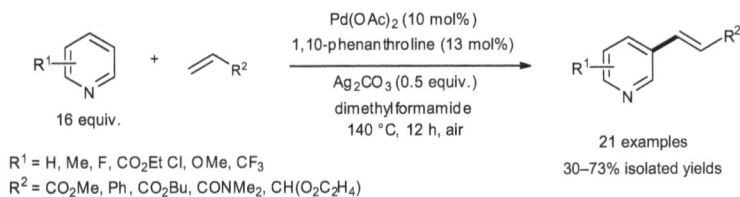

16 equiv.

R¹ = H, Me, F, CO₂Et Cl, OMe, CF₃
R² = CO₂Me, Ph, CO₂Bu, CONMe₂, CH(O₂C₂H₄)

21 examples
30–73% isolated yields

Scheme 9.7

9.2.1.2 Rhodium-Based Catalytic Systems

It is only very recently that rhodium-based catalytic systems have been described in efficient oxidative olefination reactions. Inspired by the work of Satoh and Miura on rhodium/copper-catalyzed aerobic oxidative coupling of benzoic acids with internal alkynes or acrylates (Scheme 9.11),[27] Glorius and co-workers described, in 2012, a rhodium-catalyzed directing group assisted olefination of 2-aryloxazolines under air.[28] This method, which necessitated rhodium, silver and copper metal sources, afforded the desired olefin-oxazoline products in moderate-to-good yields (Scheme 9.12).

9.2.2 Heterocoupling of Alkenes

The cross-coupling of two different alkenes constitutes a challenging process, and only rare examples refer to efficient methodologies that are able to catalyze this highly desirable transformation under aerobic conditions. In 2004, Ishii and co-workers reported the first example of cross-coupling between two acyclic alkenes to produce dienes.[29] The CDC reaction of acrylates and vinylcarboxylate used a Pd(OAc)₂/ heteropolymolybdovanadic acid (HPMV) catalytic system under an atmosphere of molecular oxygen in acetic acid to afford a mixture of *E*- and *Z*- products (Scheme 9.13). More recently, Loh and co-workers obtained higher selectivity for the direct cross-coupling of simple alkenes and acrylates to produce dienoates in good-to-excellent yields.[30] A combination of Cu(OAc)₂ and O₂ appeared to be the best oxidative conditions for this palladium-catalyzed reaction (Scheme 9.14).

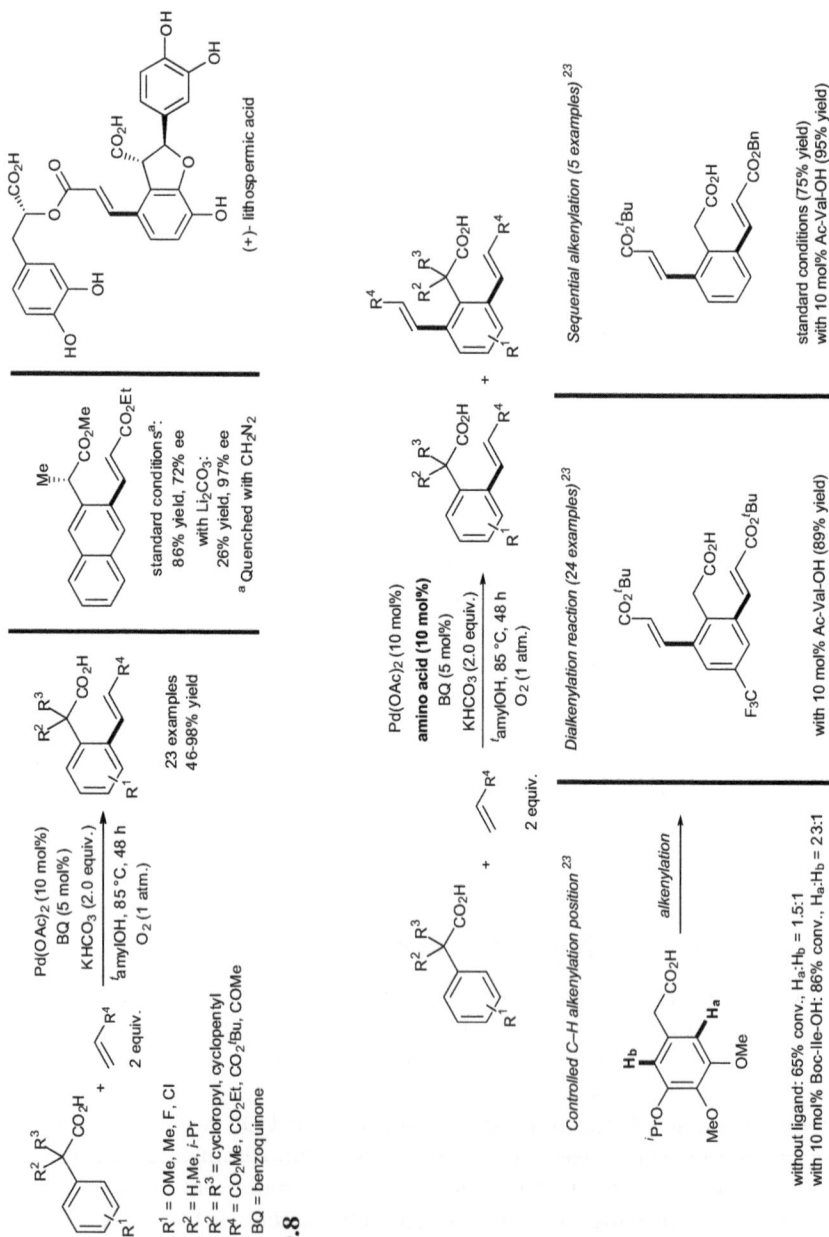

Scheme 9.8

Scheme 9.9

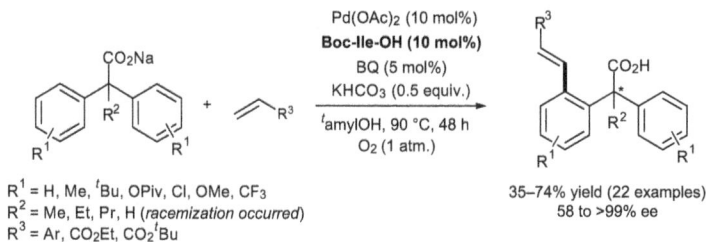

R^1 = H, Me, ^{t}Bu, OPiv, Cl, OMe, CF$_3$
R^2 = Me, Et, Pr, H (*racemization occurred*)
R^3 = Ar, CO$_2$Et, CO$_2$^{t}Bu

35–74% yield (22 examples)
58 to >99% ee

Scheme 9.10

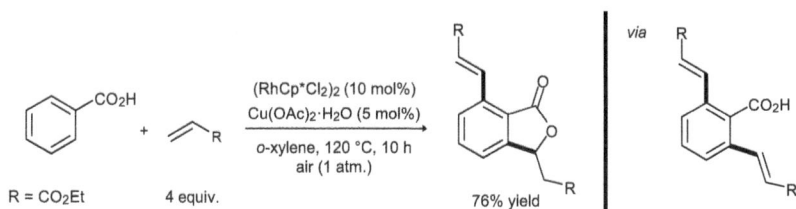

R = CO$_2$Et   4 equiv.

76% yield

Scheme 9.11

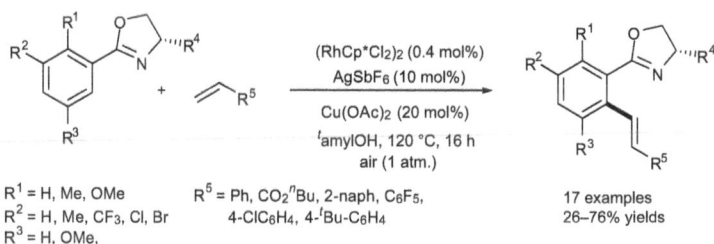

R^1 = H, Me, OMe R^5 = Ph, CO$_2$^{n}Bu, 2-naph, C$_6$F$_5$,
R^2 = H, Me, CF$_3$, Cl, Br   4-ClC$_6$H$_4$, 4-^{t}Bu-C$_6$H$_4$
R^3 = H, OMe,

17 examples
26–76% yields

Scheme 9.12

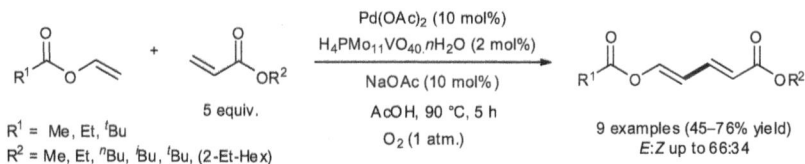

R^1 = Me, Et, ^{t}Bu
R^2 = Me, Et, ^{n}Bu, ^{i}Bu, ^{t}Bu, (2-Et-Hex)

5 equiv.

9 examples (45–76% yield)
E:Z up to 66:34

Scheme 9.13

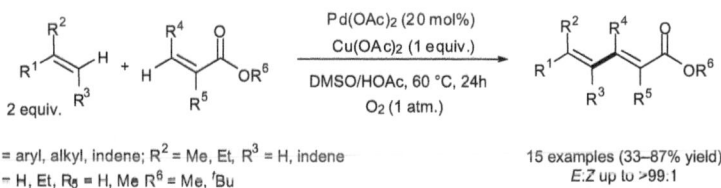

2 equiv.

R^1 = aryl, alkyl, indene; R^2 = Me, Et, R^3 = H, indene
R^4 = H, Et, R_5 = H, Me R^6 = Me, ^{t}Bu

15 examples (33–87% yield)
E:Z up to >99:1

Scheme 9.14

9.2.3 Heterocoupling of Aromatic Compounds

Biaryls are common motifs in pharmaceuticals, agrochemicals and organic materials. Although there are a myriad of strategies for the construction of the biaryl motif,[31] CDC represents the most attractive synthetic route.[32] In 2007, DeBoef and co-workers discovered a novel method for cross-coupling dissimilar arenes using molecular oxygen as a suitable terminal oxidant.[33] Indeed, the use of HPMV allowed for an increased rate of the oxidative coupling to afford selective arylation of *N*-acetylindoles and benzofurans under oxygen while minimizing their homocoupling. Moreover, the use of Cu(OAc)$_2$ as a co-oxidant permitted intramolecular reactions and cyclization of *N*-benzoylindoles (Scheme 9.15). More recently, Stahl and co-workers reported the use of 4,5-diazafluorene derivatives as ancillary ligands for the palladium-catalyzed aerobic oxidative cross-coupling of *N*-benzylsulfonyl indoles with benzenes.[34]

In 2008, Fagnou and co-workers described intramolecular oxidative carbon–carbon bond formation under air.[35] The reaction catalyzed by palladium acetate in pivalic acid (PivOH) as the solvent allowed for the efficient cyclization of a broad scope of diarylamines and diarylether substrates bearing electron-donating groups (EDG) and electron-withdrawing groups (EWG, Scheme 9.16). This contribution prompted Fujii, Ohno and co-workers to develop a one-pot N-arylation and oxidative biaryl coupling for the synthesis of a large library of carbazoles.[36]

In 2008, Buchwald and co-workers reported the *ortho*-arylation of anilides using molecular oxygen as the terminal and only oxidant.[37] The strategy, based on directing-group-assisted C–H activation in the presence of palladium catalyst afforded high efficiency, while electronic and steric effects on the arene coupling partners had a considerable effect on determining the substitution

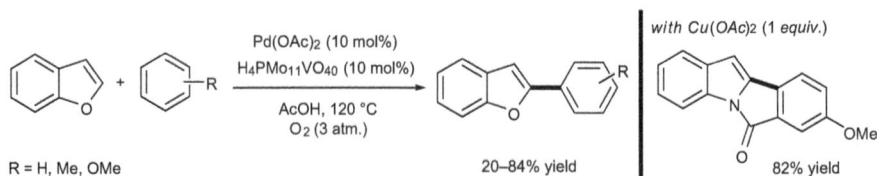

R = H, Me, OMe   20–84% yield   82% yield

Scheme 9.15

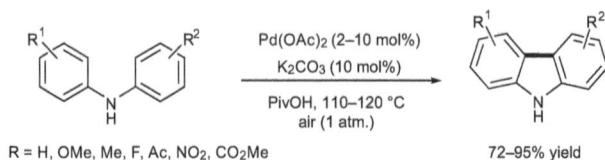

R = H, OMe, Me, F, Ac, NO$_2$, CO$_2$Me   72–95% yield

Scheme 9.16

R^1 = H, OMe, Me, F, Cl   R^2 = H, Me, OMe

59–91% yield

Scheme 9.17

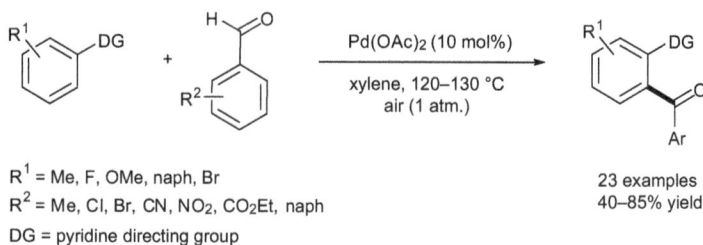

R^1 = Me, F, OMe, naph, Br
R^2 = Me, Cl, Br, CN, NO$_2$, CO$_2$Et, naph
DG = pyridine directing group

23 examples
40–85% yield

Scheme 9.18

pattern (Scheme 9.17). Using *N*-acetanilide substrates, Shi and co-workers made similar observations in their system based on copper as a co-catalyst.[38]

9.2.4 Acylation of Aromatic Compounds

The direct acylation of aromatic C–H bonds has recently emerged as a powerful method for the synthesis of aryl ketones.[39] In 2009, Cheng and co-workers described the *ortho*-acylation of pyridines which was directing-group catalyzed by palladium acetate under air (Scheme 9.18).[40] Nevertheless, a relatively limited understanding of the mechanism exists for this novel methodology, and requires more experimental evidence.[41] More recently, a copper-catalyzed aerobic intramolecular acylation reaction was reported.[42] This simple methodology led to indoline-2,3-diones in good-to-excellent isolated yields (Scheme 9.19). In the latter case, mechanistic studies suggested that the reaction did not proceed *via* a Friedel–Crafts process.

9.3 Aerobic CDC Between sp^2 and sp C–H Bonds

In recent years, the direct alkynylation of arene C–H bonds has attracted much attention as a powerful alternative to the Sonogashira reaction (Scheme 9.20). In 2010, Su and co-workers reported a copper-catalyzed direct alkynylation of polyfluoroarenes with terminal alkynes using O$_2$ as the terminal oxidant.[43] Tetrafluoroarenes were cross-coupled in moderate-to-good yields; nevertheless tri- and difluoroarenes remained unreactive. The same year, Miura and co-workers reported a similar strategy for the alkynylation of polyfluoroarenes,[44] which prompted them to evaluate their

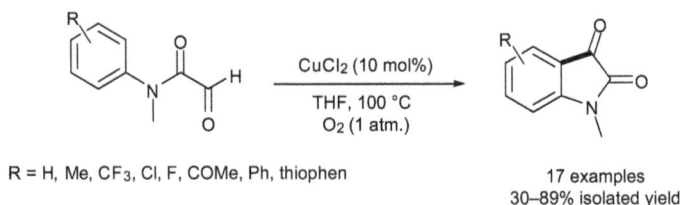

R = H, Me, CF₃, Cl, F, COMe, Ph, thiophen

17 examples
30–89% isolated yield

Scheme 9.19

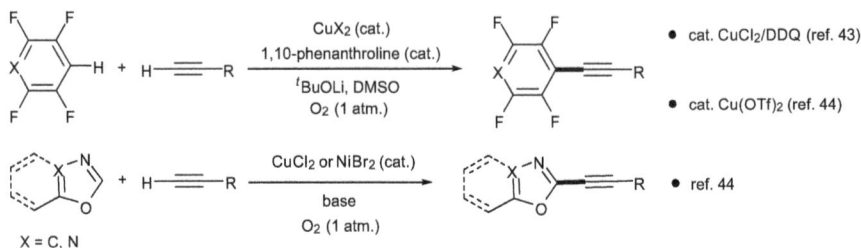

- cat. CuCl₂/DDQ (ref. 43)
- cat. Cu(OTf)₂ (ref. 44)
- ref. 44

X = C, N

Scheme 9.20

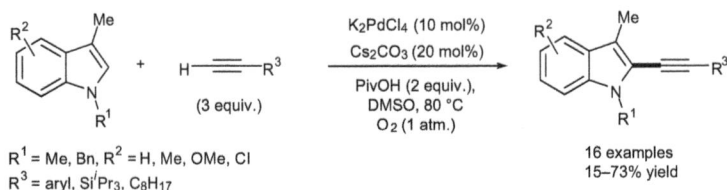

R¹ = Me, Bn, R² = H, Me, OMe, Cl
R³ = aryl, Si^{i}Pr₃, C₈H₁₇

16 examples
15–73% yield

Scheme 9.21

methodology with heterocyclic compounds such as oxazoles and oxadia-zoles.[45] It is worth noting that with benzoxazole substrates, a nickel salt demonstrated good catalytic activity in an aerobic alkynylation reaction.[43]

To date, only one report has referred to the direct alkynylation of indoles under aerobic conditions. In 2010, Li and co-workers described a palladium-catalyzed oxidative coupling of 1,3-dimethylindoles with phenylacetylenes.[46] Using a buffer system composed of 20 mol% cesium carbonate and 2 equiv. pivalic acid at 80 °C, the desired cross-coupling products were obtained in moderate-to-good yields (Scheme 9.21).

In 2011, Struder and Murarka evaluated the CDC reaction between alkynes and nitrones.[47] A catalytic system based on zinc triflate, diquinone and molecular oxygen as the terminal oxidant afforded the desired alkynylated nitrones in good-to-excellent yields (Scheme 9.22). On the other hand, Jung and co-workers reported the oxidative coupling of terminal acetylene with

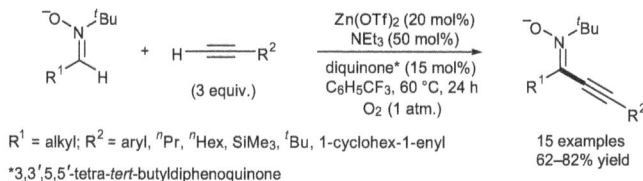

R[1] = alkyl; R[2] = aryl, [n]Pr, [n]Hex, SiMe$_3$, [t]Bu, 1-cyclohex-1-enyl

*3,3′,5,5′-tetra-*tert*-butyldiphenoquinone

15 examples
62–82% yield

Scheme 9.22

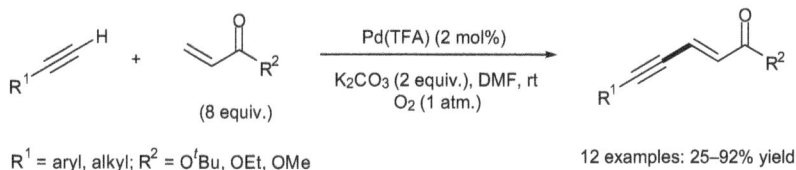

R[1] = aryl, alkyl; R[2] = O[t]Bu, OEt, OMe

12 examples: 25–92% yield

Scheme 9.23

electron-deficient olefins.[48] This simple and mild procedure with Pd(II) catalysts afforded the desired enyne products in moderate-to-excellent isolated yields (Scheme 9.23).

9.4 Aerobic CDC Between sp C–H Bonds

Oxidative acetylenic coupling to produce 1,3-diynes is an important reaction with broad applications.[49] Despite tremendous improvements in homo-coupling processes, notably the Hay procedure,[50] the construction of unsymmetrical conjugated diynes from two different terminal alkynes has only been recently reported. In 2009, Lei and co-workers described a copper/nickel-co-catalyzed aerobic cross-coupling method that tolerated a variety of functional groups such as amides, halides and free propargylic alcohols (Scheme 9.24).[51]

9.5 Aerobic CDC Between sp^3 C–H Bonds

9.5.1 sp^3 C–H Bonds Adjacent to a Nitrogen Atom

The oxidative generation of reactive iminium intermediates is an efficient strategy for the selective functionalization of sp^3 C–H bonds adjacent to nitrogen atoms in amines.[52] In 2003, Murahashi and co-workers, who evaluated the potential of ruthenium chloride to catalyze the selective α-cyanation of a broad scope of alkylamines, demonstrated the potential of the *in situ* generation of iminium cations under aerobic conditions.[53] Li and co-workers initiated the development of a general strategy using chemical oxidants to functionalize sp^3 C–H bonds adjacent to nitrogen atoms by describing efficient reactions with numerous cross-coupling partners including

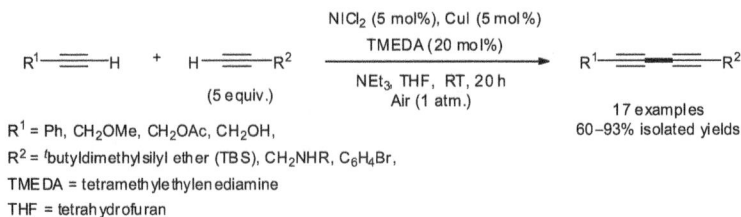

R¹ = Ph, CH₂OMe, CH₂OAc, CH₂OH,

R² = *butyldimethylsilyl ether (TBS), CH₂NHR, C₆H₄Br,

TMEDA = tetramethylethylenediamine

THF = tetrahydrofuran

Scheme 9.24

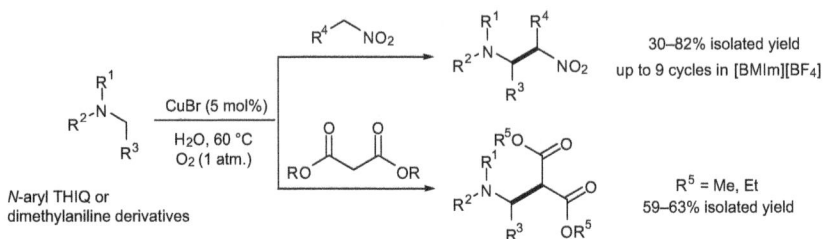

N-aryl THIQ or
dimethylaniline derivatives

Scheme 9.25

sp, sp² and sp³ C–H bonds.[1] In 2007, Baslé and Li reported the direct coupling between nitroalkanes and tertiary amines under aerobic conditions in water to generate β-nitroamines.[54] In addition, simple and highly efficient copper-catalyzed sp³ C–C bond formation was also applicable to dialkyl malonate derivatives (Scheme 9.25). Moreover, replacement of water by the ionic liquid [BMIm][BF₄] offered the possibility of both recycling the catalytic system for up to nine cycles with almost no loss of activity, and gaining a better understanding of the reaction mechanism *via* cyclic voltammetry experiments.[55] Interestingly, magnetically recoverable Fe₃O₄ nanoparticles recently demonstrated high catalytic performance in this alkylation process.[56]

Furthermore, ruthenium metal catalysts are not only applicable to the direct cyanation reaction, indeed, Xiang and co-workers reported RuCl₃-catalyzed aerobic alkylation with nitroalkanes and dialkyl malonates under one atmosphere of oxygen gas in methanol.[57] More recently, Wu and co-workers demonstrated that graphene-supported RuO₂ nanoparticles are a robust catalyst for the aerobic CDC reaction of *N*-phenyl tetrahydroisoquinoline (THIQ) derivatives with nitroalkanes in water.[58] The beneficial role of protic solvents (methanol, water) in copper-catalyzed oxidative coupling[59] was also recently observed in an efficient homogeneous gold-catalyzed aerobic alkylation with nitroalkanes and ketones.[60] Nevertheless, Prabhu and co-workers observed a better catalytic activity when performing the aerobic cyanation and alkylation coupling in absence of a solvent with a molybdenum catalyst.[61]

The aerobic CDC of an sp³ C–H bond with heterocyclic arenes was first reported in 2004 by Tsuchimoto, Shirakawa and co-workers.[62] They found

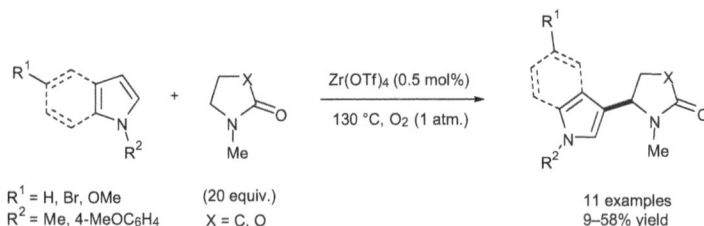

R^1 = H, Br, OMe (20 equiv.)
R^2 = Me, 4-MeOC$_6$H$_4$ X = C, O

11 examples
9–58% yield

Scheme 9.26

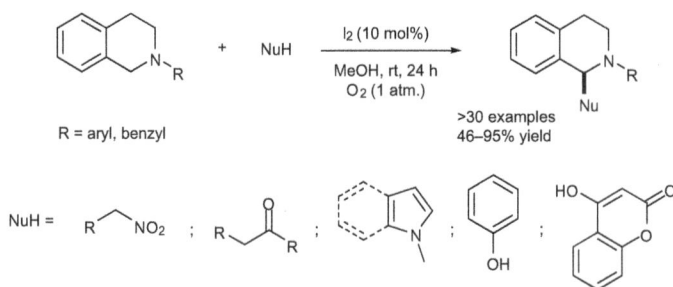

R = aryl, benzyl

>30 examples
46–95% yield

NuH =

Scheme 9.27

that of the Lewis acid and Brönsted acid catalysts they screened for the re-action, Zr(OTf)$_4$ displayed the best activity for the selective arylation of 1-methylpyrrolidin-2-one (Scheme 9.26).

More recently, it was found that the aerobic CDC reaction between *N*-heterocyclic sp^2 C–H bonds and the sp^3 C–H bonds of *N*-alkyl tertiary amines could be promoted by copper,[57,63] vanadium,[64] antimony,[65] and photo-sensitive platinum[66] catalysts. Generally, successful methodologies offered the possibility to functionalize the sp^3 C center with a variety of nucleo-philes,[57] as illustrated by the recently reported organocatalytic system based on molecular iodine (Scheme 9.27).[67]

Sunlight is a clean and inexhaustible energy source and its combination with oxygen for selective C–H functionalization has attracted considered interest in recent years (Scheme 9.28).[68] *N*-Aryl THIQs were recognized as model substrates to evaluate photocatalysts such as Au(III)-carbene,[69] Pt(II)-terpyridyl,[64] Pd(II)-porphyrin,[70] and organic eosin Y.[71,72]

9.5.2 Enolizable sp^3 C–H Bonds

In 1997, Bäckvall and co-workers described the intramolecular oxidative coupling of stabilized carbon nucleophiles with dienes.[73] The carbocycliza-tion catalyzed by Pd(OAc)$_2$ under an atmosphere of oxygen afforded the desired products in modest yields because of competing acetoxylation. In a similar fashion, Yang and co-workers developed an intramolecular oxidative coupling between β-ketoamines and alkenes.[74] Heterocycles were efficiently

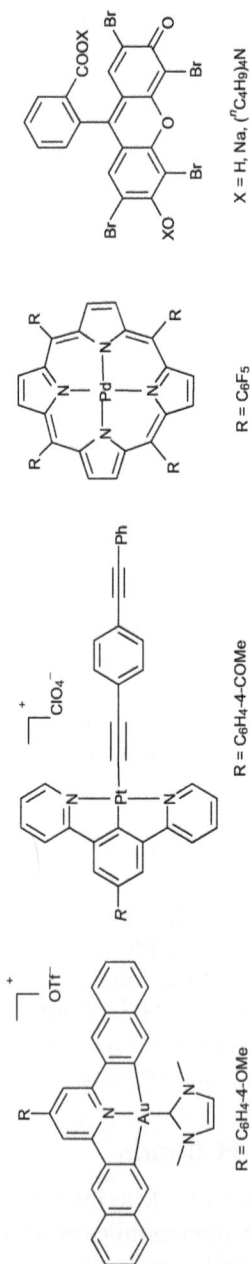

X = H, Na, (ⁿC₄H₉)₄N

R = C₆F₅

R = C₆H₄-4-COMe

R = C₆H₄-4-OMe

Scheme 9.28

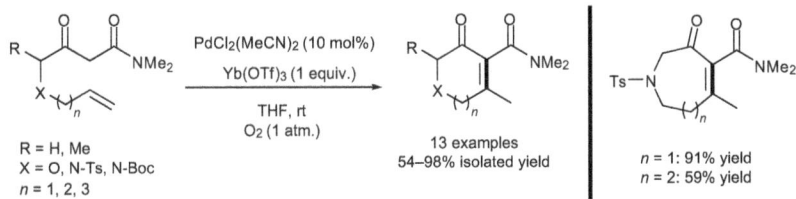

Scheme 9.29

synthesized using both $Yb(OTf)_3$ and $PdCl_2(MeCN)_2$ with molecular oxygen as the terminal oxidant. This methodology was particularly interesting, since it offered the possibility to create seven- and eight-membered ring N-heterocycles in excellent and good yields, respectively (Scheme 9.29).

In 2010, Taylor and co-workers described an intramolecular CDC reaction between an aryl sp^2 C–H bond and an enolizable sp^3 C–H bond to produce 3,3-disubstituted oxindoles.[75] The generality of this copper-catalyzed process was demonstrated with a broad substrate scope and illustrated with the efficient synthesis of a biologically active alkaloid precursor (Scheme 9.30).

9.5.3 Allylic and Benzylic sp^3 C–H Bonds

CDC involving simple allylic C–H bonds with oxygen as the terminal oxidant remains today a challenging process,[76] and to the best of our knowledge, to date, no report describes such an efficient methodology to form carbon–carbon bonds. On the other hand, benzylic sp^3 C–H bonds were recently activated and coupled with enolizable sp^3 C–H bonds.[77] Activated benzylic C–H bonds such as in xanthene, acridanes and isochromane[78] were efficiently coupled with nucleophilic carbonyl compounds with a catalytic amount of Brønsted acid in the presence of molecular oxygen (Scheme 9.31). In absence of a metal salt, auto-oxidation of the specific substrates is believed to proceed *via* peroxides and elevated partial pressures of oxygen were necessary to achieve moderate-to-good yields with challenging nucleophiles such as electron-rich arenes.[79]

9.5.4 Unactivated sp^3 C–H Bonds

Only rare examples describe efficient CDC involving unactivated sp^3 C–H bonds. In 2008, Liégault and Fagnou described an intramolecular Pd-catalyzed CDC reaction between arenes and unactivated alkanes in air.[80] Electron-poor pyrroles were alkylated in good yields with high selectivity in favor of primary sp^3 C–H bonds (Scheme 9.32).

As an alternative to the classical conjugated addition reaction, Pihko and co-workers reported, in 2012, a novel CDC reaction between sp^3 C–H bonds in the β-position of esters and indoles.[81] Soon after, the aerobic methodology was successfully extended to a variety of electron-rich arenes and phenols

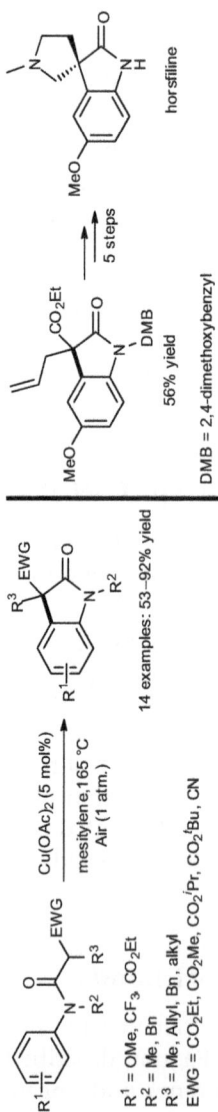

R^1 = OMe, CF$_3$, CO$_2$Et
R^2 = Me, Bn
R^3 = Me, Allyl, Bn, alkyl
EWG = CO$_2$Et, CO$_2$Me, CO$_2$^{i}Pr, CO$_2$^{t}Bu, CN

Cu(OAc)$_2$ (5 mol%)
mesitylene, 165 °C
Air (1 atm.)

14 examples: 53–92% yield

56% yield

DMB = 2,4-dimethoxybenzyl

5 steps

horsfiline

Scheme 9.30

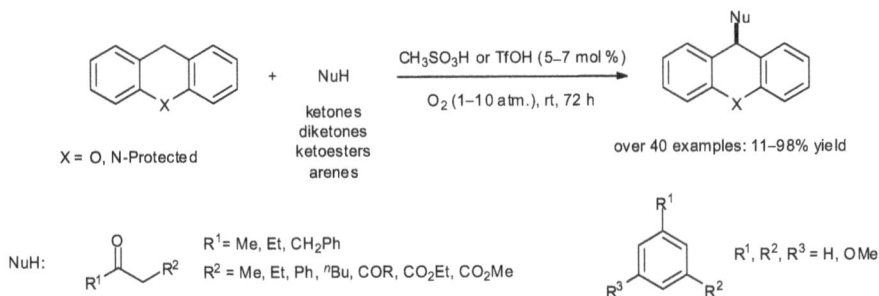

Scheme 9.31

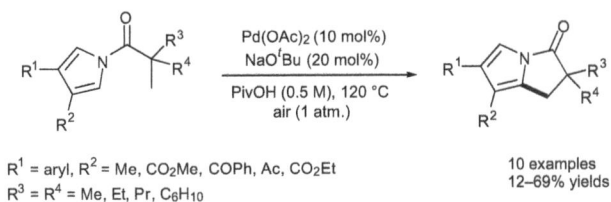

Scheme 9.32

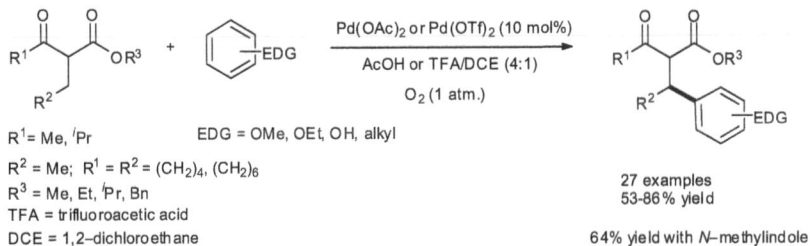

Scheme 9.33

when using Pd(OAc)$_2$ as the catalyst and diarylphosphate as the co-catalyst (Scheme 9.33).[82]

9.6 Conclusions and Outlook

Recent years have witnessed tremendous improvement in the development of novel strategies to efficiently couple different C–H bonds to form new C–C bonds. Among these successful methods, the use of molecular oxygen as the terminal oxidant (or as the only oxidant) has appeared as an ideal choice with the production of a minimum amount of waste. Nevertheless, despite the numerous recent discoveries, aerobic CDC is still in its infancy with future developments and applications expected in asymmetric catalysis, the total synthesis of biologically active molecules, and material science.

References

1. (a) C.-J. Li, *Acc. Chem. Res.*, 2009, **42**, 335; (b) C.-J. Li and Z. Li, *Pure Appl. Chem.*, 2006, **78**, 935.
2. C.-J. Li and B. M. Trost, *Proc. Natl. Acad. Sci. U. S. A.*, 2008, **105**, 13197.
3. C. S. Yeung and V. M. Dong, *Chem. Rev.*, 2011, **111**, 1215.
4. A. N. Campbell and S. S. Stahl, *Acc. Chem. Res.*, 2012, **45**, 851.
5. Y. Fujiwara, I. Moritani, S. Danno, R. Asano and S. Teranishi, *J. Am. Chem. Soc.*, 1969, **91**, 7166.
6. (a) R. S. Shue, *J. Chem. Soc. D*, 1971, 1510; (b) R. S. Shue, *J. Catal.*, 1972, **26**, 112.
7. C. Jia, T. Kitamura and Y. Fujiwara, *Acc. Chem. Res.*, 2001, **34**, 633.
8. T. Matsumoto and H. Yoshida, *Chem. Lett.*, 2000, 1064.
9. T. Matsumoto, R. A. Periana, D. J. Taube and H. Yoshida, *J. Catal.*, 2002, **206**, 272.
10. H. Weissman, X. Song and D. Milstein, *J. Am. Chem. Soc.*, 2001, **123**, 337.
11. (a) V. Ritleng, C. Sirlin and M. Pfeffer, *Chem. Rev.*, 2002, **102**, 1731; (b) E. M. Beccalli, G. Broggini, M. Martinelli and S. Sottocornola, *Chem. Rev.*, 2007, **107**, 5318.
12. J. Tsuji and H. Nagashima, *Tetrahedron*, 1984, **40**, 2699.
13. C. Jia, W. Lu, T. Kitamura and Y. Fujiwara, *Org. Lett.*, 1999, **1**, 2097.
14. M. Dams, D. E. De Vos, S. Celen and P. A. Jacobs, *Angew. Chem., Int. Ed.*, 2003, **42**, 3512.
15. (a) T. Yokota, M. Tani, S. Sakaguchi and Y. Ishii, *J. Am. Chem. Soc.*, 2003, **125**, 1476; (b) M. Tani, S. Sakaguchi and Y. Ishii, *J. Org. Chem.*, 2004, **69**, 1221; (c) T. Yamada, S. Sakaguchi and Y. Ishii, *J. Org. Chem.*, 2005, **70**, 5471.
16. E. M. Beck, N. P. Grimster, R. Hatley and M. J. Gaunt, *J. Am. Chem. Soc.*, 2006, **128**, 2528.
17. P. S. Baran and E. J. Corey, *J. Am. Chem. Soc.*, 2002, **124**, 7904.
18. E. M. Ferreira and B. M. Stoltz, *J. Am. Chem. Soc.*, 2003, **125**, 9578.
19. Y.-H. Zhang, B.-F. Shi and J.-Q. Yu, *J. Am. Chem. Soc.*, 2009, **131**, 5072.
20. M. Ye, G.-L. Gao and J.-Q. Yu, *J. Am. Chem. Soc.*, 2011, **133**, 6964.
21. M. Miura, T. Tsuda, T. Satoh, S. Pivsa-Art and M. Nomura, *J. Org. Chem.*, 1998, **63**, 5211.
22. D.-H. Wang, K. M. Engle, B.-F. Shi and J.-Q. Yu, *Science*, 2010, **327**, 315.
23. K. M. Engle, D.-H. Wang and J.-Q. Yu, *Angew. Chem., Int. Ed.*, 2010, **49**, 6169.
24. K. M. Engle, D.-H. Wang and J.-Q. Yu, *J. Am. Chem. Soc.*, 2010, **132**, 14137.
25. D.-H. Wang and J.-Q. Yu, *J. Am. Chem. Soc.*, 2011, **133**, 5767.
26. B.-F. Shi, Y.-H. Zhang, J. K. Lam, D.-H. Wang and J.-Q. Yu, *J. Am. Chem. Soc.*, 2010, **132**, 460.
27. K. Ueura, T. Satoh and M. Miura, *Org. Lett.*, 2007, **9**, 1407.
28. N. Schöder, T. Besset and F. Glorius, *Adv. Synth. Catal.*, 2012, **354**, 579.
29. Y. Hatamoto, S. Sakaguchi and Y. Ishii, *Org. Lett.*, 2004, **6**, 4623.

30. Y.-H. Xu, J. Lu and T.-P. Loh, *J. Am. Chem. Soc.*, 2009, **131**, 1372.
31. I. Cepanec, *Synthesis of Biaryls*, Elsevier, New York, 2004.
32. D. R. Stuart and K. Fagnou, *Science*, 2007, **316**, 1172.
33. T. A. Dwight, N. R. Rue, D. Charyk, R. Josselyn and B. DeBoef, *Org. Lett.*, 2007, **9**, 3137.
34. A. N. Campbell, E. B. Meyer and S. S. Stahl, *Chem. Commun.*, 2011, **47**, 10257.
35. B. Liégault, D. Lee, M. P. Huestis, D. R. Stuart and K. Fagnou, *J. Org. Chem.*, 2008, **73**, 5022.
36. T. Watanabe, S. Oishi, N. Fujii and H. Ohno, *J. Org. Chem.*, 2009, **74**, 4720.
37. G. Brasche, J. Garcia-Fortanet and S. L. Buchwald, *Org. Lett.*, 2008, **10**, 2207.
38. B.-J. Li, S.-L. Tian, Z. Fang and Z.-J. Shi, *Angew. Chem. Int. Ed.*, 2008, **47**, 1115.
39. C. Pan, X. Jia and J. Cheng, *Synthesis*, 2012, **44**, 677.
40. X. Jia, S. Zhang, W. Wang, F. Luo and J. Cheng, *Org. Lett.*, 2009, **11**, 3120.
41. O. Baslé, J. Bidange and C.-J. Li, *Adv. Synth. Catal.*, 2010, **352**, 1145.
42. B.-X. Tang, R.-J. Song, C.-Y. Wu, Y. Liu, M.-B. Zhou, W.-T. Wei, G.-B. Deng, D.-L. Yin and J.-H. Li, *J. Am. Chem. Soc.*, 2010, **132**, 8900.
43. Y. Wei, H. Zhao, J. Kan, W. Su and M. Hong, *J. Am. Chem. Soc.*, 2010, **132**, 2522.
44. N. Matsuyama, M. Kitahara, K. Hirano, T. Satoh and M. Miura, *Org. Lett.*, 2010, **12**, 2358.
45. M. Kitahara, K. Hirano, H. Tsurugi, Satoh and M. Miura, *Chem.–Eur. J.*, 2010, **16**, 1772.
46. L. Yang, L. Zhao and C.-J. Li, *Chem. Commun.*, 2010, **46**, 4184.
47. S. Murarka and A. Struder, *Org. Lett.*, 2011, **13**, 2746.
48. V. Hadi, K. S. Yoo, M. Jeong and K. W. Jung, *Tetrahedron Lett.*, 2009, **50**, 2370.
49. P. Siemsen, R. C. Livingston and F. Diederich, *Angew. Chem., Int. Ed.*, 2000, **39**, 2632.
50. A. S. Hay, *J. Org. Chem.*, 1962, **27**, 3320.
51. W. Yin, C. He, M. Chen, H. Zhang and A. Lei, *Org. Lett.*, 2009, **11**, 709.
52. L. Zhao, O. Baslé and C.-J. Li, *Proc. Natl. Acad. Sci. U. S. A.*, 2009, **106**, 4106.
53. (a) S.-I. Murahashi, N. Komiya, H. Terai and T. Nakae, *J. Am. Chem. Soc.*, 2003, **125**, 15312; (b) S.-I. Murahashi, N. Komiya and H. Terai, *Angew. Chem., Int. Ed.*, 2005, **44**, 6931; (c) S.-I. Murahashi, T. Nakae, H. Terai and N. Komiya, *J. Am. Chem. Soc.*, 2008, **130**, 11005.
54. O. Baslé and C.-J. Li, *Green Chem.*, 2007, **9**, 1047.
55. O. Baslé, N. Borduas, P. Dubois, J.-M. Chapuzet, T.-H. Chan, J. Lessard and C.-J. Li, *Chem.–Eur. J.*, 2010, **16**, 8162.
56. T. Zeng, G. Song, A. Moores and C.-J. Li, *Synlett*, 2010, **13**, 2002.

57. A. Yu, Z. Gu, D. Chen, W. He, P. Tan and J. Xiang, *Catal. Commun.*, 2009, **11**, 162.

58. Q.-Y. Meng, Q. Lui, J.-J. Zhong, H.-H. Zhang, Z.-J. Li, B. Chen, C.-H. Tung and L. Z. Wu, *Org. Lett.*, 2012, **14**, 5992.

59. (a) E. Boess, D. Sureshkumar, A. Sud, C. Wirtz, C. Fares and M. Klussmann, *J. Am. Chem. Soc.*, 2011, **133**, 8106; (b) E. Boess, C. Schmitz and M. Klussmann, *J. Am. Chem. Soc.*, 2012, **134**, 5317.

60. J. Xie, H. Li, J. Zhou, Y. Cheng and C. Zhu, *Angew. Chem., Int. Ed.*, 2012, **51**, 1252.

61. K. Alagiri and K. R. Prabhu, *Org. Biomol. Chem.*, 2012, **10**, 835.

62. T. Tsuchimoto, Y. Ozawa, R. Negoro, E. Shirakawa and Y. Kawakami, *Angew. Chem., Int. Ed.*, 2004, **43**, 4231.

63. (a) L. Huang, T. Niu, J. Wu and Y. Zhang, *J. Org. Chem.*, 2011, **76**, 1759; (b) M. Nishino, K. Hirano, T. Satoh and M. Miura, *J. Org. Chem.*, 2011, **76**, 6447.

64. K. Alagiri, G. S. R. Kumara and K. R. Prabhu, *Chem. Commun.*, 2011, **47**, 11787.

65. A. Tanoue, W.-J. Yoo and S. Kobayashi, *Adv. Synth. Cat.*, 2013, **355**, 269.

66. J.-J. Zhong, Q.-Y. Meng, G.-X. Wang, Q. Liu, B. Chen, K. Feng, C.-H. Tung and L.-Z. Wu, *Chem.–Eur. J.*, 2013, **19**, 6443.

67. J. Dhineshkumar, M. Lamani, K. Alagari and K. R. Prabhu, *Org. Lett.*, 2013, **15**, 1092.

68. A. G. Condie, J. C. Gonzalez-Gomez and C. R. J. Stephenson, *J. Am. Chem. Soc.*, 2010, **132**, 1464.

69. W.-P. To, G. S.-M. Tong, W. Lu, C. Ma, J. Liu, A. L.-F. Chow and C.-M. Che, *Angew. Chem., Int. Ed.*, 2012, **51**, 2654.

70. W.-P. To, Y. Liu, T.-C. Lau and C.-M. Che, *Chem.–Eur. J.*, 2013, **19**, 5654.

71. D. P. Hari and B. König, *Org. Lett.*, 2011, **13**, 3852.

72. Q. Liu, Y. N. Li, H.-H. Zhang, B. Chen, C.-H. Tung and L.-Z. Wu, *Chem.–Eur. J.*, 2012, **18**, 620.

73. M. Rönn, P. G. Andersson and J.-E. Bäckvall, *Tetrahedron Lett.*, 1997, **38**, 3603.

74. K.-T. Yip, J.-H. Li, O.-Y. Lee and D. Yang, *Org. Lett.*, 2005, **7**, 5717.

75. J. E. M. N. Klein, A. Pery, D. S. Pugh and R. J. K. Taylor, *Org. Lett.*, 2010, **12**, 3446.

76. C. J. Engelin and P. Fristrup, *Molecules*, 2011, **16**, 951.

77. A. Pintér, A. Sud, D. Sureshkumar and M. Klussmann, *Angew. Chem., Int. Ed.*, 2010, **49**, 5004.

78. W. J. Yoo, C. A. Correia, Y. H. Zhang and C.-J. Li, *Synlett*, 2009, 138.

79. A. Pintér and M. Klussmann, *Adv. Synth. Catal.*, 2012, **354**, 701.

80. B. Liégault and K. Fagnou, *Organometallics*, 2008, **27**, 4841.

81. M. V. Leskinen, K.-T. Yip, A. Valkonen and P. M. Pihko, *J. Am. Chem. Soc.*, 2012, **134**, 5750.

82. K.-T. Yip, R. Y. Nimje, M. V. Leskinen and P. M. Pihko, *Chem.–Eur. J.*, 2012, **18**, 12590.

CHAPTER 10

Light-Assisted Cross-Dehydrogenative-Coupling Reactions

LAURA FURST AND COREY R. J. STEPHENSON*

Department of Chemistry, University of Michigan, Ann Arbor, Michigan 48109, USA
*Email: crjsteph@bu.edu

10.1 Introduction

Cross-dehydrogenative-coupling (CDC) is an efficient method of forming C–C bonds by the functionalization of two C–H bonds. These reactions often take advantage of the activation of adjacent heteroatoms, particularly nitrogen.[1] In addition to being widely available and easy to prepare, substituted amines have favourable oxidation properties.[2] Electron-rich alkylamines are readily oxidized under photocatalytic conditions to produce reactive intermediates that can promote C–C bond-forming reactions.[3] Because radical-based methods suffer from utilizing stoichiometric amounts of powerful oxidants, toxic or pyrophoric radical initiators, and non-selective UV activation of organic molecules, chemists have employed redox active catalysts that are triggered by visible light to develop mild and environmentally friendly reaction conditions.[4] Through selective activation of photocatalysts by abundant, low-energy light, chemoselective transformations can be accomplished with fewer consequences to the environment or personal safety. This chapter will showcase the current state-of-the-art for light-assisted CDC reactions and provide a standpoint for future development.

RSC Green Chemistry No. 26
From C–H to C–C Bonds: Cross-Dehydrogenative-Coupling
Edited by Chao-Jun Li
© The Royal Society of Chemistry 2015
Published by the Royal Society of Chemistry, www.rsc.org

10.2 Visible Light Activation Mechanisms

10.2.1 Photocatalysis

Photochemical reactions operate through a number of possible mechanisms including sensitization, electron transfer, or direct bond homolysis/radical chain propagation.[5] Many methods for light-assisted CDC reactions discussed in this chapter—though not all—rely on catalytic redox cycles that function through electron-transfer processes and employ the use of: (1) a photocatalyst; (2) an oxidant or reductant (which is usually the substrate itself); and (3) a light source.[6] A representative catalytic cycle is illustrated in Figure 10.1. The cycle begins *via* visible-light excitation of the photocatalyst (PC^n) to a high-energy excited state (PC^{n*}) that can undergo a number of decay pathways, including bimolecular quenching to instigate organic reactions. The excited state species can oxidize an electron donor (**D**), such as an amine to produce an electron-rich (PC^{n-1}) species, or reduce an electron acceptor (**A**), such as oxygen or an alkyl halide to produce an electron-deficient (PC^{n+1}) species *via* single-electron-transfer (SET) processes. A second electron-transfer event re-generates the ground-state catalyst (PC^n). When an amine is employed as the electron donor, the subsequent radical cation can lead to the formation of reactive intermediates such as iminium ions or α-amino radicals that can engage in C–C cross-coupling reactions (*vide infra*).

The photocatalysts that are commonly used for CDC transformations typically fall into one of three categories:

(1) *Polypyridyl transition-metal complexes*: usually Ru and Ir-based complexes such as $Ru(bpy)_3X_n$, $Ir(ppy)_2(dtbbpy)X_n$, or $Ir(ppy)_3$ (bpy = 2,2″-bipyridine; ppy = 2-phenylpyridine; dtbbpy = 4,4″-di-*tert*-butyl-2,2″-bipyridine);
(2) *Porphyrins/organic dyes*: tetraphenylporphyrin (TPP), Rose Bengal (RB), eosin Y, and methylene blue;
(3) *Semiconductors*: TiO_2, ZnO, and CdS. These also have considerable activities in the UV range.

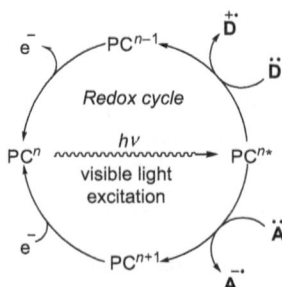

Figure 10.1 Oxidative and reductive quenching cycles for photocatalysts.

It will be demonstrated in later sections that there are several complimentary catalytic systems that can be implemented for the same transformation. For example, nitro-Mannich, Strecker-type, and alkynylation reactions can be performed under a variety of photochemical conditions. Other reactions require more specific conditions to bias pathways involving discrete intermediates, which in some cases may depend upon the use of terminal oxidants or specialized coupling partners. However, controlling the outcome of a photochemical reaction is challenging and may be influenced by certain thermodynamic or electronic properties of the molecule targeted for C–H activation (see the following section).

10.2.2 Factors Affecting C–H Activation

Chemists typically consider three properties in order to evaluate susceptibility towards C–H activation under photocatalytic conditions. These include the:

(1) redox potential (E_{red});
(2) bond dissociation energy (BDE); and
(3) pK_a

The initial step in C–H activation by visible light generally entails oxidation of the substrate by the excited-state photocatalyst. The susceptibility of a substrate, such as an amine (and in some cases, alcohols and ethers), towards oxidation can be assessed based on its corresponding redox potential $(E_{red})^7$ and the redox potential of the photocatalyst (Figure 10.2). For example, Ru(bpy)$_3^{2+}$* ($E_{red} = +0.84$ V *vs.* SCE (saturated calomel electrode))[8] can easily oxidize tertiary amines, such as triethylamine ($E_{red} = +0.73$ V *vs.* SCE),[7a,9] *N,N*-dimethylbenzylamine ($E_{red} = +0.77$ V *vs.* SCE),[7a] or 4-methoxy-*N,N*-diphenylaniline ($E_{red} = +0.76$ V *vs.* SCE).[10] This comparison is usually applied qualitatively, since redox potentials are influenced by subtle changes in solvent, pH, viscosity, and temperature, among other factors. Thus, reported values do not always provide accurate predictions for practical redox

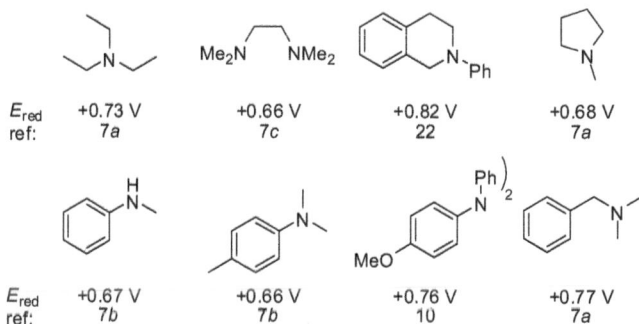

E_{red} ref:	+0.73 V 7a	+0.66 V 7c	+0.82 V 22	+0.68 V 7a

E_{red} ref:	+0.67 V 7b	+0.66 V 7b	+0.76 V 10	+0.77 V 7a

Figure 10.2 Redox potentials (*vs.* SCE) of common amines.

behavior. In general, electron-rich amines will have larger reduction potentials than electron-poor amines.

Upon oxidation, the subsequent radical cation can decompose in a number of different ways to generate reactive intermediates (Scheme 10.1). The first possibility involves direct H-atom abstraction of the α-C–H bond of the oxidized amine (**I**) to generate an iminium ion (**II**), which is susceptible to nucleophilic attack *via* polar reaction mechanisms (pathway a). Deprotonation of **I** may also form a carbon-centered radical species (**III**) that can react with typical radical traps, such as olefins or arenes (pathway b). Generation of the iminium ion may also occur indirectly through oxidation of **III** *via* SET to the photocatalyst or another oxidant (pathway c). Finally, radical cation **I** can undergo non-productive pathways such as back-electron transfer with the reduced photocatalyst (PC^{n-1}) to re-generate the neutral amine and PC^n (pathway d).

The increased propensity of α-amino hydrogen atoms to undergo abstraction can be rationalized by a comparison of BDEs between the neutral and oxidized species. Studies by Wayner *et al.* have shown that upon oxidation to the radical cation, the BDE of the α-amino C–H bond in trimethylamine (Me_3N) is lowered by as much as 30 kcal mol^{-1}.[11] This dramatic decrease reflects the thermodynamic driving force of the radical species to form a more stable iminium ion intermediate (Scheme 10.2).

Oxidation of organic molecules can also affect the acidity of C–H bonds. Whereas H-atom abstraction forms ionic species such as iminium ions,

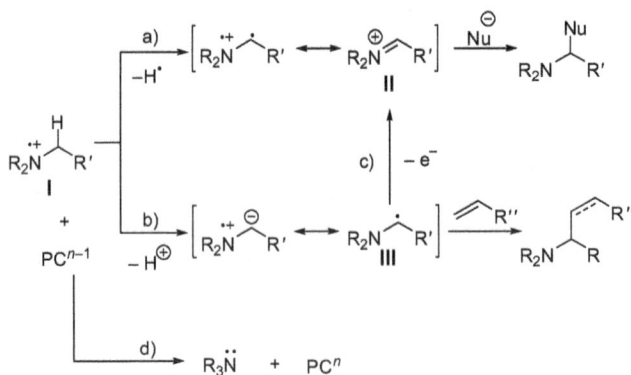

Scheme 10.1   Decomposition pathways of amine radical cations.

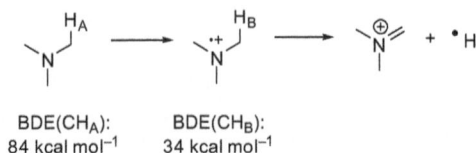

BDE(CH_A):
84 kcal mol^{-1}

BDE(CH_B):
34 kcal mol^{-1}

Scheme 10.2   Hydrogen-atom abstraction of the Me_3N radical cation.

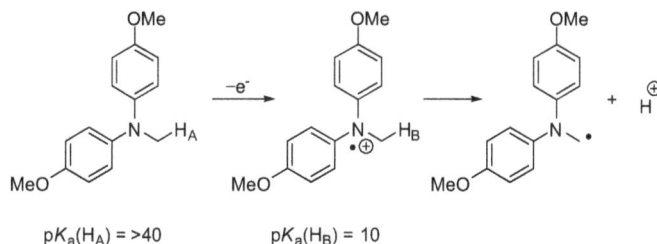

Scheme 10.3 Deprotonation of the di-*para*-anisylmethylamine radical cation.

deprotonation forms a neutral radical intermediates. Nicholas and Arnold established a correlation between the pK_a of the α-C–H bond of oxidized amines to both the BDE(CH$_A$) and E_{red} of the neutral amine.[12] Applications of this equation have demonstrated that the pK_a values of amine radical cations are substantially lowered relative to neutral amines. For example, calculations by Dinnocenzo *et al.* have shown that the di-*para*-anisylmethylamine radical cation is approximately 10 pK_a units in acetonitrile, which is dramatically more acidic than the un-oxidized starting material (Scheme 10.3).[13]

While understanding the factors that influence the activation of C–H bonds, certain on-going challenges limit the generalization of light-assisted CDC reactions. Substrates must have a suitable E_{red} for oxidation and a subsequently weakened C–H bond that can be cleaved. As such, many organic compounds do not meet these criteria. For instance, secondary or non-benzylic amines typically perform poorly in CDC reactions or not at all. Indeed, many examples shown below involve substrates with similar structures such as activated benzylic or cyclic tertiary amines. However, such transformations exhibit broad reactivity with respect to the coupling partners, produce highly functionalized compounds with great atom economy, and usually involve mild reaction conditions.

10.3 Photochemical CDC Reactions Using Transition Metal Complexes

10.3.1 Reactions of Ionic Intermediates

10.3.1.1 Aerobic Cycles

In 2010, Stephenson and co-workers[14] studied CDC reactions of tertiary amines with a focus on visible-light activation using polypyridyl transition metal complexes as electron-transfer agents.[15] Such complexes are readily available, easy to synthesize, and amenable to low catalyst loadings. Furthermore, their photophysical properties[16] have been studied extensively and their redox behavior with tertiary amines is well known. Electron-rich amines are generally utilized in photocatalysis as reducing agents for a wide range of transformations. Rather than acting only as sacrificial electron or

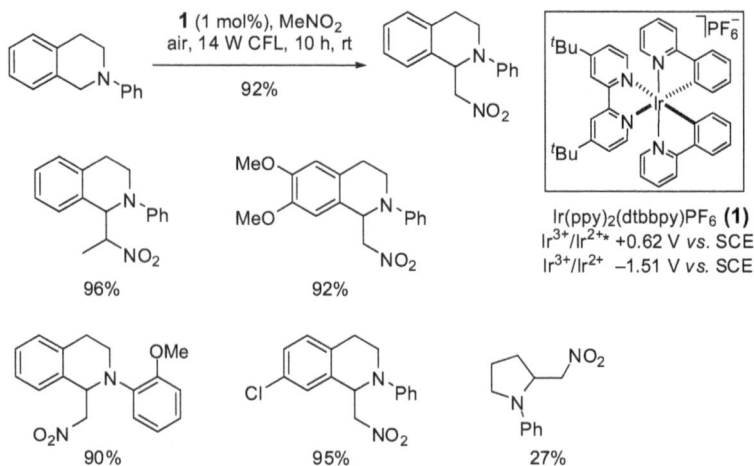

Scheme 10.4 CDC nitro-Mannich reactions.

hydrogen-atom donors, their unique reactivity can be exploited for chemo-selective C–H bond functionalization. Stephenson and co-workers initially demonstrated the use of photoredox catalysis for nitro-Mannich (or aza-Henry) reactions of *N*-aryl tetrahydroisoquinolines (THIQs) with nitroalkanes. A reaction mixture containing a THIQ, a nitroalkane, and Ir(ppy)$_2$(dtbbpy)PF$_6$ (**1**)[17] was irradiated with a 14 W compact fluorescent light (CFL) bulb under an aerobic atmosphere to produce nitro-Mannich products in yields >90% (Scheme 10.4). The CDC reaction of non-benzylic *N*-phenylpyrrolidine also occurred, though oxidation was more sluggish (72 h) and exhibited poorer conversion. A slow background reaction was observed in the absence of **1** over a period of 7 days; however, in the presence of a photocatalyst, reactions were typically complete in fewer than 24 h with excellent yields.

A representative mechanism for the oxidative nitro-Mannich reaction of THIQs is shown in Scheme 10.5. Oxidation of the THIQ with excited state Ir^{3+}* gives the amine radical cation (**Ia**) and Ir^{2+}. Catalyst turnover is presumed to occur by Ir^{2+}-mediated reduction of either adventitious oxygen or the nitroalkane to the corresponding radical anion species. This radical anion can abstract the α-H atom of the amine radical cation, resulting in the formation of the iminium ion (**IIa**), which is then trapped by the nitroalkane. Alternatively, deprotonation of **Ia** can occur to form α-amino radical **IIIa**, which is then oxidized to form **IIa**. Stephenson and co-workers were able to rule out the role of singlet oxygen in the predominate reaction pathway by demonstrating that photosensitizers were ineffective in promoting this transformation.

The encouraging results with photoredox catalysis sparked an explosion of additional investigations of visible-light-promoted oxidation of sp^3 carbons. Many studies explored use of related polypyridyl transition metal based

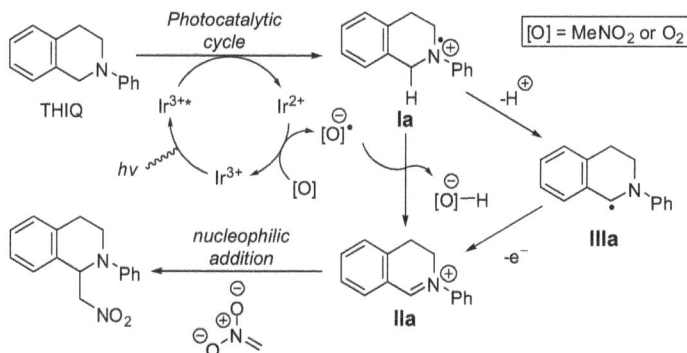

Scheme 10.5 Stephenson's proposed mechanism for oxidative nitro-Mannich reactions.

photocatalysts with different combinations of ligands and metal centers. Rueping *et al.* studied a wide range of intermolecular CDC reactions of benzylic and tertiary amines by tuning reaction conditions for each desired nucleophile. For Mannich-type additions, THIQ iminium ions were generated using the photocatalyst Ru(bpy)$_3$PF$_6$ (**2**)[18] in acetonitrile, and were trapped by acetone derivatives aided by L-proline as an organocatalyst (Scheme 10.6). The authors found that rapid generation of the iminium ion was counterproductive given the slower rate of enamine formation/addition, which could lead to undesired degradation of the iminium ion. This problem was ameliorated by using lower intensity light sources such as a 5 W CFL rather than LEDs or UV light, the use of which resulted in decreased yields. In a related experiment, Xia and co-workers carried out Mannich-type additions utilizing Ru(bpy)$_3$Cl$_2$ (**3**)[19] and silyl enol ethers as nucleophiles in methanol (Scheme 10.6). Interestingly, the authors found that a slower reaction rate was detrimental due to the gradual formation of undesired sideproducts and that by speeding up the reaction, specifically by the use of blue LEDs, the yields improved substantially. The seemingly contradictory observations by Rueping and Xia suggest it is possible that light may also be influencing the reactivity of the nucleophile.[20]

Rueping *et al.* also developed separate conditions for the addition of phosphates,[21] cyanide,[22] and acetylides.[23] Cyanations (or Strecker-type reactions) of THIQ derivatives were achieved using Ir(tbppy)$_2$(bpy)PF$_6$ (tbppy = 2-(4-*tert*-butyl-phenyl)-pyridine; **4**) in acetonitrile and tertiary benzylic amines with potassium cyanide, in which the addition of acetic acid was crucial for HCN liberation and solubility. Product yields range from 81–97% (Scheme 10.7). In addition to THIQ derivatives, substituted anilines also underwent coupling, with alkyl C–H bonds oxidized in preference to benzylic C–H bonds.

While the majority of transition metal photocatalysts are based upon ruthenium and iridium complexes, similar intermolecular cyanation reactions can also be accomplished using gold-based transition metal photocatalysts.

Rueping *et al.*

2 (1 mol%)
L-Proline (10 mol%)

MeCN, air, 5 W CFL
24–48 h

$R^1/R^2 =$
Ph/Me, 95%
Ph/Et, 47%
p-Cl-C$_6$H$_4$/Me, 83%
o-Tol/Me, 61%

Xia *et al.*

3 (5 mol%)
MeOH, air

Blue LEDs, 4–13 h

$R^1/R^2 =$
Ph/Ph, 96%
Ph/*o*-Cl-C$_6$H$_4$, 75%
m-Me-C$_6$H$_4$/*p*-Cl-C$_6$H$_4$, 87%
CHCO$_2$Et/Ph, 92%

Ru(bpy)$_3$X$_2$

X = PF$_6$ **(2)**   Ru^{2+*}/Ru$^+$ +0.84 V *vs.* SCE
X = Cl **(3)**   Ru^{2+}/Ru$^+$ −1.31 V *vs.* SCE

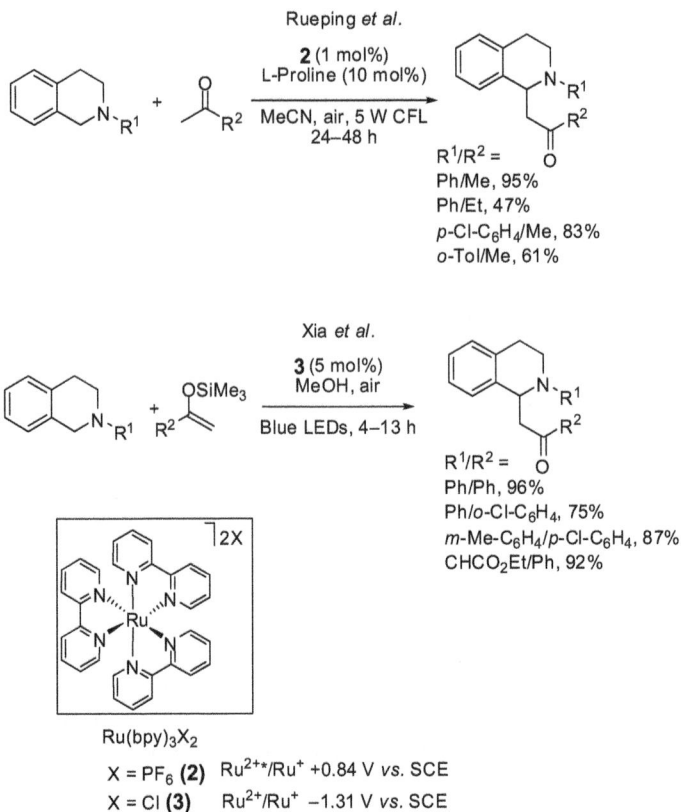

Scheme 10.6   CDC Mannich reactions.

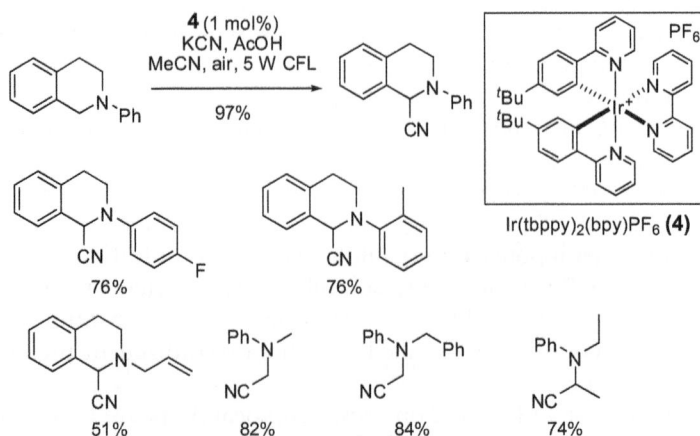

4 (1 mol%)
KCN, AcOH
MeCN, air, 5 W CFL

97%

Ir(tbppy)$_2$(bpy)PF$_6$ **(4)**

76%

76%

51%   82%   84%   74%

Scheme 10.7   Oxidative cyanations of amines.

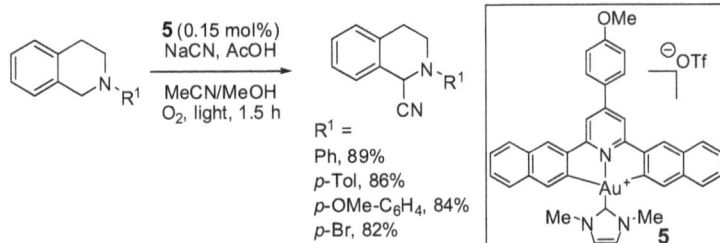

Scheme 10.8 Oxidative cyanations of amines using gold catalyst **5**.

Che and co-workers demonstrated the use of the pyridyl Au(III) complex **5** bound to a *N*-heterocyclic carbene ligand (Scheme 10.8) for the oxidation of tertiary amines.[24] Using this photocatalyst in only 0.15 mol%, cyanations of THIQs were achieved in 82–92% yields.

For CDC reactions of amines with alkynes, the photocatalyst Ru(b-py)$_2$(dtbbpy)(PF$_6$)$_2$ (**6**) in dichloromethane was shown to be more efficient for the oxidation of THIQ derivatives (Scheme 10.9).[23] The addition of terminal alkynes was accomplished in the presence of (MeCN)$_4$CuPF$_6$ as a co-catalyst to form a nucleophilic copper acetylide intermediate. This dual catalytic system enabled the addition of a wide range of alkynes in yields ranging from 53 to 88%. In the case of a *para*-(*tert*-butyl)phenyl substituted alkyne, the use of Cu(I) was ineffective, however the addition of the Ag(I) salt Ag(O$_2$CCF$_3$) led to a 77% yield of the desired product. Upon changing the irradiation source from a 5 W CFL to blue LEDs, the use of Ag(O$_2$CCF$_3$) as a co-catalyst instead led to decomposition.

Direct arylations of C–H bonds have also been reported under photocatalytic conditions. Li and co-workers reported photo-oxidations of α-amino carbonyl compounds at elevated temperatures.[25] In the presence of **3** and 1 atm. O$_2$ in dichloromethane at 40 °C, amino ketones and esters were oxidized to the corresponding iminium ion and trapped with indole nucleophiles to give products in 50–71% yields (Scheme 10.10). The reaction tolerates indole *N*-substitution with alkyl and benzyl groups; however the addition of *N*-acyl indole was unsuccessful. Notably, similar arylations of α-amino carbonyl compounds using copper-based CDC methods lead to over-oxidation. Wu and co-workers demonstrated that Pt(II) terpyridyl complex **7** is effective for photochemical CDC arylations of THIQ derivatives.[26] A variety of substituted indoles underwent addition in the presence of **7** and FeSO$_4$ in DMF (*N,N*-dimethylformamide) under ambient air and visible-light irradiation using blue LEDs. In the absence of FeSO$_4$ under aerobic conditions, amide **8** was formed as a side-product (Scheme 10.11). As complex **7** has previously shown to be a photosensitizer for oxygen,[27] the authors used electron paramagnetic resonance experiments to support a mechanism involving the superoxide radical anion (O$_2$•⁻) rather than singlet oxygen (¹O$_2$) as the key reactive intermediate. Thus, it was presumed that under an oxygen atmosphere, O$_2$•⁻ could react with amino radical **IIIa** to form

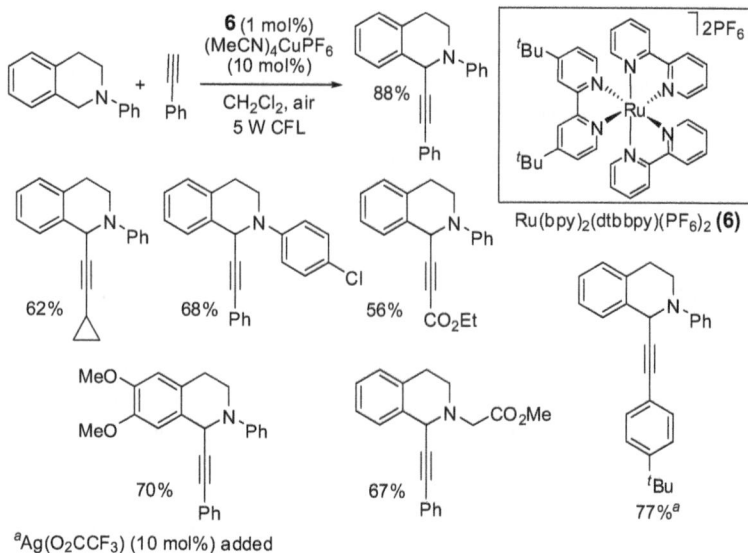

Scheme 10.9   CDC alkynylation reactions.

amide **8**. The addition of FeSO$_4$ to sequester O$_2$$^{\bullet-}$ led to improved yields and no observed formation of **8**.

Wang and co-workers discovered that photo-oxidation of dialkylamines results in concomitant fragmentation.[28] Photo-oxidation of tetra-methylethanediamine (TMEDA) using photocatalyst **3** in acetonitrile under an atmosphere of oxygen forms amine radical cation **Ib** (Scheme 10.12). With a β-N atom present, C–C bond cleavage can occur in preference to α-H atom abstraction to generate iminium ion **IIb** as well as α-amino radical **IIIb**. The authors were able to trap **IIb** with a variety of nitroalkanes. Dialkylamines such as dipiperidinylethane and dimorpholinoethane undergo similar transformations. However, upon oxidation, β-alkoxyamines such as 2-(dimethylamino)ethyl benzoate favor α-H atom abstraction over β-C–C fragmentation to give products such as **9**. The authors were also able to detect amino radical **IIIb** *via* free-radical polymerization of 2-hydroxyethylacrylate with TMEDA under the photocatalytic conditions.

In addition to iminium ions, other electrophilic species can be generated using photocatalysis. For example, a 1,3-dipole can be generated from the *in situ* oxidation of amines substituted with an appropriate electron-withdrawing group (EWG). This modification extends the reactivity profile of photogenerated intermediates to include the formation of C–C bonds with less polar reactive partners such as olefins and alkynes (Scheme 10.13).

The research groups of Xiao and Rueping applied this concept towards [3 + 2] dipolar cycloadditions using amino acetate derivatives that form azomethine ylides under photocatalysis. Xiao and co-workers performed aerobic oxidations of ethyl 2-(3,4-dihydroisoquinolin-2(1*H*)-yl) acetates in the

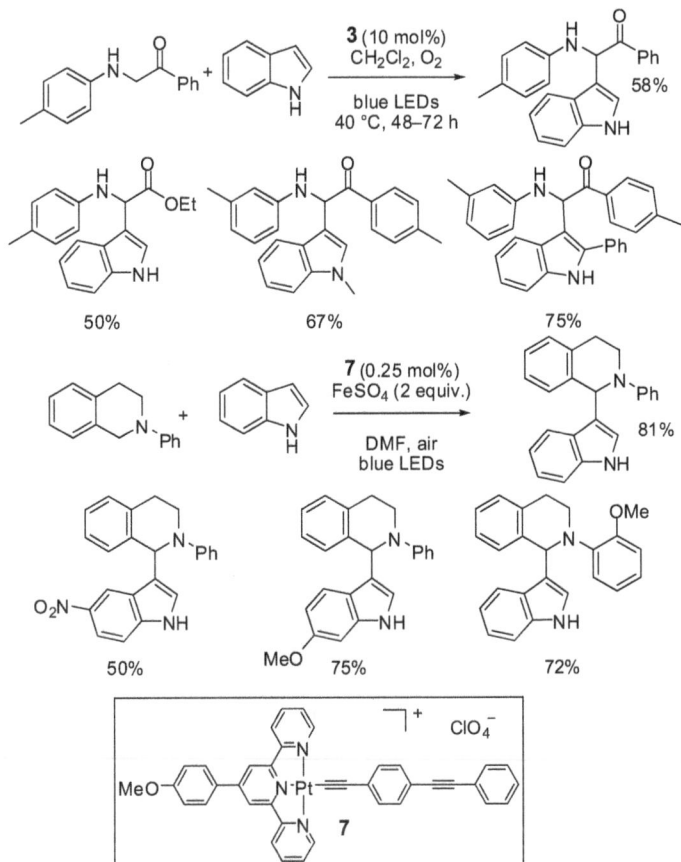

Scheme 10.10 CDC reactions of amino ketones, THIQs and indoles.

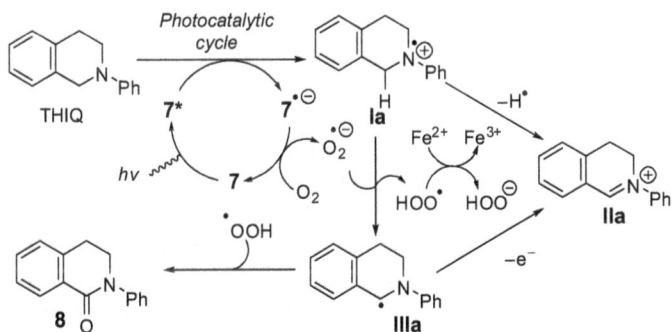

Scheme 10.11 Wu's proposed mechanism for amine oxidation with 7.

presence of **3** and underwent cycloadditions with dipolarophiles such as *N*-substituted maleimides (**10** and **11**, Scheme 10.14), olefins (**12** and **14**), maleic anhydrides (**13**), or alkynes (**14**).[29] A mixture of oxidation products

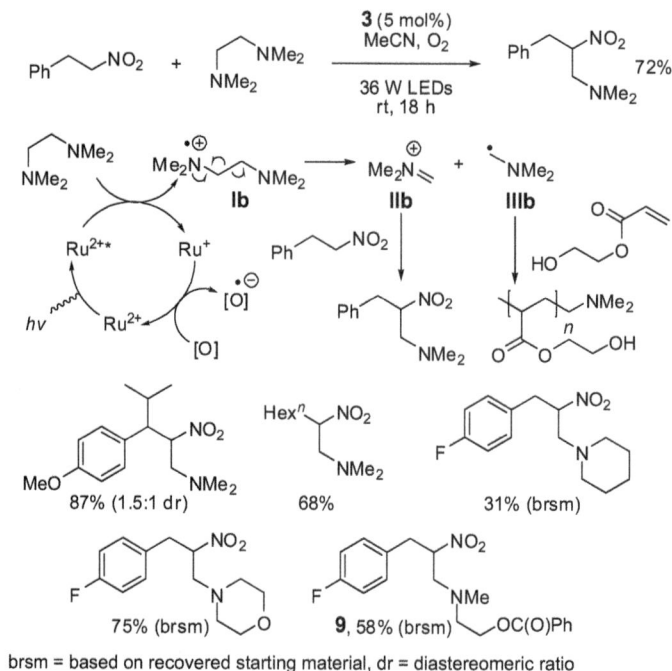

Scheme 10.12   Nitro-Mannich/fragmentation reactions with diamines.

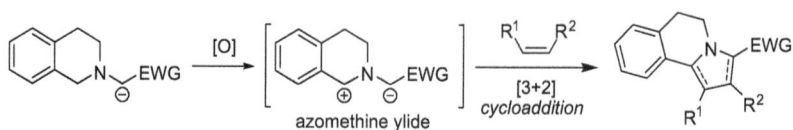

Scheme 10.13   Formation of azomethine ylides using photocatalysis.

was formed in some cases, with the minor product being the fully saturated tetrahydropyrrolo[2,1-*a*]isoquinoline. Hence, the authors found that subsequent addition of *N*-bromosuccinimide (NBS) cleanly affords dihydropyrrolo[2,1-*a*]isoquinolines such as **9–13** in 51–94% yields. Rueping *et al.* studied related cycloadditions of THIQ derivatives with maleimides using the photocatalyst Ru(dtbbpy)$_3$(PF$_6$)$_2$ **15** under air and light irradiation (Scheme 10.15).[30] By using *N*-malonyl rather than *N*-acetate THIQ derivatives, the corresponding tetrahydropyrrol[2,1-*a*]isoquinolines **16** are produced exclusively with good-to-moderate *exo/endo* (*syn,anti/syn,syn*) selectivities and moderate yields.

   Polypyridyl metal catalysts are soluble in a range of organic and inorganic solvents, lending ease and efficiency to experimental operation, but also making it difficult to recover catalysts from crude reaction mixtures for possible re-use. Lin and co-workers aimed to tackle this particular challenge by developing heterogeneous photocatalysts for CDC reactions

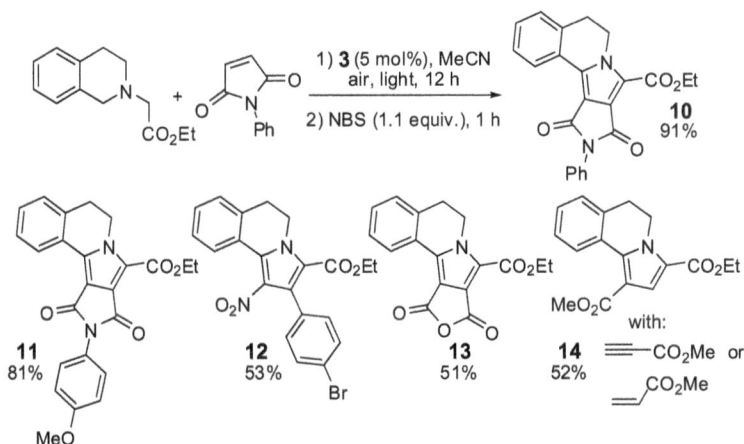

Scheme 10.14 Xiao's photocatalytic cycloadditions of amines and olefins.

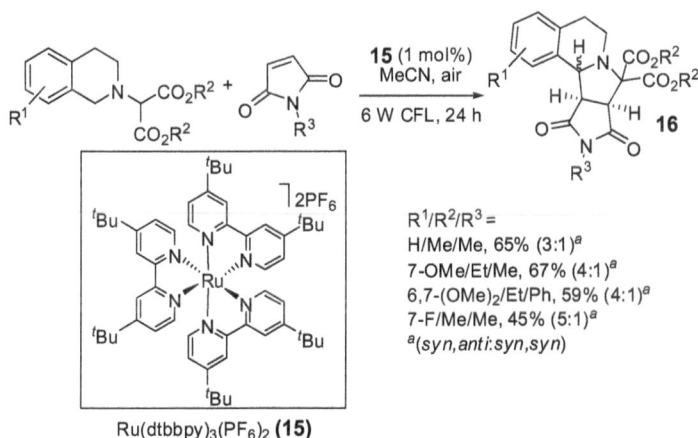

R^1/R^2/R^3 =
H/Me/Me, 65% (3:1)^a
7-OMe/Et/Me, 67% (4:1)^a
6,7-(OMe)$_2$/Et/Ph, 59% (4:1)^a
7-F/Me/Me, 45% (5:1)^a
^{a}(*syn,anti:syn,syn*)

Ru(dtbbpy)$_3$(PF$_6$)$_2$ **(15)**

Scheme 10.15 Rueping's photocatalytic cycloadditions of THIQ derivatives with maleimides.

(Figure 10.3).[30–32] The aim of this approach is to minimize wasteful disposal of expensive rare earth metal complexes by instead utilizing recoverable photocatalysts that could be recycled in subsequent reactions without loss of activity. Lin and co-workers developed solid-supported photocatalysts by embedding chromophores into cross-linked polymers. The authors converted Ru(bpy)$_3$$^{2+}$- and Ir(ppy)$_2(bpy)^+$-based complexes into polymers by linking each chromophore with substituted tetraphenylmethane units to generate both porous and non-porous cross-linked polymers (PCP and CP, respectively) that are photochemically active. Using 0.2 to 1 mol% of **Ru-PCP** and **Ir-PCP**, nitro-Mannich reactions of THIQs were accomplished in the presence of nitroalkane solvents, air, and a 26 W CFL (Scheme 10.16).[31]

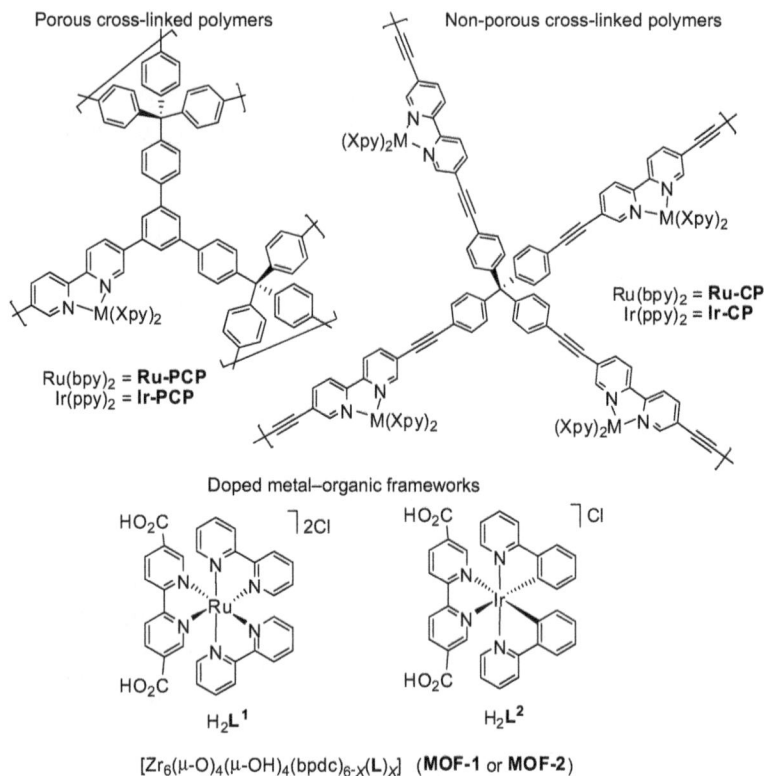

Porous cross-linked polymers

$(Xpy)_2M$

Ru(bpy)$_2$ = **Ru-PCP**
Ir(ppy)$_2$ = **Ir-PCP**

Non-porous cross-linked polymers

Ru(bpy)$_2$ = **Ru-CP**
Ir(ppy)$_2$ = **Ir-CP**

$(Xpy)_2M$

Doped metal–organic frameworks

HO$_2$C   2Cl

HO$_2$C

H$_2$L^1

HO$_2$C   Cl

HO$_2$C

H$_2$L^2

$[Zr_6(\mu\text{-O})_4(\mu\text{-OH})_4(bpdc)_{6-x}(L)_x]$ (**MOF-1** or **MOF-2**)

Figure 10.3   Solid-supported photocatalysts.

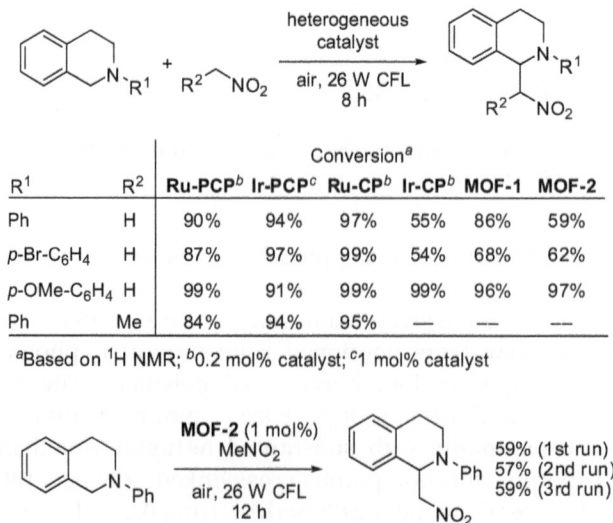

heterogeneous
catalyst
———————→
air, 26 W CFL
8 h

R^1	R^2	Conversion^a					
		Ru-PCP^b	Ir-PCP^c	Ru-CP^b	Ir-CP^b	MOF-1	MOF-2
Ph	H	90%	94%	97%	55%	86%	59%
p-Br-C$_6$H$_4$	H	87%	97%	99%	54%	68%	62%
p-OMe-C$_6$H$_4$	H	99%	91%	99%	99%	96%	97%
Ph	Me	84%	94%	95%	—	—	—

^{a}Based on ^{1}H NMR; ^{b}0.2 mol% catalyst; ^{c}1 mol% catalyst

MOF-2 (1 mol%)
MeNO$_2$
———————→
air, 26 W CFL
12 h

59% (1st run)
57% (2nd run)
59% (3rd run)

Scheme 10.16   Heterogeneous photocatalysis for CDC nitro-Mannich reactions.

Upon filtration of the catalysts from the reaction mixture, conversions between 84–97% were reported and were typically higher than those corresponding to the monomeric metal complex. Results were more varied among **Ru-CP** and **Ir-CP**, in which **Ru-CP** greatly out-performed **Ir-CP** for substrates that were more electron-deficient.[32] Control experiments to support the heterogeneity of these systems employed the supernatant solution for subsequent reactions and resulted in only minimal photocatalytic activity (<12% conversion).

Another approach undertaken by Lin and co-workers towards heterogeneous catalysis involved doping metal–organic frameworks (MOFs) with the metal complex chromophores (Figure 10.3). The authors incorporated Ru and Ir complexes H_2L^1 and H_2L^2, respectively, into porous $Zr_6O_4(OH)_4(bpdc)_6$ (bpdc = *para*-biphenyldicarboxylic acid) frameworks to produce **MOF-1** and **MOF-2**, respectively, which were used in nitro-Mannich reactions of THIQ derivatives (Scheme 10.16).[33] Conversions were generally lower for **MOF-1** and **MOF-2** than for cross-linked polymers or their monomeric counterparts; however, these catalysts could be readily filtered off and re-used for subsequent reactions. For example, *N*-phenyl THIQ was subjected three times to the photocatalyzed nitro-Mannich reaction in the presence of **MOF-2**, which provided consistent conversions (57–59%) for each run.

10.3.1.2 Anaerobic Cycles

The photocatalytic cycles discussed so far utilize oxygen as the stoichiometric oxidant *via* exposure of the reaction mixture to ambient air or solution saturation with an O_2 gas balloon. Oxygen may be implicated in a number of mechanistic events including amine oxidation, H-atom abstraction, and catalyst turnover. Also, a number of reactive oxygen-derived intermediates such as peroxide, superoxide anion, or singlet oxygen may be involved. The use of oxygen has many practical advantages including low cost and availability as well as operational ease relative to inert reaction conditions. However, with the many mechanistic roles of oxygen, there is also a greater potential for side-product formation arising from α-amino radical **III** (Scheme 10.17) both in its absence and presence, including dimerization (**IV**) and O_2 coupling (**V**), respectively.

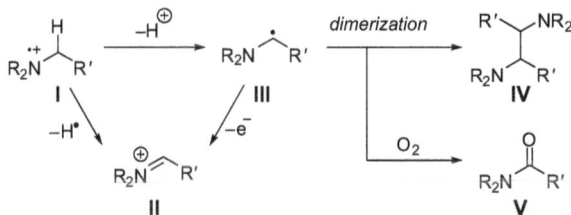

Scheme 10.17 Potential side-products in photochemical CDC reactions.

Researchers have developed a variety of aerobic conditions for photo-generation and trapping of iminium ions that are fine-tuned to each desired nucleophile. This includes the use of different solvents, light sources, additives, and photocatalysts. Another approach towards biasing the re-activity of iminium ion **II** over α-amino radical **III** is to replace oxygen as the terminal oxidant. Stephenson and co-workers found that bromotri-chloromethane (BrCCl$_3$) was a suitable oxidant for catalyst turnover, and it allowed for the addition of several different types of nucleophiles to oxidized THIQ derivatives.[34] Using photocatalyst Ru(bpy)$_3$Cl$_2$ (**3**) in DMF in the presence of BrCCl$_3$ and blue LEDs, full conversion to the iminium ion was observed in 3 h. Then, different nucleophiles were added to the reaction mixture (Scheme 10.18). The authors also found that removing the reaction

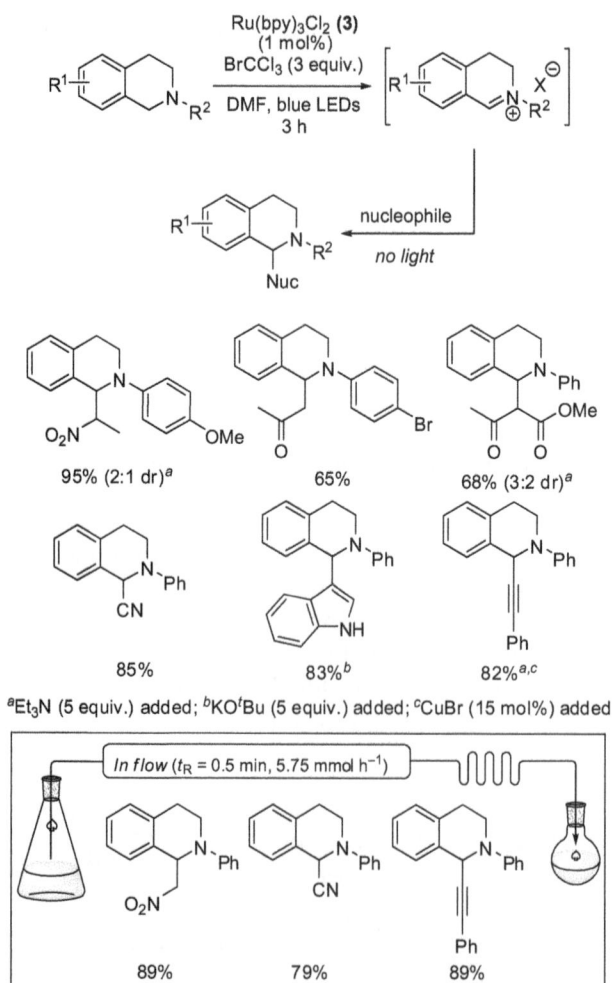

95% (2:1 dr)^a 65%   68% (3:2 dr)^a

85%   83%^{b} 82%^{a,c}

^{a}Et$_3$N (5 equiv.) added; ^{b}KO^{t}Bu (5 equiv.) added; ^{c}CuBr (15 mol%) added

In flow (t_R = 0.5 min, 5.75 mmol h^{-1})

89%   79%   89%

Scheme 10.18   Photochemical CDC reactions of THIQs with **3** and BrCCl$_3$ in batch (top) and flow (bottom).

mixture from light upon generation of the iminium ion helped to minimize side-product formation. A broad range of nucleophiles including nitroalkanes, allylsilanes, silyl enol ethers, 1,3-dicarbonyls, cyanide and indoles were accommodated using this method. Acetylides could also be added with the aid of 15 mol% CuBr *via* dual catalytic cycles.

Stephenson and co-workers subsequently applied their conditions in flow to demonstrate a practical means for conducting photochemical reactions on a large scale.[35] A mixture containing 0.5 mol% 3, BrCCl$_3$, DMF, and 1.0 g *N*-phenyl THIQ was pumped through a reactor consisting of coiled perfluoroalkoxy tubing (479 μL) that was exposed to blue LEDs. The mixture was then diverted into a flask containing the desired nucleophile. Products were formed in 79–89% yields with a residence time (t_R) of 5 min. In flow, a higher rate of conversion to the iminium ion was observed with a turnover rate of 5.75 mmol h^{-1}. Batch reactions in contrast required 3 h for full conversion, corresponding to a turnover rate of 0.081 mmol h^{-1}. This remarkable difference exemplifies the utility and efficiency of flow chemistry.

The gradual formation of CHCl$_3$ over time with iminium ion generation supported a mechanism involving H-atom abstraction by the trichloromethyl radical (CCl$_3$•, Scheme 10.19). Bromotrichloromethane was initially employed to oxidize Ru$^+$ to Ru^{2+}, but it has also been known to oxidize Ru^{2+}* to Ru^{3+}. Thus, both of these pathways are possible and the amine substrate can act as a reductant for either Ru^{3+} or Ru^{2+}*. The CCl$_3$• formed in the cycle can directly abstract the α-C–H of amine radical cation **I**, providing iminium ion **II** as well as CHCl$_3$. A radical chain mechanism that involves oxidation of α-amino radical **III** by BrCCl$_3$ *via* either electron or atom transfer to generate **II** may be operative.

Stephenson and co-workers also explored the persulfate anion (S$_2$O$_8$$^{2-}$) as an oxidant for light-assisted CDC reactions of *N*-alkylamides.[36] *N,N*-Dialkylamides such as DMF, dimethylacetamide, and *N*-methylpyrrolidinone could be oxidized using a combination of Ru(bpy)$_3$Cl$_2$ (3) and ammonium persulfate [(NH$_4$)$_2$S$_2$O$_8$] in the presence of blue LEDs at 25–30 °C (Scheme 10.20). Upon generation of *N*-acyliminium ion **IIc**, a variety of electron-rich arenes reacted to form α-arylated products in yields ranging from 23 to 89%. The reaction also proceeded *via* thermolysis of S$_2$O$_8$$^{2-}$ at

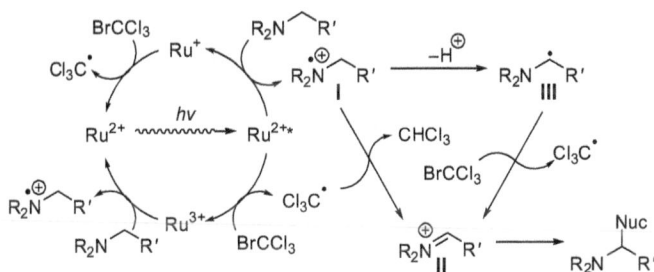

Scheme 10.19 Proposed mechanism for CDC reactions with **3** and BrCCl$_3$.

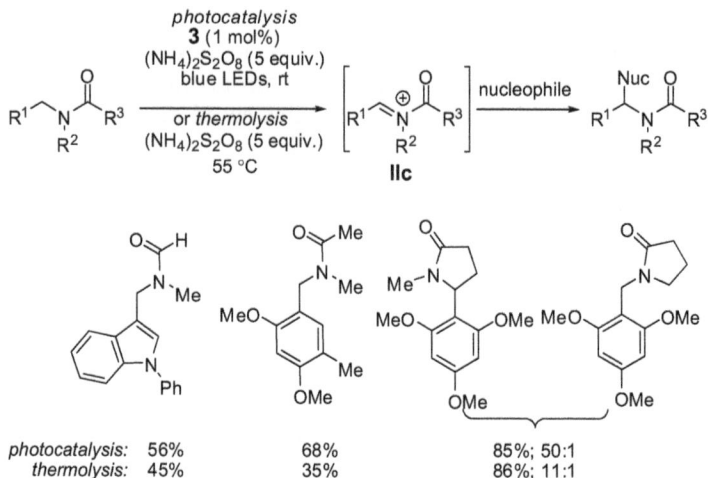

Scheme 10.20 Photochemical CDC arylation reactions with **3** and persulfate.

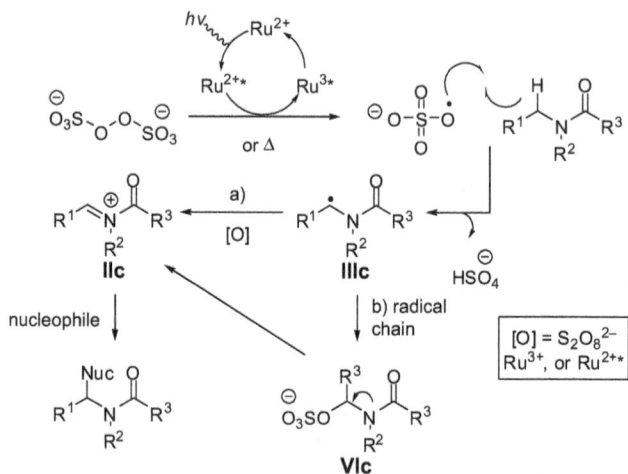

Scheme 10.21 Stephenson's proposed mechanism for amine α-arylation.

55 °C in the absence of the photocatalyst and light. However, yields and regioselectivities were generally better under photocatalytic conditions.

Unlike previous examples, the proposed mechanism for the Friedel–Crafts amidoalkylation does not involve initial oxidation of the amide substrate *via* a Ru^{2+}/Ru^{+} redox cycle (see Scheme 10.19). Rather, given the strong oxidizing ability of $S_2O_8^{2-}$ and the relatively weaker reducing ability of amides compared with amines, the authors proposed a predominantly Ru^{2+}/Ru^{3+} redox cycle (Scheme 10.21). Visible light excitation of Ru^{2+} forms $Ru^{2+}*$, which then reduces persulfate to the sulfate radical anion ($SO_4^{\bullet-}$). This species can directly abstract the α-C–H of the amide to form α-amido radical

IIIc. From here, there are two plausible pathways to generate the *N*-acyliminium intermediate **IIc.** One possibility involves single-electron oxidation by Ru^{3+}, Ru^{2+}*, or S$_2$O$_8$$^{2-}$ under photocatalytic conditions or just S$_2$O$_8$$^{2-}$ under thermolytic conditions (pathway a). Alternatively, a radical chain mechanism may occur to form oxyalkylamide **VIc,** which eliminates sulfate to provide **IIc** (pathway b). Lastly, interception by the nucleophile forms the product.

Rovis *et al.* also explored alternative oxidants to effect unique photochemical CDC reactions of amines. They aimed to achieve enantioselective α-acylation of THIQ substrates by combining chiral *N*-heterocyclic carbene (NHC) catalysis with visible light photoredox catalysis.[37] In their studies, the authors found the addition of a weak co-oxidant, such as *meta*-dinitrobenzene (*m*DNB), to be crucial for high product yields and efficient catalytic turnover, whereas oxidants such as BrCCl$_3$ appeared to be incompatible with the NHC catalyst. Using 1 mol% photocatalyst **3** and 10 mol% of NHC catalyst **17** in CH$_2$Cl$_2$ with *m*DNB and an aldehyde, α-amino ketones could form in 51–94% yields in up to 99% enantiomeric excess (ee) upon irradiation with blue LEDs (Scheme 10.22). Interestingly, greatly decreased yields of **18** were observed upon exposure of the reaction mixture to atmospheric oxygen (46%) as well as upon vigorous degassing (75%), indicating that oxygen may play an important role in mediating the reaction at low concentrations.

The authors proposed dual catalytic cycles for this transformation (Scheme 10.23). The photocatalytic cycle begins with oxidation of Ru^{2+}* by *m*DNB to form Ru^{3+}, which oxidizes the amine to radical cation **Ia.** This

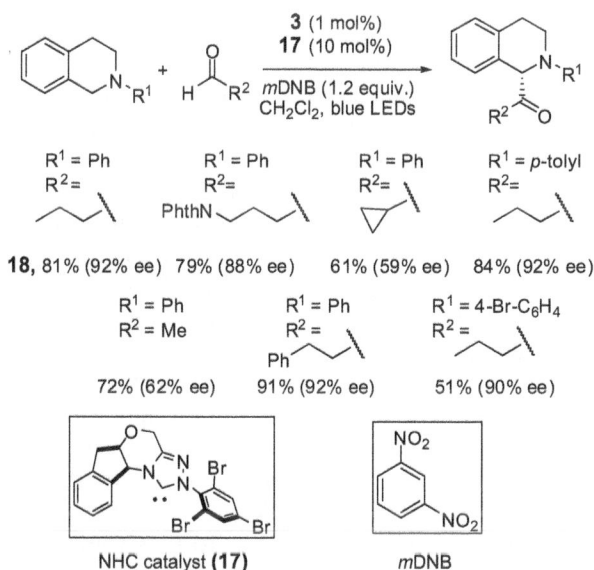

Scheme 10.22 Photochemical CDC reactions of THIQs and aldehydes.

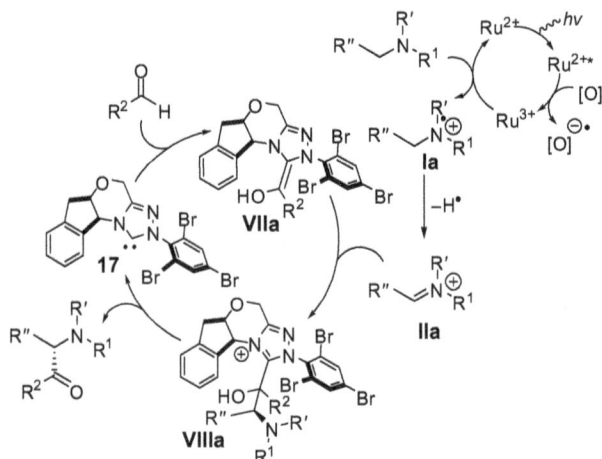

Scheme 10.23 Dual catalytic cycle for asymmetric amine acylation with 3 and 17.

species undergoes H-atom abstraction to generate iminium ion **IIa**. Simultaneously, the organocatalytic cycle begins with NHC catalyst **17** reacting with the aldehyde to generate Breslow intermediate **VIIa**. This attacks iminium ion **IIa** to form amino alcohol **VIIIa**, followed by elimination of NHC **17** to provide the chiral α-amino ketone product.

10.3.2 Reactions of Radical Intermediates

The photo-oxidation of substrates such as amines generates a number of reactive intermediates including amine radical cations, iminium ions, or α-amino radicals. The previous section discussed photogeneration of polar intermediates that can be intercepted by nucleophiles. Another type of light-assisted CDC reaction involves reactions of *in situ* generated α-amino radicals. These intermediates have distinct reactivities and can be intercepted by radical acceptors such as arenes and olefins. However, under photocatalysis they are readily converted to the iminium ion and trapped by polar species. Nucleophilic addition partners are added exogenously or formed as by-products in the catalytic cycle. For instance, when commonly employed oxidants like oxygen, BrCCl$_3$, or persulfate are reduced, superoxide anion, bromide, or hydrogen sulfate and sulfate, respectively can be liberated. Those species can react to form the iminium ion directly or transient α-substituted amine intermediates that are essentially iminium ion equivalents (*e.g.*, **VIc**, Scheme 10.21). The iminium ion is then trapped by another polar species to form stable and isolable products. However, the reaction conditions can be engineered appropriately to bias transformations of the α-amino radical. This is generally accomplished in the absence of external oxidants or the use of a base to promote deprotonation of the amine radical cation.

MacMillan *et al.* discovered through extensive reaction screening that C–H bonds of tertiary amines can couple with sp^2 C–CN bonds of electron-deficient arenes in the presence of *fac*-Ir(ppy)$_3$, NaOAc, and visible light.[38] Reiser and co-workers later demonstrated conjugate additions of photo-generated α-amino radicals with electron-deficient olefins.[39] Using either photocatalyst **1** or **3** in MeCN, THIQ derivatives were oxidized and underwent conjugate additions in the presence of α,β-unsaturated ketones, aldehydes, nitriles, and lactones in moderate-to-good yields, with catalyst **1** providing better yields in most cases (Scheme 10.24).

Under these conditions, amine radical cation **Ia** is deprotonated to form α-amino radical **IIIa**, which reacts with the olefin to form radical adduct **IXa**. Since no external oxidant is present, the adduct radical presumably turns over Ru$^+$ to Ru^{2+} and is reduced to the anion. Protonation of the enolate forms the product. For intramolecular additions, dehydrogenation occurs rather than reduction to provide 5,6-dihydroindolo[2,1-*a*]-THIQs such as **18**.

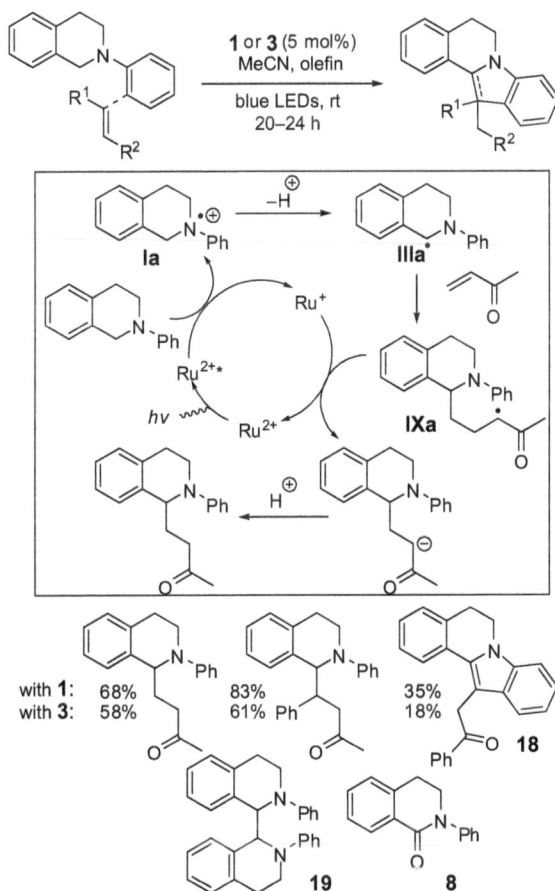

Scheme 10.24 Photochemical CDC reactions of THIQ α-amino radicals with conjugated olefins.

The authors provided evidence in support of amino radical **IIIa** as an intermediate by performing the reaction in the absence of the olefin and observing the formation of dimer **19**, formed from the radical combination of **IIIa** with itself. Furthermore, under an oxygen atmosphere in the absence of an alkene radical, **IIIa** was intercepted by molecular oxygen to form amide **8** in 42% yield.

Nishibayashi and co-workers developed similar conditions for light-assisted conjugate additions of α-amino radicals.[40] A variety of *N*-alkyl and -aryl tertiary amines were oxidized in the presence of the photocatalyst Ir(ppy)$_2$(dtbbpy)BF$_4$ (**20**) and visible light in *N*-methylpyrrolidine (NMP) and reacted with trisubstituted electron-deficient olefins to give alkylated amines in excellent yields (Scheme 10.25). The authors conducted radical clock experiments with cyclopropyl substituted olefins to provide support for a radical mechanism. They were also able to rule out a radical chain mechanism by conducting the reaction during intervals of illumination and darkness, during which time no reaction progression was observed. In addition, the use of radical initiators did not lead to conversion of starting material. This implies that formation of the amino radical is not propagated by H-atom abstraction by the radical adduct. Therefore, these results indicate that catalyst turnover occurs by reduction of the adduct radical.

Later studies revealed the influence of oxygen in the mechanism of photocatalytic reactions of α-amino radicals. Rueping and co-workers investigated intermolecular conjugate additions of dimethylaniline-derived radicals both in the absence and presence of oxygen using the photocatalyst Ir(ppy)$_2$(bpy)PF$_6$ (**21**) in MeCN (Scheme 10.26).[41] Under oxygen-free conditions, *N,N*-dimethyl-4-methylaniline underwent conjugate addition to 2-benzylidenemalononitrile to yield **22** in 91% yield *via* reduction/protonation of adduct radical **IXd**. When the reaction was open to air, cyclization product **23** was formed in 68% yield. Oxygen was presumed to be responsible for catalyst turnover rather than **IXd**, producing the superoxide radical anion. This allows reversible cyclization of **IXd** to **Xd**, which is converted to **23** by the superoxide radical anion.

The authors applied this concept to intramolecular visible-light-assisted CDC reactions of olefin-tethered anilines. Irradiation of **24** with visible light

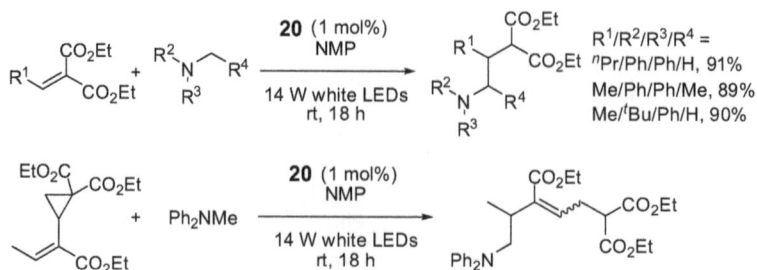

Scheme 10.25 Reactions of α-amino radicals using photocatalyst **20**.

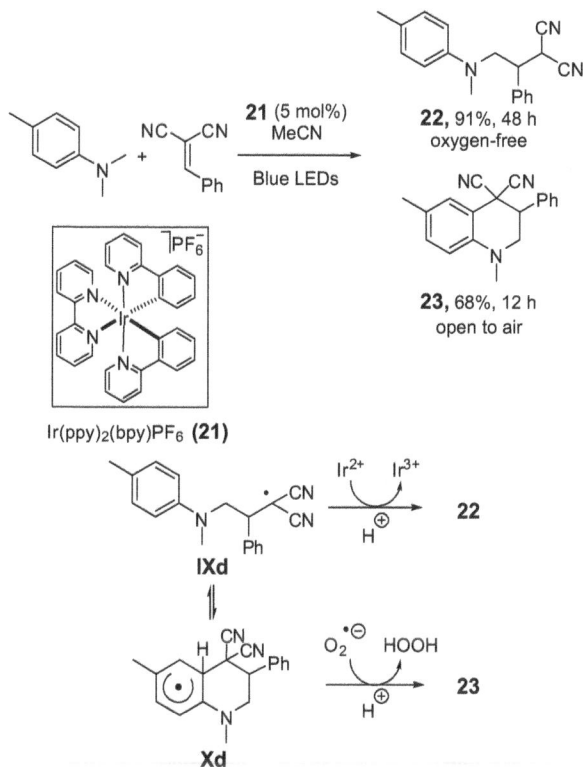

Scheme 10.26 Influence of oxygen on radical addition/cyclization reactions.

using $Ir(ppy)_2(dtbbpy)PF_6$ (**1**) in chloroform under an oxygen balloon produced the unexpected indole aldehyde product **25** in 63% yield (Scheme 10.27). The authors proposed the involvement of oxygen to explain the concomitant C–C fragmentation. One mechanistic possibility is outlined in Scheme 10.27. After conjugate addition of the α-amino radical, radical adduct **IXe** can be reduced and protonated to form indoline **Xe**. Under these conditions, **Xe** can undergo a subsequent oxidation to produce indole **26**. The indole can engage in another photocatalytic cycle to produce indolyl radical cation **XIe**, which can be deprotonated to produce α-carbonyl radical **XIIe**. This intermediate can be intercepted by molecular oxygen to form peroxy radical **XIIIe**. Cyclization gives intermediate **XIVe**, which can fragment to produce observed the carboxaldehyde product **25** and acetic acid.

10.4 Photochemical CDC Reactions Using Porphyrins/Organic Dyes

The benefit of using transition metal photocatalysts is often selective activation over organic molecules, which typically do not absorb light in the visible range. However, a broad range of organic dyes and porphyrins with

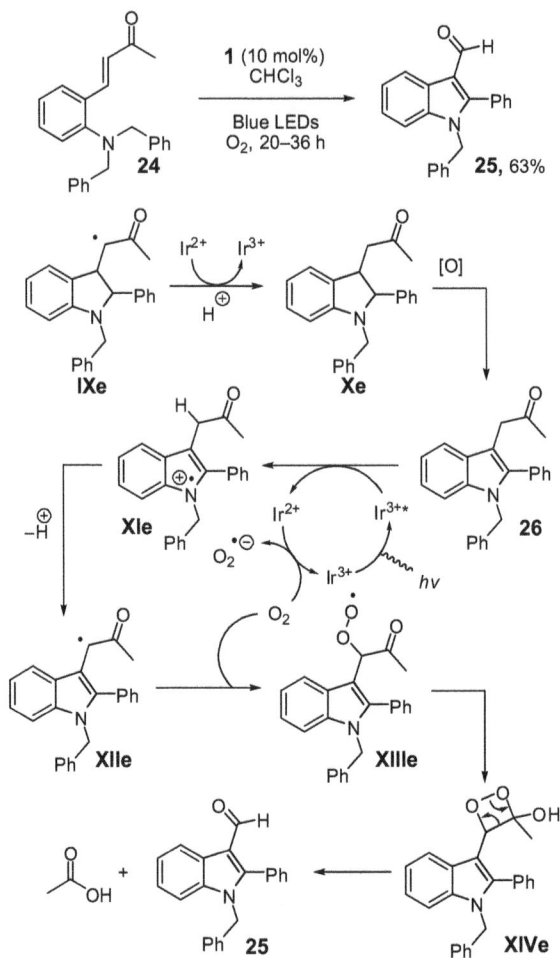

Scheme 10.27 Intramolecular CDC reaction/fragmentation of amino radicals.

variable absorption properties can be utilized as electron-transfer agents or as photosensitizers.[42] Though more commonly known for their use in dye-sensitized solar cells,[43] these highly conjugated molecules can be selectively activated by visible light and have suitably long excited-state lifetimes to mediate organic redox transformations (Figure 10.4). Organic dyes are typically less expensive than metal-based reagents and can be structurally altered with relative ease to change their electronic properties (*e.g.*, eosin Y *vs.* fluorescein).

10.4.1 Sensitization

Many of the photocatalysts discussed so far are capable of excited-state oxygen sensitization in addition to electron transfer. Hence, the oxidation of

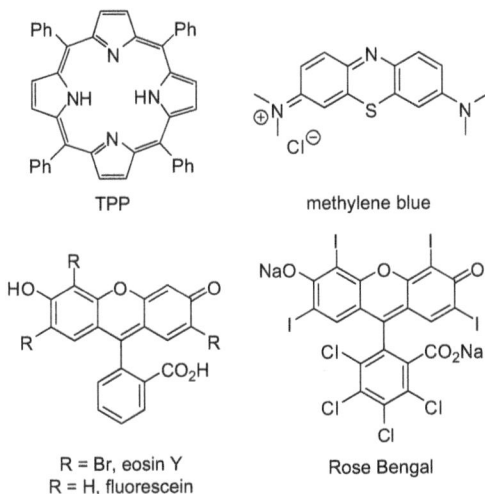

Figure 10.4 Dyes commonly used in photocatalysis.

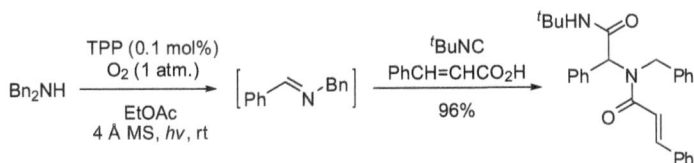

Scheme 10.28 Che's pioneering photochemical CDC reactions *via* sensitization. MS = molecular sieves.

substrates can be promoted by singlet oxygen ($^1O_2$) rather than through redox cycles between catalysts and organic molecules. Dyes have long been recognized as efficient oxygen photosensitizers and have been utilized for photo-oxidations of organic molecules.[44] Che and co-workers reported, in 2009, one of the first examples of photochemical CDC reactions which utilized the organic dye TPP. The authors demonstrated CDC reactions of benzylic amines *via* singlet oxygen ($^1O_2$) mediated generation of iminium ions that could participate in Ugi-type multi-component reactions in excellent yields (Scheme 10.28).[45] The authors employed TPP as an oxygen photosensitizer that was activated by a powerful 300 W tungsten lamp. Although these types of light sources have wide-ranging spectral emission between 200 and 3000 nm, as well as considerable thermal energy production, this nevertheless provided the foundation for use of dyes in visible-light-promoted organic transformations.

The electronic properties of porphyrin-based photocatalysts may be altered by modifying the substituents on the porphyrin backbone or by coordination to a metal. Che and co-workers reported the synthesis of Pt(II)- and Pd(II)-bound porphyrins that were effective for visible-light-mediated CDC reactions of THIQ derivatives.[46] In particular, complex **PdF$_{20}$TPP** has

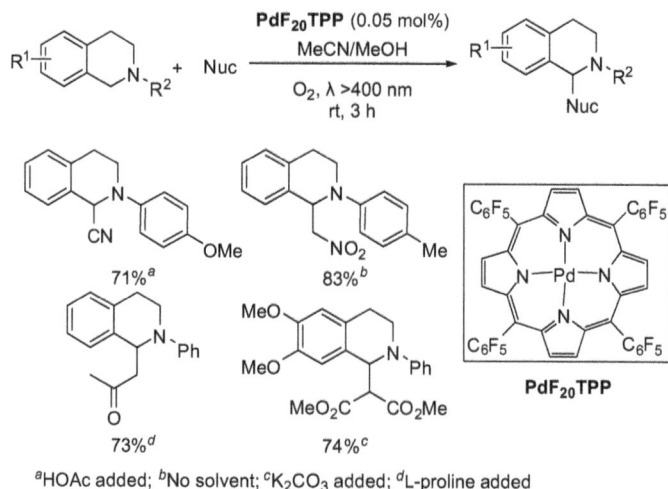

Scheme 10.29   CDC reactions of THIQs using Pd-bound porphyrin sensitizers.

photophysical properties comparable to Ru(bpy)$_3$ as well as substantial absorption in the visible range. Using this catalyst at very low loadings (0.05 to 0.01 mol%) on the gram-scale under an atmosphere of oxygen, a wide range of nucleophiles could be added to oxidized THIQ derivatives to give products in yields typically between 70 and 99% (Scheme 10.29). In addition, the authors demonstrated intramolecular cyclization of tethered nucleophiles. A variety of analytical indicators were used to distinguish a mechanism involving $^1O_2$ *vs.* photo-induced electron transfer. Emission quenching of excited-state PdF$_{20}$TPP* was observed in the presence of *N*-phenyl THIQ, but the corresponding reduced species, [PdF$_{20}$TPPP]$^-$ was not detected with nanosecond time-resolved absorption measurements. The authors noted the possibility of rapid back electron transfer that did not allow for the observation of this species. Moreover, using a comparison of redox potentials, the authors determined that some successful CDC reactions were too thermodynamically unfavorable for electron-transfer processes to occur. This supported a pathway principally involving oxidation by $^1O_2$.

10.4.2   Electron Transfer

Dyes can also be utilized as electron-transfer agents. Tan and co-workers utilized RB for the photogeneration of iminium ion **II** under aerobic conditions (Scheme 10.30).[47] To best match the absorption maxima of the photocatalyst, green LEDs were used as the light source to generate RB*, which can accept an electron from the amine substrate to form RB$^{\bullet-}$ and amine radical cation **I** *en route* to iminium ion **II**. Nitro-Mannich and Mannich-type addition reactions occurred in good yields (68–95%) in the presence of nitroalkanes or ketones as the solvent, respectively. For

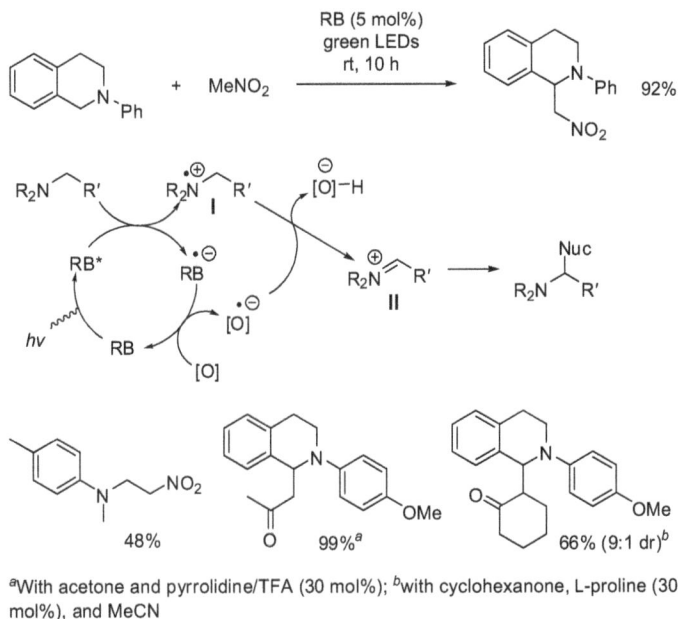

Scheme 10.30 Photochemical Mannich-type CDC reactions with RB.

Mannich-type reactions, an organocatalyst such as pyrrolidine/trifluoroacetic acid (TFA) or L-proline was employed to assist in the additions of ketones.

In addition to RB, eosin Y could be used for similar transformations of amines. König *et al.* utilized eosin Y and green LEDs to oxidize THIQ derivatives. The transient iminium ion was then trapped with a variety of nucleophiles including nitroalkanes, malonates, and dialkylphosphonates (Scheme 10.31).[48] The addition of malonitrile, however, led to α-cyanation rather than formation of the β-malonitrile, a result which has also been observed under non-photochemical oxidative conditions.[1]

Soon after, Wu and co-workers conducted mechanistic studies on eosin Y mediated CDC reactions.[49] In their report they utilized an eosin Y tetrabutylammonium salt (TBA-EY) and a 500 W Hanovia mercury lamp (filtered for $\lambda > 450$ nm) to study Mannich and nitro-Mannich reactions of THIQ derivatives with malonates, acetone/L-proline derivatives, and nitroalkanes under an O_2 atmosphere (Scheme 10.32). The authors then conducted thorough mechanistic investigations using a combination of analytical techniques. Using flash photolysis, they were able to observe generation of the triplet excited state of eosin Y (EY*) from visible light irradiation. The radical anion species (EY•−) was also observed with absorption spectroscopy upon introduction of *N*-phenyl THIQ. This supports a mechanistic pathway involving electron transfer from the amine to EY* to generate EY•− and the amine radical cation **Ia** *en route* to iminium ion **IIa**. The authors also utilized electron spin resonance (ESR) spectroscopy to gain insight into the role of

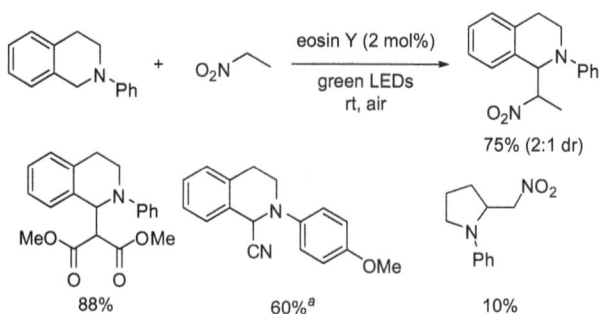

Scheme 10.31 Photochemical CDC reactions with eosin Y.

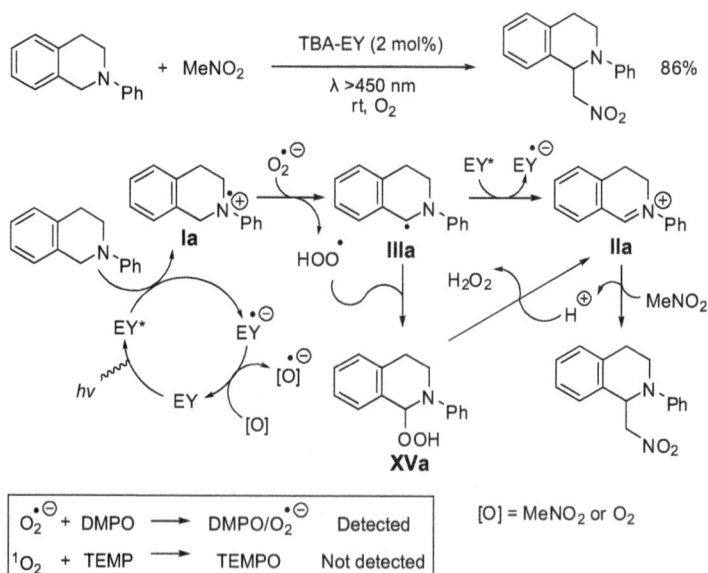

Scheme 10.32 Wu's proposed mechanism for eosin Y-promoted CDC reactions.

oxygen in the redox process, specifically whether the superoxide radical anion ($O_2^{\bullet-}$) or singlet oxygen ($^1O_2$) was the key reactive intermediate. Using 5,5-dimethyl-1-pyrroline-*N*-oxide (DMPO) and 2,2,6,6-tetramethylpiperadine (TEMP) to scavenge $O_2^{\bullet-}$ and $^1O_2$, respectively, they were able to detect a strong ESR signal for a DMPO/$O_2^{\bullet-}$ adduct with no signal for TEMPO resulting from reaction of TEMP with $^1O_2$. This supports a superoxide radical anion mediated oxidation of **Ia** to radical anion **IIIa** as the predominant pathway. Furthermore, they purported the formation of peroxide intermediate **XVa** for the continued generation of iminium ion **IIa** in the dark.

10.5 Photocatalysis Using Semiconductors

As discussed in previous sections, the use of heterogeneous photocatalysts allows for conservation of resources due to the recoverability and recyclability of catalysts. One type of heterogeneous system mentioned in Section 10.2.1 involves organometallic chromophores that have been engineered into polymeric networks. However, there are a number of other materials, such as semiconductors, that can be activated by visible light to induce organic redox transformations. Semiconductors are particularly attractive materials for photocatalysis given their wide availability and variable electronic properties.[50] Furthermore, they are generally cheap, virtually non-toxic, and environmentally friendly. The photocatalytic activity of semiconductors operates *via* the formation of electron–hole pairs that can participate in bimolecular electron-transfer processes (Figure 10.5). These pairs are formed as a consequence of photo-excitation of the valence band (VB) electrons into the conductance band (CB) with the incident wavelength being the same or a higher energy corresponding to the band gap (VB→CB). After migration of the separated electron–hole pair towards the surface of the semiconductor, electron transfer can then occur with acceptors (**A**) or donors (**D**). The high surface area available for reactivity on semiconductors provides a distinct advantage over other photocatalysts by increasing reaction efficiency and potentially stabilizing charged intermediates.

Metal oxide semiconductors such as TiO_2 have long been utilized in photocatalysis for a variety of applications including water-splitting and degradation of organic-based pollutants.[51] Given their large conductance band gaps (3.2 eV), which correspond to large redox potentials, photochemical activation usually occurs in the UV range, although structural and chemical modifications can be made to improve visible light absorption. Despite this, Rueping *et al.* showed that unmodified TiO_2 and ZnO nanoparticles were sufficiently active to mediate aerobic CDC reactions of THIQ derivatives in the presence of visible light.[52] Mannich and nitro-Mannich-type reactions as well as cyanations were achieved using Aeroxide P25 TiO_2 nanoparticles, whereas phosphorylations were accomplished in higher

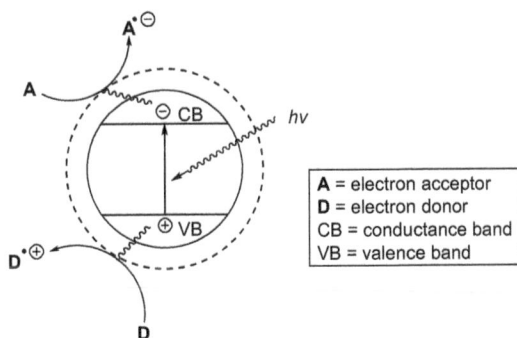

Figure 10.5 Light-promoted electron transfer processes of semiconductors.

yields using ZnO (Scheme 10.33). Furthermore, both catalysts could be re-covered and re-used in consecutive reactions without loss in reactivity.

Cadmium sulfide, with its smaller band gap of 2.4 eV, has more activity in the visible light range than TiO_2 and ZnO, and has been better studied, al-though not extensively, in visible-light-promoted organic transformations.[53] König *et al.* demonstrated that CdS is a suitable photocatalyst for CDC re-actions of THIQ derivatives. Mannich-type reactions were achieved by em-ploying a 5 mg mL^{-1} mixture of CdS in the desired ketone using proline as an organocatalyst (Scheme 10.34).[54] Reactions were run under a balloon of O_2 and irradiated by high-power LEDs at 460 nm. Products were formed in 79 to >99% yields.

König and co-workers also studied CdS-mediated coupling of benzylic alcohols, ethers, and amines. Using high-power blue LEDs, the authors were

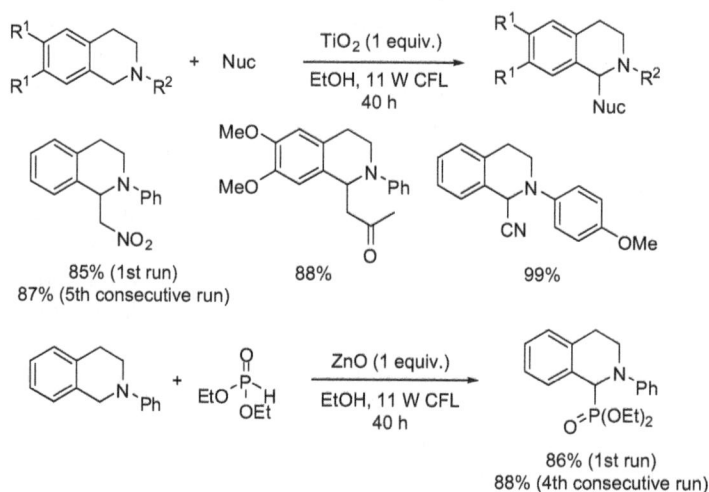

Scheme 10.33   Photochemical CDC reactions using metal oxides.

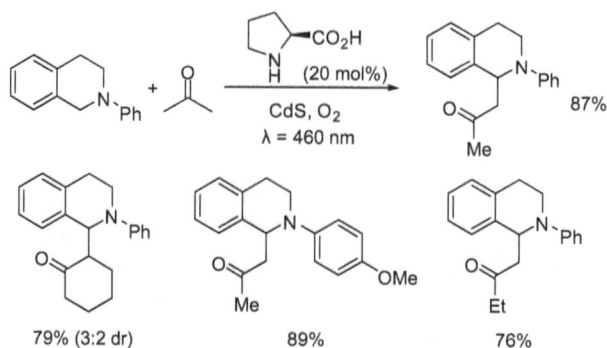

Scheme 10.34   Addition of ketones to THIQs using CdS with proline organocatalyst.

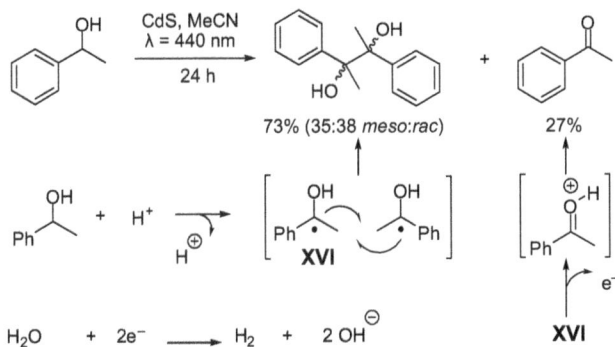

Scheme 10.35 Photodimerization of benzylic alcohols using CdS.

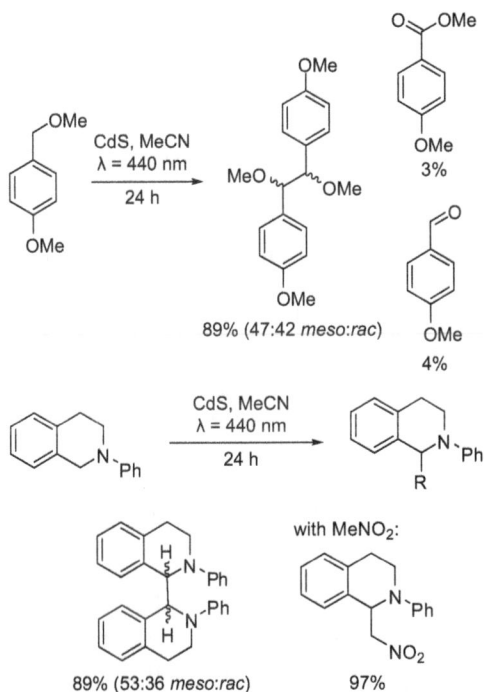

Scheme 10.36 Photochemical CDC reactions of benzylic ethers and amines using CdS.

able to dimerize α-methyl benzyl alcohol in the presence of CdS in MeCN under anaerobic conditions to give of a mixture of diols in 73% yield along with acetophenone as a side-product in 27% yield (Scheme 10.35).[55] The key intermediate is likely benzylic radical **XVI** generated from oxidation of the alcohol at the semiconductor hole (h^+). Radical combination forms the diol product. Alternatively, radical **XVI** can be oxidized by either electron injection into the conductance band of the semiconductor or by oxidation at

h^+ to produce acetophenone. The authors also observed the formation of H_2 by gas chromatography measurements due to the reduction of water, of which a small amount is found at the semiconductor monolayer.

Alcohols, benzylic ethers and amines similarly undergo dimerization under photocatalytic conditions. *para*-Methoxybenzyl alcohol methyl ether dimerizes to produce a mixture of bis-ethers in 89% yield and a trace amount of ester and aldehyde side-products (Scheme 10.36). Under these conditions *N*-phenyl THIQ dimerizes in 89% yield, presumably through the analogous α-amino radical intermediate. However, in the presence of a large excess of nitromethane, the oxidative nitro-Mannich reaction can outcompete dimerization to give a 97% yield of the heterocoupling product.

10.6 Conclusions and Outlook

Light-assisted CDC reactions represent mild, efficient, and environmentally benign methods for direct C–C bond formation through the C–H oxidation/functionalization of organic molecules. *In situ* photo-oxidation provides ionic or radical species that can react with diverse functional groups. A variety of catalytic systems have been developed for promoting CDC reactions triggered by low-energy visible light. Photocatalysts based on transition metal complexes, organic dyes, or semiconductors have comparable oxidative abilities, but each system has its own distinct advantages. Polypyridyl transition metal catalysts are numerous, encompass a wide range of redox potentials, and can be used in low catalyst loadings. Furthermore, they can be engineered for heterogeneous catalysis to maximize resource conservation. Photocatalysts based on organic dyes and porphyrins are diverse with respect to structure and redox properties and can also be used at low catalyst loadings. Additionally, they are less-expensive and offer metal-free alternatives for similar experiments. Semiconductors are not as well studied for visible light CDC reactions, but researchers have demonstrated their abilities and advantages. These materials are cheap and widely available, have large surface areas to make them efficient redox mediators, and can also be re-used for sequential photochemical reactions.

Photocatalysis can operate through different mechanisms, including electron transfer and sensitization, and can generate intermediates that allow for a variety of transformations. The generation of iminium ions enables diverse functionalization and many conditions exist that allow the addition of a wide variety of nucleophiles including nitroalkanes, cyanide, enolates, and electron-rich arenes. Alternatively, with modified reaction conditions, radical intermediates can be generated and trapped with electron-deficient alkenes and arenes. Redox cycles may be fine-tuned with the choice of terminal oxidant, such as O_2, $BrCCl_3$, $S_2O_8{}^{2-}$, *m*DNB, *etc.* Often, photoredox catalysis can be applied in parallel with other processes. For example, nucleophilic additions to photogenerated iminium ions have been aided by organocatalysis and copper catalysis.

Alcohols, ethers, and particularly amines, are good substrates for C–H functionalization since they form a number of stable intermediates upon oxidation. This behavior is reflected by their relatively low redox potentials, which enable them to interact with excited-state photocatalysts and/or singlet oxygen and/or other oxidants. With sufficiently activated substrates, the oxidized intermediates can undergo C–H activation through deprotonation or H-atom abstraction. Given the criteria, the current limiting factor for light-assisted CDC reactions is the substrate scope. Reactions usually involve benzylic, tertiary, or cyclic amines, alcohols or ethers. Conversely, the scope of functionalization is broad and atom-economical. As the field of photocatalysis continues to grow, there will be opportunities for improving catalytic systems and altering redox properties to achieve C–H functionalizations of less-activated organic molecules. The use of visible light to promote organic transformations will undoubtedly increase, and current progress indicates a promising future for continued research.

References

1. C. Li, *Acc. Chem. Res.*, 2009, **42**, 335.
2. K. R. Campos, *Chem. Soc. Rev.*, 2007, **36**, 1069.
3. L. Shi and W. Xia, *Chem. Soc. Rev.*, 2012, **41**, 7687.
4. (a) G. Ciamician, *Science*, 1912, **36**, 385; (b) A Albini and M. Fagnoni, *Green Chem.*, 2004, **6**, 1.
5. G. J. Kavarnos and N. J. Turro, *Chem. Rev.*, 1986, **86**, 401.
6. J. Xuan and W.-J. Xiao, *Angew. Chem., Int. Ed.*, 2012, **51**, 6828.
7. (a) J. R. L. Smith and D. Masheder, *J. Chem. Soc., Perkin Trans. 2*, 1976, 47; (b) L. I. Lagutskaya and Y. I. Bellis, *Theor. Exp. Chem.*, 1973, **6**, 456; (c) J. R. L. Smith, *J. Chem. Soc., Perkin Trans. 2*, 1977, 1732.
8. A. Juris, V. Balzani, F. Barigelletti, S. Campagna, P. Belser and A. von Zelewsky, *Coord. Chem. Rev.*, 1988, **84**, 85.
9. Y. L. Chow, W. C. Danen, S. F. Nelsen and D. H. Rosenblatt, *Chem. Rev.*, 1978, **78**, 243.
10. K. Sreenath, C. V. Suneesh, K. R. Gopidas and R. A. Flowers, *J. Phys. Chem. A*, 2009, **113**, 6477.
11. D. D. M. Wayner, J. J. Dannenberg and D. Griller, *Chem. Phys. Lett.*, 1986, **131**, 189.
12. D. E. P. Nicholas and R. Arnold, *Can. J. Chem.*, 1982, **60**, 2165.
13. J. P. Dinnocenzo and T. E. Banach, *J. Am. Chem. Soc.*, 1989, **111**, 8646.
14. A. G. Condie, J. C. Gonzalez-Gomez and C. R. J. Stephenson, *J. Am. Chem. Soc.*, 2010, **132**, 1464.
15. (a) C. K. Prier, D. A. Rankic and D. W. C. MacMillan, *Chem. Rev.*, 2013, **113**, 5322; (b) J. M. R. Narayanam and C. R. J. Stephenson, *Chem. Soc. Rev.*, 2011, **40**, 102; (c) F. Teply, *Collect. Czech. Chem. Commun.*, 2011, **76**, 859.

16. (a) L. Flamigni, A. Barieri, C. Sabatini, B. Ventura and F. Barigelletti, *Top. Curr. Chem.*, 2007, **281**, 143; (b) S. Campagna, F. Puntoriero, F. Nastasi, G. Bergamini and V. Balzani, *Top. Curr. Chem.*, 2007, **280**, 117.

17. J. D. Slinker, A. A. Gorodetsky, M. S. Lowry, J. Wang, S. Parker, R. Rohl, S. Bernhard and G. G. Malliaras, *J. Am. Chem. Soc.*, 2004, **126**, 2763.

18. M. Rueping, C. Vila, R. M. Koenigs, K. Poscharny and D. C. Fabry, *Chem. Commun.*, 2011, **47**, 2360.

19. G. Zhao, C. Yang, L. Guo, H. Sun, C. Chem and W. Xia, *Chem. Commun.*, 2012, **48**, 2337.

20. R. Rathore and J. K. Kochi, *Tetrahedron Lett.*, 1994, **35**, 8577.

21. M. Rueping, S. Zhu and R. M. Koenigs, *Chem. Commun.*, 2011, **47**, 8679.

22. M. Rueping, S. Zhu and R. M. Koenigs, *Chem. Commun.*, 2011, **47**, 12709.

23. M. Rueping, R. M. Koenigs, K. Poscharny, D. C. Fabry, D. Leonori and C. Vila, *Chem.–Eur. J.*, 2012, **18**, 5170.

24. W.-P. To, G. S.-M. Tong, W. Lu, C. Ma, J. Liu, A. L.-F. Chow and C.-M. Che, *Angew. Chem., Int. Ed.*, 2012, **51**, 2654.

25. Z.-Q. Wang, M. Hu, X.-C. Huang, L.-B. Gong, Y.-X. Xie and J.-H. Li, *J. Org. Chem.*, 2012, **77**, 8705.

26. J.-J. Zhong, Q.-Y. Meng, G.-X. Wang, Q. Liu, B. Chen, K. Feng, C.-H. Tung and L.-Z. Wu, *Chem.–Eur. J.*, 2013, **19**, 6443.

27. X.-H. Li, L.-Z. Wu, L.-P. Zhang, C.-H. Tung and C.-M. Che, *Chem. Commun.*, 2001, **37**, 2280.

28. S. Cai, X. Zhao, X. Wang, Q. Liu, Z. Li and D. Z. Wang, *Angew. Chem., Int. Ed.*, 2012, **51**, 8050.

29. Y.-Q. Zhou, L.-Q. Lu, L. Fu, N.-J. Chang, J. Rong, J.-R. Chen and W.-J. Xiao, *Angew. Chem., Int. Ed.*, 2011, **50**, 7171.

30. M. Rueping, D. Leonori and T. Poisson, *Chem. Commun.*, 2011, **47**, 9615.

31. Z. Xie, C. Wang, K. E. deKrafft and W. Lin, *J. Am. Chem. Soc.*, 2011, **133**, 2056.

32. C. Wang, Z. Xie, K. E. deKrafft and W. Lin, *ACS Appl. Mater. Interfaces*, 2012, **4**, 2288.

33. C. Wang, Z. Xie, K. E. deKrafft and W. Lin, *J. Am. Chem. Soc.*, 2011, **133**, 13445.

34. D. B. Freeman, L. Furst, A. G. Condie and C. R. J. Stephenson, *Org. Lett.*, 2012, **14**, 94.

35. J. W. Tucker, Y. Zhang, T. F. Jamison and C. R. J. Stephenson, *Angew. Chem., Int. Ed.*, 2012, **51**, 4144.

36. C. Dai, F. Meschini, J. M. R. Narayanam and C. R. J. Stephenson, *J. Org. Chem.*, 2012, **77**, 4425.

37. D. DiRocco and T. Rovis, *J. Am. Chem. Soc.*, 2012, **134**, 8094.

38. A. McNally, C. K. Prier and D. W. C. MacMillan, *Science*, 2011, **334**, 1114.

39. P. Kohls, D. Jadhav, G. Pandey and O. Reiser, *Org. Lett.*, 2012, **14**, 672.

40. Y. Miyake, K. Nakajima and Y. Nishibayashi, *J. Am. Chem. Soc.*, 2012, **134**, 3338.

41. S. Zhu, A. Das, L. Bui, H. Zhou, D. P. Curran and M. Rueping, *J. Am. Chem. Soc.*, 2013, **135**, 1823.
42. P. Esser, B. Pohlmann and H.-D. Scharf, *Angew. Chem., Int. Ed.*, 1994, **33**, 2009.
43. A. Mishra, M. K. R. Fischer and P. Bäuerle, *Angew. Chem., Int. Ed.*, 2009, **48**, 2474.
44. M. C. DeRosa and R. T. J. Crutchley, *Coord. Chem. Rev.*, 2002, **233–234**, 351.
45. G. Jiang, J. Chen, J.-S. Huang and C.-M. Che, *Org. Lett.*, 2009, **11**, 4568.
46. W.-P. To, Y. Liu, T.-C. Lau and C.-M. Che, *Chem.–Eur. J.*, 2013, **19**, 5654.
47. Y. Pan, C. W. Kee, L. Chen and C.-H. Tan, *Green Chem.*, 2011, **13**, 2682.
48. D. P. Hari and B. König, *Org. Lett.*, 2011, **13**, 3852.
49. Q. Liu, Y.-N. Li, H.-H. Zhang, B. Chen, C.-H. Tung and L.-Z. Wu, *Chem.–Eur. J.*, 2012, **18**, 620.
50. H. Kisch, *Angew. Chem., Int. Ed.*, 2013, **52**, 812.
51. K. Nakata and A. Fujishima, *J. Photochem. Photobiol., C*, 2012, **13**, 169.
52. M. Rueping, J. Zoller, D. C. Fabry, K. Poscharny, R. M. Koenigs, T. E. Weirich and J. Mayer, *Chem.–Eur. J.*, 2012, **18**, 3478.
53. Y. Zhang, N. Zhang, Z.-R. Tang and Y.-J. Xu, *Chem. Sci.*, 2012, **3**, 2812.
54. M. Cherevatskaya, M. Neumann, S. Füldner, C. Harlander, S. Kummel, S. Dankesreiter, A. Pfitzner, K. Zeitler and B. König, *Angew. Chem., Int. Ed.*, 2012, **51**, 4062.
55. T. Mitkina, C. Stanglmair, W. Setzer, M. Gruber, H. Kisch and B. König, *Org. Biomol. Chem.*, 2012, **10**, 3556.

CHAPTER 11

Mechanisms of Cross-Dehydrogenative-Coupling Reactions

ALTHEA S.-K. TSANG, SOO J. PARK AND MATTHEW H. TODD*

School of Chemistry, The University of Sydney, NSW 2006, Australia
*Email: matthew.todd@sydney.edu.au

11.1   Introduction and Scope

For the purposes of this chapter, a cross-dehydrogenative-coupling (CDC) reaction will be defined as the oxidative coupling of species that are nominally related to electrophiles (*proelectrophiles*) with those nominally associated with being nucleophiles (*pronucleophiles*), see Scheme 11.1.

An example of the former is an amine (commonly *N*-phenyl tetrahydroisoquinoline, **1**), which may be oxidized in a CDC reaction to give an imine, and an example of the latter is a malonate (*e.g.*, **2**) which may react as a carbon-based nucleophile upon tautomerization. The distinguishing feature of CDC reactions is that such species may be mixed in one pot, unfunctionalized, along with an oxidant. Of the multiple possible chemical pathways that can be envisaged in this melting pot, one predominates with sometimes striking selectivity, and it is this aspect of CDC reactions that has captured the attention of the research community.[1–3] In this chapter we are concerned with the mechanisms of such reactions.

To maintain a manageable focus we will ignore couplings where there is no clear pronucleophile/electrophile, and this means we will ignore the various mechanisms involved in reactions usually described as involving

RSC Green Chemistry No. 26
From C–H to C–C Bonds: Cross-Dehydrogenative-Coupling
Edited by Chao-Jun Li

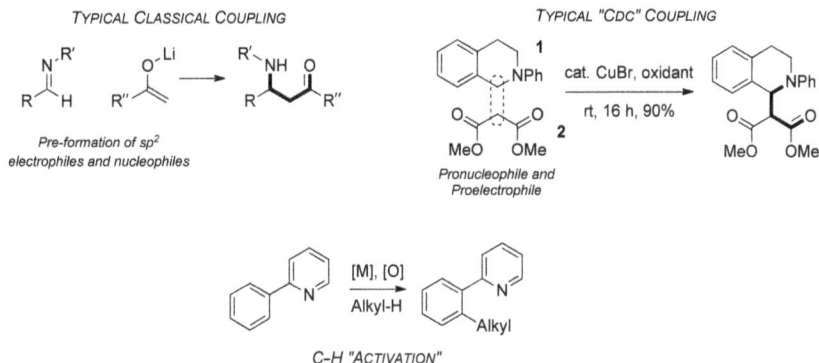

Scheme 11.1 Scope of reactions covered and comparison with C–H activation.

C–H *activation*[4] (rather than the occasionally used term in the CDC field, C–H *functionalization*) but which are sometimes grouped with CDC reactions.[5] The mechanistic distinction is subtle, and recent articles may be consulted for comparison of the mechanisms[3,6–11] but to a first approximation the mechanisms not covered are those involving insertion of a metal atom into a C–H bond.

We will focus on reagent-mediated oxidations and will therefore not cover electrochemical oxidations, though significant work has been done on the relevant reactions.[12–18] This means we will not be covering Yoshida's "cation pool" method for the electrochemical generation of cations to which nucleophiles may be added,[19] though some of the mechanistic details will overlap. Since a major benefit of CDC reactions is their one-pot nature we will deprioritize stepwise processes, for example *N*-halogenations that can be used in Hofmann–Löffler-type cyclizations.[20]

We must also deprioritize simple oxidation reactions where there is no added pronucleophile, such as oxidations at mildly activated positions without attendant coupling processes, since a major advantage of CDC reactions is that we can combine pronucleophiles and proelectrophiles; clearly some of these studies provide mechanistic information as to the composite steps.[21–23] While much of the related chemistry is relevant to *N*-demethylation reactions,[24] oxidative dehalogenations[25] or other processes[26] we do not cover those studies here since they are not aimed at (constructive) synthesis, which is the primary focus of CDC reactions.

We are not able to cover the enormous associated literature of electron transfer, but clearly the interested reader can follow the references cited herein to access the most relevant papers. There is a great deal of literature on the oxidations of organic compounds, for example, and how such reactions can be mimics of enzymatic oxidation processes.[27,28] A frequent participant in the mechanisms described herein is the radical ion, a fundamentally important species in organic chemistry (that is perhaps underappreciated by the synthetic community) and the subject of a recent review.[29]

The processes must be oxidative for inclusion. Metals such as copper and iron can catalyze the addition of alkynes to amines in a CDC process that is presumed to involve an intermediate iminium ion and formation of metal acetylides as the reactive nucleophiles.[30–34] The oxidative component of this reaction is the generation of the iminium ion. The corresponding non-oxidative process, the aldehyde–alkyne–amine (A^3) coupling, will not be a focus here and has been recently reviewed.[35] Interesting intramolecular processes such as cyclizations involving hydride transfers that are overall redox neutral must be similarly excluded.[36–38]

The motivation for this chapter is that while there has been significant progress in exploring the scope of CDC methodology there has been less work revealing the mechanistic details of the reactions. Clearly such detail is necessary if we are to make progress towards being predictive about where such chemistry might be applied and towards explanations as to why certain reactions 'fail'. Besides differences in oxidant, there are clear differences in the catalysts that are employed across the spectrum of CDC reactions and one of the messages of this chapter is that the mechanisms are likely to be substrate-specific, which can make for interesting reading but confusing science. Thus while the subject matter is arranged according to oxidant system, substrate will be seen to influence reaction outcome strongly. For example the oxidative functionalization of an sp^3 carbon atom adjacent to an oxygen atom is regarded as more difficult than the corresponding process at the position adjacent to a nitrogen atom, due to the higher oxidation potential of the C–H bond in the former case, and clearly this will influence the outcomes of the reactions with these substrates under otherwise similar conditions.

This chapter will take a broad look at those reports where there is *experimental* evidence for reaction mechanisms and will deprioritize those papers where a mechanism is putative, suggested, or supported by one or two possibly ambiguous observations. In many studies, preliminary mechanisms are investigated through the addition of radical traps, yet such results need to be treated with caution, as described below. An example is the employment of TEMPO (2,2,6,6-tetramethylpiperidine-1-oxyl) as a radical trap, awkward in this field because TEMPO is itself an oxidant or can be used as a co-oxidant.[39–41]

Changing the oxidant could change the reaction outcome as well as the mechanism. In early work on one-pot oxidative bond-forming reactions that is seldom cited, it was shown that modification of the oxidant from chlorine dioxide (ClO_2) to $Hg(OAc)_2$ gave rise to different products (**4** *vs.* **5**, from **3**, Scheme 11.2) that likely arise from different mechanisms.[42] There is evidence that the ClO_2-mediated reaction proceeds *via* an initial slow single-electron transfer (SET) to form a radical cation that is converted quickly to an iminium ion in a mechanism that we will see repeatedly below.[43–45] The mercuric acetate process is thought (but not proven) to proceed by co-ordination of the nitrogen center to the metal followed by base-effected elimination to the iminium ion, with the latter being shown to form (without

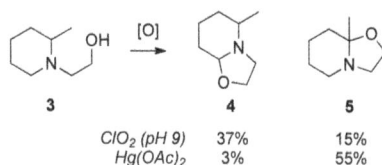

Scheme 11.2 Variation of product, and possibly mechanism, with oxidant.

necessarily being part of the reaction pathway) *via* crystallization of picrate salts.[46] The putatively different mechanisms, arising from different reagents (and reaction conditions) manifest themselves as the significantly different product distributions shown.

We will look at thermal CDC processes first, both metal-catalyzed and metal-free, then briefly at photochemical CDC methods and so-called "SOMO (Singly Occupied Molecular Orbital)-activation" mechanisms and finally we will note an important stereo-electronic feature of some of these reactions that is likely responsible for their ability to produce very well-defined reaction outcomes.

11.2 Thermal Metal-Mediated CDC Reactions

The most common form of CDC reaction employs a metal catalyst. These will be considered first.

11.2.1 Ruthenium Catalysis

Ruthenium salts were used in early reports of CDC reactions. As can be seen from the examples discussed below, the suggested roles of the catalyst were to oxidize the substrate, to complex to an iminium intermediate, to activate the nucleophile, or to facilitate nucleophilic addition. The use of a stoichiometric oxidant was necessary, with peroxides or molecular oxygen being common. The suggested role of the oxidant ranges from the conversion of the amine to the iminium ion or the re-generation of the catalyst.

Murahashi showed that the oxidation of tertiary (aryl) amines (such as **6**, Scheme 11.3) by *tert*-butyl peroxide (TBHP) in the presence of a ruthenium catalyst gave the corresponding α-substituted peroxy compounds (*e.g.*, **7**) in a one-pot process.[47] The perhaps more expected *N*-oxide was not observed, and this intermediate might have led to alternative outcomes such as Polonovski reactions[48] resulting in C–N bond cleavage, rather than the constructive bond formations sought here. Simple acidic hydrolysis could be used to complete a demethylation reaction but formal CDC processes were shown to be possible with the inclusion of nucleophiles such as methanol[49] (using RuCl$_3$ and H$_2$O$_2$) or cyanide.[50–52] These reports, and the original, contain the results of mechanistic studies that furnish Hammett plots and inter-/intramolecular deuterium isotope effects. The results show an apparent influence on the rate of reaction (judged by the quantity of product

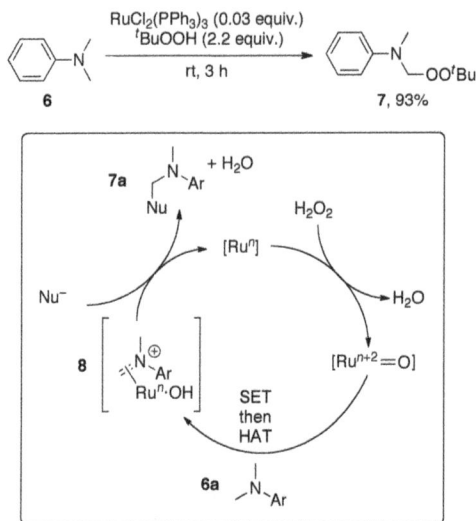

Scheme 11.3 Ruthenium-catalyzed coupling reactions reported by Murahashi, and a composite proposed mechanism with added nucleophile.

formed using NMR spectroscopy) by the electronic contribution of substituents on the ring, suggesting cationic character (negative ρ values observed) in an intermediate involved in the rate-determining step, and clearly positive kinetic isotope effects (KIEs, *ca.* 3–4). In the initial reports the isotope data were taken as a suggestion of radical character in the C–H bond-breaking event, suggestive of initial SET followed by H-atom transfer (HAT) to give the iminium ion. In the later report the data were taken as indicative of more ionic character in the C–H bond-breaking event, meaning the steps were proposed to be the other way around, and more in line with the literature mechanism of P450 enzymes.[53] The latter suggestion of the order (electron transfer followed by atom transfer) would appear to match the data better and fits with later mechanistic suggestions by others (*vide infra*) in different reactions. The ruthenium was proposed to interact with the oxidant and the resulting metal oxide effects electron- and atom transfer resulting in an iminium ion complex (**8**) where (presumably) the atom transfer is rate limiting. Several authors (see below) have suggested that the iminium ion intermediate may be in equilibrium with a 'trapped' intermediate arising from attack of, for example, the peroxide, and the Ru-catalyzed case is no exception.[54] Regardless, the putative iminium ion complex is able to react with the nucleophile to give the product and re-generate the catalyst. In later reports, where oxygen was used as the oxidant,[50,52] a slightly modified catalytic cycle was proposed with an initial co-ordination to Ru by the amine followed by the oxidative steps (where the Ru is oxidized while co-ordinating to the iminium ion) that resulted in the formation of H_2O_2, which then feeds into the cycle shown, and this accounted for the observed stoichiometry of the reaction of $1:2$ O_2:substrate. Murahashi's work is pioneering in many respects,[55] and brings to light many of the recurrent questions relevant to a mechanistic understanding of the field: how do we prove the occurrence of reactive intermediates on the reaction pathway, and on which step in the multi-step reaction pathway are the data providing insight?

11.2.2 Iron Catalysis

Early work by Miura identified an iron/O_2 system capable of oxidative cyanation adjacent to a nitrogen atom (Scheme 11.4), along with proposals for the mechanism of the reaction that did not explicitly involve the metal center.[56,57] The reactions studied typically gave mixtures of products, most often (alongside product **7b**) the formamide **7c** and starting material. From a synthetic point of view this is undesirable, but this does allow insights into the mechanism as the conditions are changed, and such insights are relevant to later studies in the CDC field. For example, the product distribution is tightly dependent on the iron salt employed. In the earlier paper of the two, focusing on the oxidation of the starting material, it was found that addition of the radical trap BHT (2,6-di-*tert*-butyl-4-methyl phenol) inhibited the oxidation when one iron salt was used ([Fe(salen)]OAc) but had no effect on the reaction catalyzed by a different iron salt (FeCl$_3$) which is a cautionary tale against drawing conclusions about a mechanism for a given metal center from single mechanistic experiments. Intramolecular kinetic isotope data were obtained for the substrate **6b** (*N*-methyl-*N*-trideuteromethylaniline) undergoing oxidation with different catalysts, quantified by means of measurement of product distributions between **9** and **9a**. It was found the

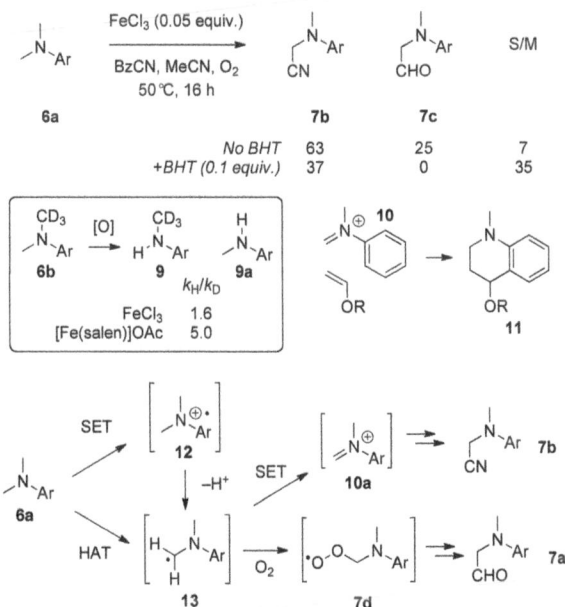

Scheme 11.4 Miura's early oxidative cyanation of *N*,*N*-dimethylanilines (BzCN = benzoyl cyanide).

intramolecular KIE depended on the catalyst employed, again implying a strong correlation between catalyst and mechanism (or at least the identity of the rate-limiting step) for a given starting material/product combination. The presence of the iminium ion **10** was suggested by its trapping when a dienophile (a vinyl ether) was added to give the corresponding 6-membered ring product **11**. In one experiment catalyzed by $FeCl_3$ in the presence of BHT the formation of the formamide product was suppressed while the cyanation product was still formed (albeit in reduced yield). This was taken to imply that different mechanisms (shown in the schematic) may be operating in the same one-pot process, with BHT essentially shutting down those involving free radicals (*i.e.*, **7d** or the α-aminyl radical **13**) while not so much influencing those involving cations or radical cations (*i.e.*, **10a** or **12** respectively). Similar conclusions were drawn from studies on catalyst systems employing $CoCl_2$ and oxygen where the substantial intramolecular KIE suggested a rate limiting loss of a hydrogen atom.[58]

Clearly the feature of this chemistry that makes it distinct from other iron-catalyzed processes (*e.g.*, Gif, Fenton) is that the reactions construct new synthetically useful (generally carbon–carbon) bonds during the oxidative metal-catalyzed reaction. More recent work has illustrated this well, for example in the coupling of benzylic positions with 1,3-dicarbonyl compounds with Fe(II) catalysis that was proposed (but not shown) to operate *via* a radical mechanism rather than one involving an intermediate carbocation.[59] An interesting iron-catalyzed oxidative coupling between an aromatic ring and a β-ketoester leading to benzofurans (**14**, Scheme 11.5) was proposed on the basis of KIE and cross-over experiments to proceed *via* a radical-based oxidative step initially, then a cyclization step in which the role of the iron switched to Lewis acid.[60] The proposal for the radical pathway was justified based on limited recent precedence of related reactions rather than *via* direct experiment.

Shirakawa recently proposed a mechanism for an iron-catalyzed CDC reaction between aromatic nucleophiles and *N*-methylamides that involved the production of a naked C-based radical (**16**) that is then oxidized by the metal (Scheme 11.6A).[61] Thus iron acts as a means to oxidize the radical to the iminium ion (**17**) – a case of HAT before SET. The evidence to support this is an experiment when the iron is left out: dimers of the putative radicals are formed (**15b**) rather than the CDC reaction product **15a** arising from attack of the nucleophile. However it is not clear whether this is the required order of events, and the presence of the C-based radical was not experimentally or

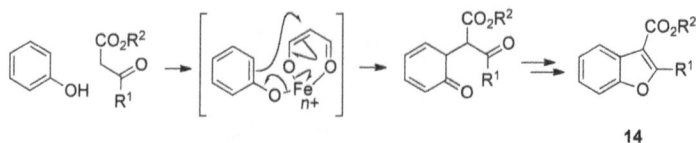

14

Scheme 11.5   Li's Fe-catalyzed synthesis of benzofurans, proceeding *via* an initial coupling between phenol and a β-ketoester.

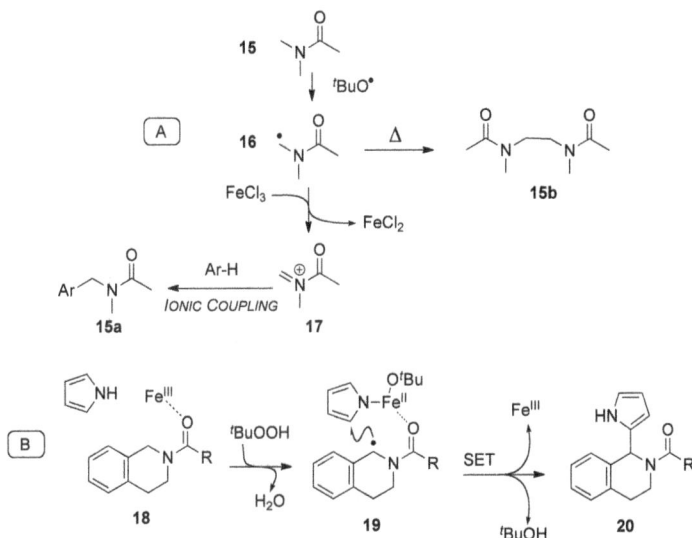

Scheme 11.6 Two mechanistic hypotheses for iron-catalyzed CDC reactions.

computationally established. A similar mechanism has been proposed by others.[62] Standing in contrast is the mechanism of an iron-catalyzed reaction proposed by Schnürch (coupling of indoles to isoquinoline amides to give **20**, Scheme 11.6B),[63] where a proposed C-based radical (**19**, formed from **18**) actually performs the coupling with the aromatic ring directly, followed by SET to give the product, supported by the deleterious effect of TEMPO on the reaction. This mechanism is partially related to that for Klussmann's Cu-catalyzed reaction described below which also invokes a naked C-based radical adjacent to nitrogen, but which is in that case rapidly trapped by the peroxy radical *in situ*, with subsequent steps being ionic in nature.

A different possibility, that of removal of an electron from nitrogen by Fe(III) to give an amine radical cation, followed by loss of the H atom as originally described by Miura (above), was suggested by Doyle in a recently reported iron-catalyzed CDC that used oxygen, rather than peroxide, as the terminal oxidant; the argument for this route was based on the known redox potentials of Fe(II)/(III) and amine/radical cation pairs.[64] This possibility was examined in some detail, though for the case of Rh-catalyzed CDC reactions employing aqueous *tert*-butyl hydroperoxide (T-HYDRO) as an oxidant.[65,66] The iminium ion intermediate may react with the solvent methanol (kinetically) but prefers to exist as a peroxy hemi-aminal, supporting the "reservoir hypothesis" of the iminium ion, mentioned further below. The product is a thermodynamic sink. Competition experiments, reaction monitoring and Hammett analysis led to the conclusion that SET was part of the rate-determining step. Kinetic and product isotope effects gave an interesting picture across a number of catalysts studied (Rh-, Ru-, Cu-, Fe- and Co-based) with some apparent dependence of the mechanism on the catalyst employed, but this was shown to be due to competing pathways arising from dissolved oxygen, which when corrected for, revealed that the catalysts studied appeared to be operating by the same mechanism and this has interesting implications for the field. The alternatives by which the amine radical cation undergoes irreversible C–H bond cleavage were either direct HAT, or proton transfer (PT) followed by SET. Proton transfer would generate an α-amino radical (of the kind described repeatedly in this chapter) and to test whether such a species was present, radical-trap substituents were employed (Scheme 11.7). In fact, the Rh-catalyzed reaction proceeded without substituent ring opening using substrate 21, implying either HAT as a mechanism, or SET that is faster than the ring opening, but giving traces of ring-opened products when the more sensitive radical trap 22 was employed. These results suggested proton transfer followed by a very fast SET step to complete the formation of the iminium ion, but there could be a mixture of PT/SET and direct HAT operating. The mechanistic picture that results involves the metal only in the generation of *tert*-butyl peroxy radicals[67] and these then abstract an electron from the amine substrate in a reversible process. Proton transfer is irreversible to give an α-amino radical from which SET (to a *tert*-butyl oxy radical) is fast, giving the iminium ion which is able to participate in 'trapping' equilibria with various nucleophiles in solution until the product forms as the thermodynamic end point.

These extensive studies reveal a mechanistic picture confirming many of the suggestions made by Miura (Scheme 11.4). The implication that the mechanisms of reactions previously thought to operate by different processes are actually related is an important one, particularly given the reduced role of the metal, and what that implies for the prognosis of catalytic, asymmetric versions of CDC reactions.[68,69] Indeed, the role of the metal in all of the processes described in this section varies, with perhaps a general chronological trend being the lessening of its importance in electron transport. On the basis of several mechanistic experiments, Zhiping Li recently proposed a similar catalytic pathway for an iron-mediated oxidative

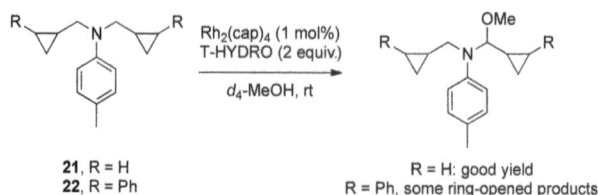

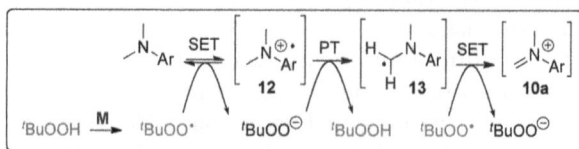

Scheme 11.7 Doyle's composite mechanism applicable to diverse metal-catalyzed CDC reactions, with a limited role for the metal (cap = caprolactamate).

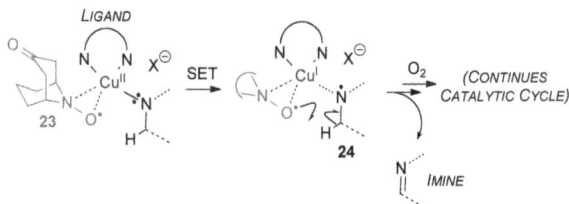

Scheme 11.8 Part of a catalytic cycle involving the *N*-oxyl radical mediated oxidation of amines to imines in which the metal is intimately involved in electron transport.

conversion of amines to amides,[70] while a copper-mediated CDC involving nitrones and ethers/amines proceeded in a similar fashion but with a role in electron transport ascribed to copper.[71] In an interesting recent report of a new catalytic *N*-oxyl radical (**23**, Scheme 11.8) able to effect the mild oxidation of amines to imines, a significant role was ascribed to the metal, playing an active role in electron transport.[72] The involvement of the aminyl radical **24** was proposed on the basis of radical clock experiments similar to those employed by Doyle. It was noted that the starting material amine is actually an unproductive inhibitor of the catalytic cycle *via* co-ordination to the copper center, but the amine ligand may be encouraged off the metal by the addition of dimethylaminopyridine to the reaction mixture, providing oxygen itself with the opportunity to co-ordinate to the metal to complete the catalytic cycle. We will return to further experimental analyses of copper-mediated reactions below.

11.2.3 Vanadium Catalysis and a Comment on Radical Traps

In the case of a vanadium-catalyzed CDC reaction[73] a Murahashi-style amine–metal co-ordination[50] was proposed to assist in the metal-mediated SET/HAT events that lead to the iminium ion–metal complex; subsequently the metal is re-oxidized by the terminal oxidant, but there was no experimental validation of this hypothesis. A similar mechanism involving amine–metal co-ordination and metal oxidation while complexed to the iminium ion was proposed for a gold-catalyzed CDC reaction.[74]

Where molybdenum ($MoO_2(acac)_2$) or vanadium (V_2O_5) salts were used as catalysts with O_2 (Scheme 11.9), a metal–nitrogen interaction (in **25**) was proposed to facilitated simple elimination to give the iminium ion **26**, to which nucleophilic addition occurred to give the product **27**.[75] The rationale for this mechanism is that the *N*-oxides (which can be isolated) do not lead to CDC products under standard reaction conditions. The stoichiometric oxidant (O_2) was thought to re-generate the catalyst, rather than oxidize the amine to the iminium ion, but no evidence for this was provided. When $VO(acac)_2$ was used with ^{t}BuOOH as the oxidant, the *N*-oxides could not be observed in the reaction but when they were separately made, they did *not* lead to the CDC reaction product; coupled with the inhibition of the reaction in the presence of BHT, the conclusion was that the iminium ion formation

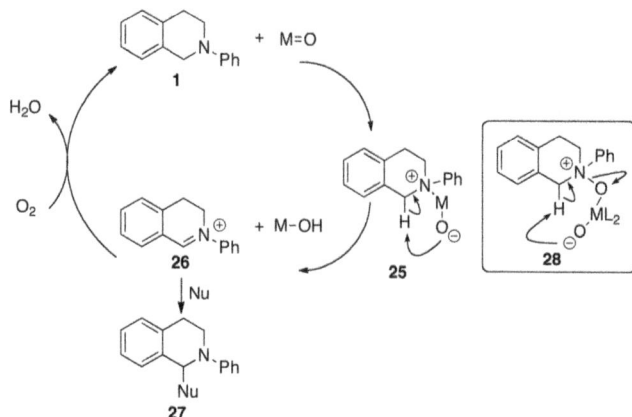

Scheme 11.9 Proposed mechanism of molybdenum- and vanadium-catalyzed CDC reactions.

was radical in nature. In stark contrast, for the related catalyst system of VO(acac)$_2$/mCPBA (acac = acetylacetonate, mCPBA = $meta$-chloroperoxy-benzoic acid) the mechanism was proposed to involve the formation of the N-oxide [which then co-ordinated to the metal (**28**, inset)] because this compound *could* be isolated in the absence of other reagents and *did* give rise to reaction products.[76] Co-ordination to the metal gave rise to polar elimination to the iminium ion in the same way as above. Thus even with the same metals and related substrates, the mechanisms appear to be following subtly different paths.

The conclusions about polar/radical reaction mechanisms above bring up the issue of how these conclusions were reached. Given the possibility of radical intermediates along the reaction pathway, there has been widespread use of radical traps as a means of probing the mechanism of CDC reactions. The addition of BHT as a trap has been popular. If the reaction to which the trap was added still proceeded upon addition of BHT, then the conclusion was generally that radicals are not involved and an ionic mechanism is probably operating, *e.g.*, for vanadium/O$_2$-[75,77] or iron/BuOOH-[78] catalyzed reactions. Other reactions gave lower yields in the presence of BHT, which was taken as evidence of a radical pathway, *e.g.*, vanadium/O$_2$-,[73] vana-dium/^{t}BuOOH,[79] cobalt(II)-,[58,80] gold-,[74] iron/^{t}BuOOH[63] or zirconium/O$_2$[81] -catalyzed reactions (where in the last case TEMPO was used as the radical scavenger). These results need to be treated with caution. Clearly a mech-anism could involve both radical and ionic steps, for example in a CDC re-action where the formation of a reactive intermediate involves SET but where the subsequent coupling of the iminium ion is polar ionic in nature. Indeed Miura proposed this in his original work with Fe/O$_2$ oxidative cyanations, where it was suggested that BHT acted selectively on free radical-based processes but left others relatively untouched (above).[57] Clearly also there are underlying assumptions when using a radical trap, for example that the radical, if formed, is available for intermolecular reaction with the trap, and that it is reactive enough to do so, giving a stable product. It would be wise not to place too much faith in our radical traps and not to over-conclude from simplistic experiments. In our own work, described below, we observed a CDC reaction to be uninhibited by the addition of TEMPO and also concluded the reaction is ionic in nature, yet subsequent unpublished studies on the early stages of the reaction suggest it is much more likely the overall CDC process is a mixture of radical and ionic steps, presumably where the radical stages are either too fast or sterically inaccessible to the scavenger (or both). Li also showed that a reaction, proposed to involve radical ion intermediates, proceeded upon addition of these radical traps.[70] BHT itself also contains a benzylic (and hence technically oxidizable) center. It stands to reason that any reduction in product yield or reduced reactivity could be due to competing reactions through the introduction of BHT as another substrate!

11.2.4 Copper Catalysis

Li originally proposed mechanistic pathways for copper-catalyzed CDC reactions. The oxidant (usually *tert*-butyl peroxide or oxygen) was thought to oxidize the substrate **6a** to the corresponding cation, with the resulting copper complex co-ordinating to the iminium π-bond giving **8a** and facilitating addition of the nucleophile (also activated by copper) to give the product **7a** (Scheme 11.10).[30,82–86] It was suggested that *tert*-butyl peroxide

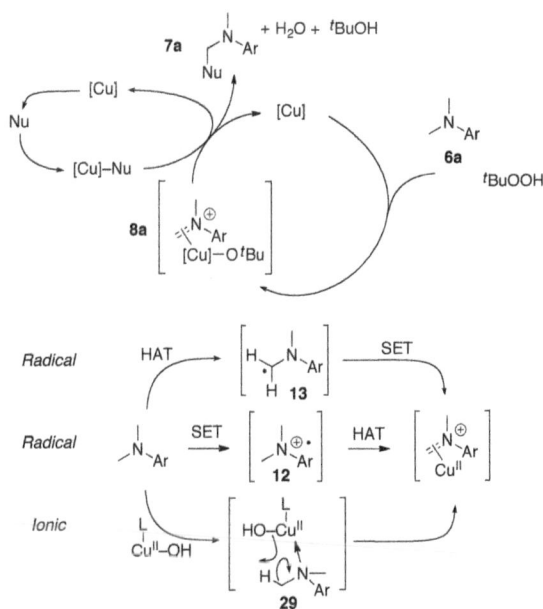

Scheme 11.10 Li's proposed mechanisms for copper-mediated CDC reactions.

adducts could be reactive intermediates. In one case where molecular oxygen was used as the oxidant, it was found that half an equivalent of O_2 gave one equivalent of coupled product,[87] (as had been observed previously by Miura in a Co(II)-catalyzed oxidation[80]) but in another[85] it was found that un-intended oxidation of the nucleophile led to higher than expected O_2 consumption, which is a side-reaction of continual danger in the CDC field. In the copper-catalyzed coupling of the allylic positions of hydrocarbons with active methylene compounds, a π-allyl metal complex was postulated as an intermediate on the basis of a deuterium labeling experiment.[88] Generally in these earlier reports on copper-catalyzed CDC reactions, no direct evidence for the intermediates was reported. The more basic question was also left unaddressed, as it often is in reports of CDC reactions: why was it frequently the case that Cu(I) salts performed better than Cu(II) salts when the catalyst is present alongside an excess of oxidant?

Radical *vs.* ionic mechanisms were described as possible routes to the formation of the iminium ion complex[89] using similar mechanisms to those we have already encountered in this chapter. The radical processes were thought to involve HAT and SET in either order, and the ionic process was thought to involve elimination of water from a nitrogen–amine co-ordination complex (29). A copper-catalyzed CDC reaction between N-phenyl tetrahydroisoquinoline (THIQ) and nitromethane was found to proceed in 70% yield when BHT was added, which was taken as evidence for the ionic mechanism.

In one copper-catalyzed CDC reaction between a functionalized amino acid (30) and phenyl boronic acid it was found that the role of the copper was likely just in the formation of the imine (31), to which the aryl group was delivered intramolecularly to give the product 34 (Scheme 11.11).[90] The key to the reaction was thought to be the co-ordination by an iminol intermediate (32) to the boron (to give 33), activating the phenyl for delivery unassisted by the catalyst (in contrast to a related coupling between amino acids and malonates).[91] This conclusion was derived from two observations: (1) substrate specificity: the reaction using the same substrate but where the

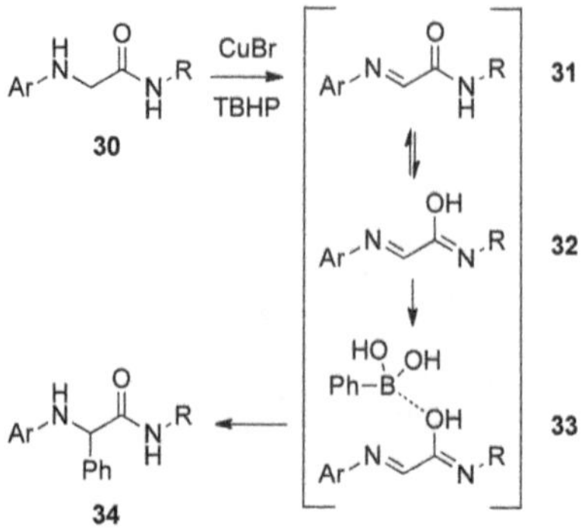

Scheme 11.11   Metal-free delivery of nucleophiles after metal-mediated oxidation.

secondary amide was replaced by a tertiary amide or an ester (*i.e.*, unable to form the iminol) did not give product formation; and (2) pre-formation and isolation of the imine and its subsequent reaction with boronic acid in the absence of metal catalyst gave a high yield of the expected product.

In some reactions regular ketones are used as pronucleophiles, raising the question of how, under the conditions of the reaction, the nucleophile is actually formed. In one CuI/O$_2$ CDC reaction it was suggested that the beneficial effect of adding several equivalents of acetic acid to the reaction arose from the acid promoting enolization of the ketone and hence a faster trapping of the intermediate iminium ion.[86]

Klussmann and co-workers have performed the most detailed dissection of a copper-mediated CDC reaction, in this case between *N*-phenyl THIQ (**1**) and various nucleophiles. The earlier of the studies is notable for its structural characterization of the iminium ion intermediate and its explanation of the beneficial role of methanol as an additive.[92,93] The reaction in question (Scheme 11.12) when run with half an equivalent of catalyst and no nucleophile generated the salt **26a** consisting of the expected iminium ion and a dichlorocuprate counterion and which was characterized by X-ray crystallography; it is clear from the structure that in this case there is an ionic interaction rather than the often-supposed π-iminium complex. A 2D NMR exchange spectroscopy experiment provided evidence for dynamic equilibria between the iminium ion and compounds **27a** and **27b**. Running the oxidation with CuBr gave a related salt where (Cu$_2$Br$_4$)$^{2-}$ was the

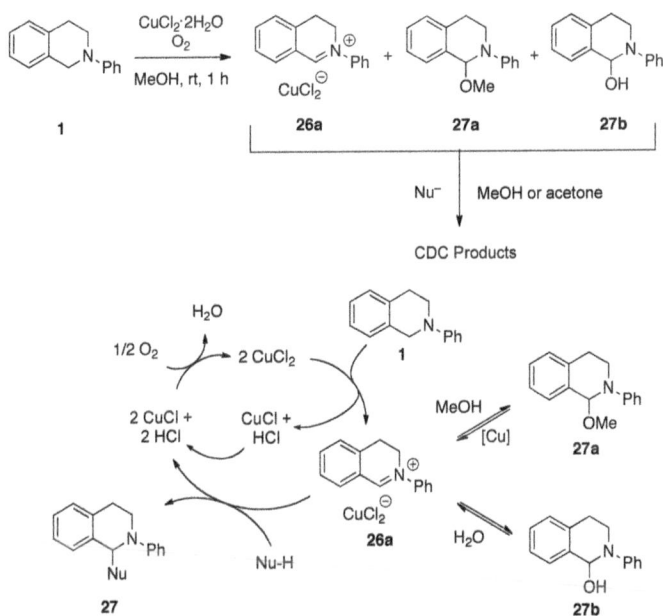

Scheme 11.12 Reaction, reactive intermediates and proposed catalytic cycle for a copper-catalyzed CDC reaction.

counterion. The reactivity of the iminium ion with added nucleophiles was found to be heavily dependent on the counterion, with NMR studies suggesting that weakly co-ordinating counterions gave smooth reactions to CDC products. The isolated methanol adduct **27a** reacts slowly with the added nucleophile, apparently requiring the copper catalyst for a reaction to occur. This suggests copper plays the additional role of Lewis acid in mediating an off-cycle equilibrium between the methanol adduct and the iminium ion, and led Klussmann to propose the catalytic cycle shown (modified by subsequent studies described below). The copper cycles between oxidation states I and II and is the species that interacts with the terminal oxidant. An interesting corollary of the proposed cycle is the role of methanol in helping to trap (and in effect stabilize) the reactive iminium ion. If such a compound were formed to too great an extent the CDC reaction would never proceed to completion, with the methanol being responsible for creating an unreactive reservoir of aminal side-product, but the presence of the Lewis acid ensures the presence of a low level of iminium ion that can proceed to the expected CDC product through reaction with the added nucleophile. It is also interesting that the attachment of methanol to produce **27a** liberates a copper center that can then re-enter the catalytic cycle of the oxidation. This important influence of solvent on the rate of oxidation was indeed demonstrated by comparison of reactions in methanol and acetone, with the former showing the expected rate acceleration, as well as a cleaner reaction that could be due to this trapping of the iminium ion as a reasonably stable adduct that additionally might prevent its involvement in unwanted side-reactions that could explain the generally cleaner reactions observed in this solvent. (The sensitivity to solvent of the formation of methoxy- *vs.* peroxy-trapped intermediates in a Cu-catalyzed CDC reaction had been noted by others.[94]) The presence of acetic acid was found to be beneficial in cyanations by both Klussmann[95] and Murahashi,[50] has been proposed to liberate oxonium ions from trapped O-linked dimers,[96] and in a cobalt-based system investigated by Miura the addition of acetic anhydride was found to accelerate oxygen uptake.[80]

The reaction above ($CuCl_2 \cdot 2H_2O/O_2$) was compared in detail with the Li catalytic system (CuBr/TBHP) in a clear demonstration that the mechanism of a CDC reaction is dependent on the catalyst employed.[95] If the mechanistic picture was as simple as nucleophiles reacting with iminium ions, then it should be the case that a given nucleophile reacts with a given iminium ion were it to form. In fact, because of the complexity of CDC processes the picture is not so straightforward, even for one reaction mediated by the same metal center. Using the $CuCl_2$ system, and employing a range of nucleophiles with known nucleophilicity parameters[97,98] a lower limit could be established for nucleophiles that would react under standard conditions, which was found to be approximately $N = 3.8$, or between a trimethylsilyl enolate and 2-methylfuran. This allowed a corresponding electrophilicity parameter to be calculated for the iminium ion (about -8 to -9), a result that must be treated with care given that the iminium ion may be present in only trace amounts.

Interestingly, when *N*-phenyl THIQ and copper chloride were mixed in the absence of oxidant, the iminium ion was formed immediately, which means the copper is able to oxidize the amine, and implies that the role of the oxidant is to re-oxidize the copper. By the use of a mono-deuterated analog of the starting material (**35**, Scheme 11.13), and the observation of a primary KIE value of 4.5 as determined by product ratios, the rate-determining step appears to involve H-atom transfer and it was supposed that this follows a more rapid initial SET step.

The Li catalytic system of CuBr and peroxide behaves differently, with a lesser role for copper in line with the Doyle analysis above. The nucleophile 2-naphthol is found to be reactive in this catalyst system while it is ineffective in the CuCl$_2$-catalyzed reaction. Analysis of the reaction mixture by NMR spectroscopy did not reveal the presence of any iminium ion but rather the formation of appreciable quantities of the peroxy adduct **27c**. Formation of this compound was found to be relatively insensitive to the addition of methanesulfonic acid, suggesting a radical pathway for its formation. This conclusion was supported by a slowing of the formation of the peroxy adduct in the presence of BHT, though these experiments are another cautionary example of the use of BHT since the additive by no means completely suppressed this reaction pathway. It was postulated that while the formation of the peroxy adduct was radical based (a rate-limiting abstraction of a hydrogen atom was proposed that was supported by a primary KIE of 3.4

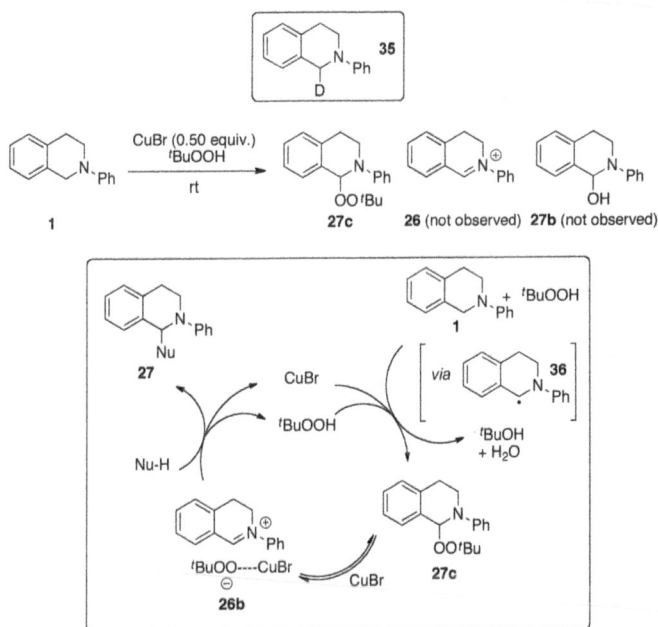

Scheme 11.13 Copper(I) bromide-catalyzed coupling of nucleophiles involving radical and polar steps.

when using the same mono-deuterated substrate as before), the peroxy compound reacts with added nucleophile through a simple ionic mechanism that is acid-mediated and operates *via* the iminium ion; peroxy intermediates as traps able to release iminium ions have been proposed by others.[47,50,54,66,82,99] The explanation of differing reactivities of the nucleophiles in the two different reaction types was suggested then to arise from differences in the pH of the reaction solutions. The mechanism shown for the $CuCl_2$ system involves the formation of HCl, while the CuBr system does not. Experimentally the pH of the solutions were found to be 5.3–6.1 and 6.3–7.2 respectively, neatly explaining why the CDC reaction involving nitromethane as a nucleophile is so much faster in the more basic CuBr case.

These copper-catalyzed CDC reactions were recently subjected to computational analysis.[100] Of the possibilities assayed for the Cu/O_2 reagent system, the calculations best supported an initial SET followed by proton-coupled electron transfer, intimately mediated by the copper complex in oxidation state II. A second possible pathway was found, in which oxygen mediates the proton and electron loss directly to generate the iminium ion and the copper center remains in oxidation state I. These pathways clearly showed H atom loss was rate-determining and gave good theoretical predictions of the experimentally found KIE values. For the reaction analyzed (involving nitromethane) copper was suggested to play a role in the formation of the active (azinic) form of the nucleophile. For the peroxide-mediated pathway the picture was mixed, with several possible pathways operational [including one based on Cu(III)] which would account for the experimental KIE value lying between the calculated extremes required by the various alternatives. The KIE aside, the radical pathway proposed by Klussmann was seen as the most likely.

Schnürch *et al.* compared their work with iron catalysis (above) to a related copper-catalyzed system and suggested a different mechanism was operating.[101] The reaction was again the coupling of indoles to protected THIQs **37** to give **38** which could be effected with either $Cu(NO_3)_2 \cdot 3H_2O$ or $Fe(NO_3)_3 \cdot 9H_2O$ as the catalyst with 'BuOOH as the oxidant (Scheme 11.14). Addition of TEMPO as a radical trap gave different results: the iron-catalyzed reaction was completely suppressed while the copper-catalyzed reaction proceeded with undiminished product yield. The authors proposed that the Cu(II)- and Fe(III)-catalyzed reactions occurred *via* two different mechanisms, ionic and radical in nature. The radical mechanism was described above while the proposed ionic mechanism involved nitrogen co-ordination to the metal in a manner not entirely dissimilar to the mechanisms proposed for vanadium catalysis (Scheme 11.9).

In an interesting recent report of a copper-catalyzed intramolecular CDC reaction, cross-over experiments and experiments with enantio-enriched starting materials suggested that the iminium ion formed along the reaction pathway (**39**, Scheme 11.15) was labile with respect to detachment of the imine.[102] The quinone methide **40** was trapped and characterized. The CDC product could possibly be formed by a cycloaddition reaction on **40**, to give

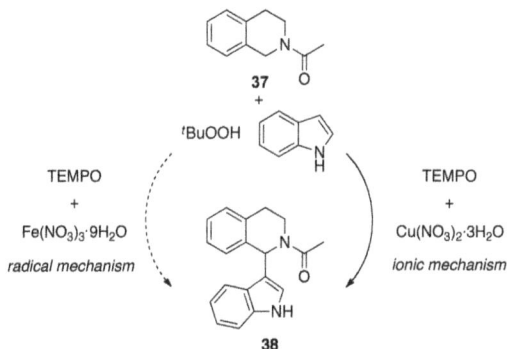

Scheme 11.14 Iron(III)- and copper(II)-catalyzed coupling of indoles to THIQs.

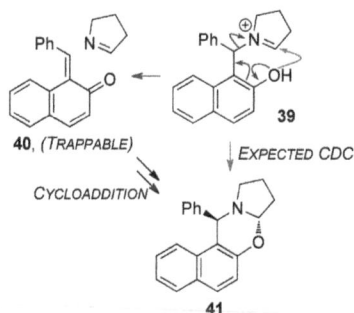

Scheme 11.15 Frangible iminium ion pathway operating in parallel with the expected CDC reaction pathway in an intramolecular copper-catalyzed coupling.

the same stereochemical outcome as the direct reaction (**39** to **41**). That loss of stereochemical integrity of enantio-enriched starting materials was not complete, implied that the pathways may co-exist.

Results from our laboratory showed that metal catalysis was necessary for the DDQ (2,3-dichloro-5,6-dicyano-1,4-benzoquinone)-mediated coupling of a difficult CDC reaction, the arylation of isochroman (Scheme 11.16).[103] Formation of the oxocarbenium ion (**45**) was implied by the result of replacement of anisole (**42**) with ethanol, giving the relevant ethoxy-substituted product, with the implication from the reaction conditions being that the oxocarbenium ion formation was rapid under ambient conditions. However, there was no reaction between the oxocarbenium and anisole, despite these compounds being expected to react smoothly according to the simple Mayr descriptions of electro- and nucleophilicity. A metal catalyst was found to be needed for the coupling of the aromatic ring, with Cu(II) performing best. It was striking to observe that catalysis occurred with redox-active metals such as copper and zinc but not with nickel, for which no Ni(I) complexes are known.[104] The proposed redox cycling parallels proposed by Klussmann for the copper-mediated process employed other oxidants,

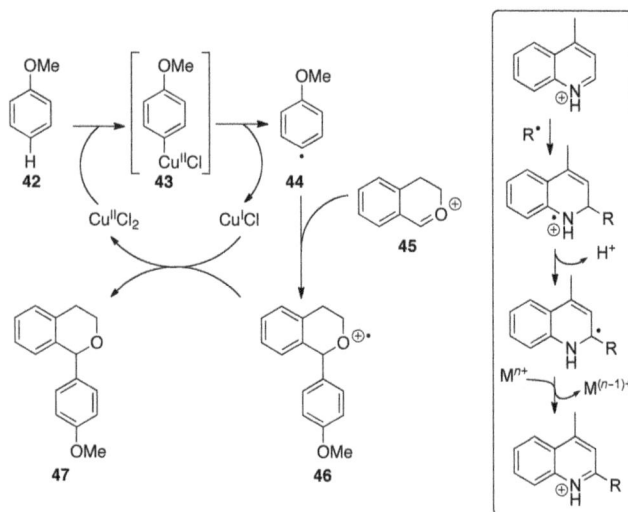

Scheme 11.16 Proposed mechanism of the Cu(ɪɪ)-catalyzed DDQ-mediated arylation of isochroman, and comparison with a Minisci radical–
heteroaromatic ring coupling.

although in that case the metal center is proposed to interact directly
through the terminal oxidant. Such mechanisms impart a significantly different role to the metal beyond possible simple Lewis acid catalysis (seen for
example in another CDC reaction involving isochroman where isolated
'dimer' compounds were shown to react to give CDC products when in the
presence of a Brønsted acid).[96] A possibility in the arylation reaction, which
has not been experimentally verified, is a Minisci-type mechanism[105] (inset)
involving a carbon-based radical coupling to an activated heterocycle which,
in the typical Minisci case, involves a protonated heteroatom, for which the
oxocarbenium is a surrogate (but which gains no energetic benefit from re-
aromatization). An intermediate solid appears in this arylation reaction
which by solid-state NMR spectroscopy (unpublished results) appears to be
the DDQ-bridged dimeric structure where the dihydroquinone has reacted
with two oxocarbenium ions. Interestingly when this solid is mixed with the
catalyst system without further oxidant, but with TEMPO, no further reaction
occurs, which lends some weight to the idea, but does not prove, that the
coupling with the aromatic ring is a radical-mediated process. It is possible
that the copper participates in a redox cycle, giving **43** *via* insertion into the
C–H bond at the *para*-position, leading to formation of a reactive radical **44**.
Electron transfer as part of the redox cycle gives the CDC product **47** (*via* **46**)
and re-generates the copper catalyst. The 'naked' aryl radical is highly un-
usual, however, and requires proper validation, but the insertion of Cu
atoms into aromatic C–H bonds (or an agostic interaction leading to
chemical exchange) has limited precedent when accompanied by stabilizing
ligands.[106,107] A Pd-catalyzed CDC reaction that may be operating *via* a
Minisci-type mechanism has also been reported.[108]

11.3 Thermal Metal-Free CDC Reactions

Over the years there has been significant interest in metal-free oxidation processes for the usual reasons of toxicity/cost. Much of the early work focused on the basic behavior of such compounds in amine oxidations rather than on our focus of CDC-type processes, but oxidants such as chlorine dioxide (ClO_2) have been used in intermolecular cyanations (**47** to **48/49**, Scheme 11.17).[42] There have been more recent reports of CDC reactions mediated by reagents such as KI/TBHP where the role of the iodide was proposed to be either: (1) a precursor to molecular iodine oxidation of the amine to the iminium ion;[109] or (2) the conversion of the peroxide reagent to a *tert*-butyl oxide free radical that abstracts a H atom from the carbon adjacent to the nitrogen (part of an amide in this case) to give an α-amino radical that is oxidized by a molecule of iodine to give the iminium ion.[110] In a more recent case, employing iodine and hydrogen peroxide it was proposed that *in situ* generated HOI is responsible for oxidation of the amine to the iminium ion.[111] A CDC-based cyanation reaction was recently reported where the oxidation was effected by the tropylium ion.[112] An interesting oxidative coupling between indoles and methyl ketones was recently published that is effected by iodine, but the mechanism is likely to be significantly different to those described here, relying initially on a rate-determining α-iodination reaction as the initial step.[113,114]

More recently Klussmann reported an auto-oxidative coupling between benzylic positions (in *e.g.*, xanthenes, **50**, Scheme 11.18) and C-nucleophiles where preliminary mechanistic experiments have suggested the involvement of peroxide intermediates such as **51** that may couple simply under the influence of a strong Brønsted acid.[115–117]

The great majority of the more recent literature on metal-free CDC reactions has involved DDQ, so a short detour will be taken to cover mechanistic aspects of this reagent. There has been vigorous debate for decades about whether DDQ operates with a radical[118] or an ionic[119–123] mechanism,

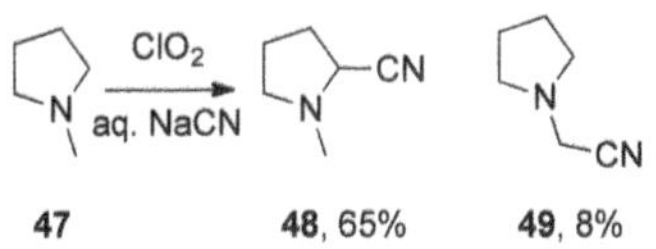

47　　　　**48**, 65%　　**49**, 8%

Scheme 11.17　An example of ClO_2-mediated oxidative cyanation.

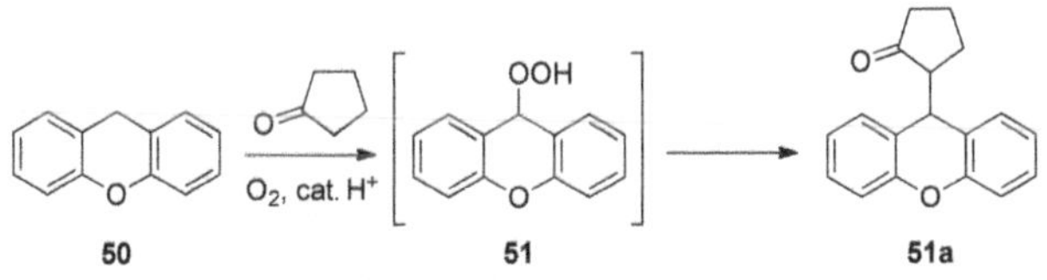

50　　　　　**51**　　　　　**51a**

Scheme 11.18　Auto-oxidative CDC coupling effected by a Brønsted acid and air.

or whether the mechanism is intractably substrate/conditions dependent (Scheme 11.19). Hydride transfer is generally thought to be the slow step that is followed by rapid proton transfer to complete the reaction. An alternative is that hydride abstraction consists of an initial H-atom abstraction step to give radical intermediates, followed by SET to give ionic intermediates and finally proton transfer. Evidence for the presence of ionic intermediates in DDQ oxidations was obtained upon isolation of the tropylium salt **54** from the oxidation of cycloheptatriene.[119]

These possibilities were followed by the formulation of others. For example Rüchardt and co-workers described the possible mechanistic routes for quinone oxidation of hydroarenes with radical pathways (SET/HAT), ionic pathways (direct hydride transfer) and even pericyclic pathways.[124] The conclusions were that initial H-atom transfer appeared more likely than hydride transfer.

The most striking visual feature of DDQ-mediated reactions is the vibrant color produced, often attributed to charge-transfer complexes arising from an interaction between electron donors and acceptors that lowers the activation barrier for subsequent reaction.[125–127] For example in an early report of dehydrogenation reactions of phenols by DDQ[128] Becker proposed that the formation of charge-transfer complex **56** (Scheme 11.20) preceded any further transformation of the substrate by the benzoquinone.[129]

Following these early proposals on the general nature of DDQ-mediated reactions, more recent research has provided a much deeper level of understanding of such processes. Fukuzumi and Kochi and others have

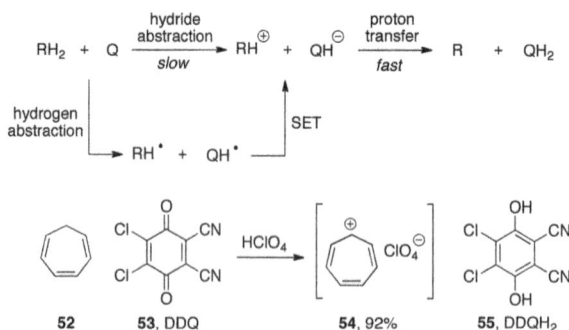

Scheme 11.19   Early mechanisms for DDQ oxidations, and evidence of ionic intermediates.

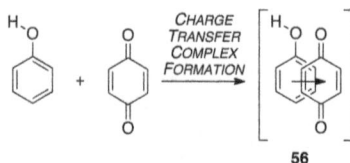

Scheme 11.20   Charge-transfer complex **56** of phenol and benzoquinone.

advanced a particularly strong case for the theory that the oxidation occurs by an electron-transfer process which involves intermediate charge-transfer complexes rather than *via* a hydride transfer mechanism.[130–137] Yet there remain continuing disagreements about the details of the mechanism,[138,139] and it would appear that (inevitably) the reaction mechanisms and nature of electron transport are highly dependent on both substrate and even on re-action conditions/solvent.[140–142] It was not until recently that direct meas-urements were made on CDC processes.

In understanding quinone-mediated oxidations through measurement it should be noted that quinones can form quinhydrones, a 1:1 molecular complex of the quinone and the oxidized hydroquinone form. The complex **57** was assembled through π–π stacking interactions of the quinone and the hydroquinone in an alkaline environment, with stabilization *via* hydrogen bonding between the quinone carbonyl and the hydroquinone hydroxyl groups (Scheme 11.21).[143] The complex is known to be a weak acid,[144] with the loss of the phenolic protons giving the semiquinone radical **58** a species that can be observed spectroscopically.

Early examples of DDQ-mediated oxidative bond-forming reactions were carried out by Mukaiyama[145] and Xu[146,147] and mechanisms were proposed but not experimentally investigated. Some of these reactions were performed stepwise, however, to avoid cross-reaction between nucleophile and oxidant, precluding them from being considered CDC reactions in the strict sense of the term. Stepwise reactions of this type are still being performed.[148,149] Mechanistic investigation in these early reactions was light, beyond the observation that LiClO$_4$ was frequently found to give superior results, taken as evidence for the formation of a stabilized salt in the intermediate.

We identified an efficient DDQ-mediated CDC reaction between *N*-phenyl THIQ and nitromethane that provides access to a number of rare chiral vicinal diamines (*via* **27d**, Scheme 11.22).[150,151] In the course of this work we observed NMR spectroscopic evidence for the long-supposed iminium ion and noted the precipitation of a solid in the reaction mixture which when isolated and then dissolved led to the reaction product. Subsequent focus on this solid using elemental analysis led to its suggested identity being **26c**, where an iminium ion paired with the anion of the DDQ reaction product (the dihydroquinone) and this salt precipitated with another molecule of the protonated dihydroquinone (DDQH$_2$).[152] The solid was successfully

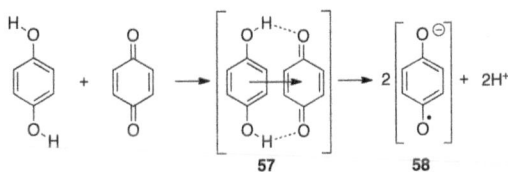

Scheme 11.21 Formation of quinhydrone **57** from benzoquinone and its hydro-quinone. Subsequent collapse of the quinhydrone may give the corresponding semiquinone radical **58**.

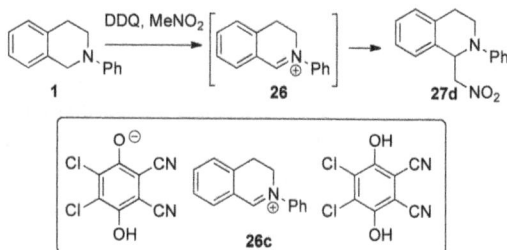

Scheme 11.22   Formation of an intermediate iminium ion that was characterized and reacted further.

characterized with solid-state NMR spectroscopy and X-ray crystallography, supporting the structure shown and providing the first experimental evidence for the intermediacy of the iminium ion in a CDC process. This was published around the same time Klussmann identified the iminium intermediate in a copper-mediated CDC reaction (described above). Synthetically it was interesting that the iminium ion was so well trapped in this solid that it could be isolated, washed and studied as a solid under ambient conditions, allowing us to trial various nucleophiles over a year later and without the need for additional oxidant. This illustrated that we could employ nucleophiles that were incompatible with the one-step CDC reaction presumably because they reacted irreversibly with oxidant. Chiral counterion catalysis has since given enantiomerically enriched products using this strategy.[153,154] The mass balance was such that half the starting material was captured in the solid, with the remainder in the supernatant. Interestingly the structure provides evidence that the DDQH⁻ does not react with the iminium ion to form a C–O bond at the benzylic position, but instead remains as a salt.

Floreancig has developed an aerobic oxidative cyclization reaction (**59** to **60**, Scheme 11.23) that is initiated by SET, and mediated by a reagent system involving the catalytic oxidant *N*-methylquinolinium hexafluorophosphate (NMQPF₆), oxygen, light, and toluene as a sensitizer.[155,156] The radical cation leads to a weakening of the adjacent bonds (in **61/61a**) and in this case selective cleavage of the benzylic bond to produce an iminium ion (**62**) that is trapped by the internal nucleophile. An interesting stereo-electronic influence on the bond cleavage was highlighted in a comparison of the oxidative cleavage of the two related substrates **63** and **66**. The former, acyclic structure has at least two possible bonds that can align well with the relevant partially filled orbital on the nitrogen atom (oxidation is assumed to lead to a SOMO on the carbamate) and in this case an appreciable amount of each corresponding product is formed because the lone pair on the pendant methoxy group can overlap well with the σ* orbital of the C–C bond to be broken (inset). In the case of the cyclic compound, the oxazolidine **66**, the only orbital overlap possible is that with the benzylic C–C bond (inset, where the cleavage is then stabilized by the aromatic π-system, as was the case in

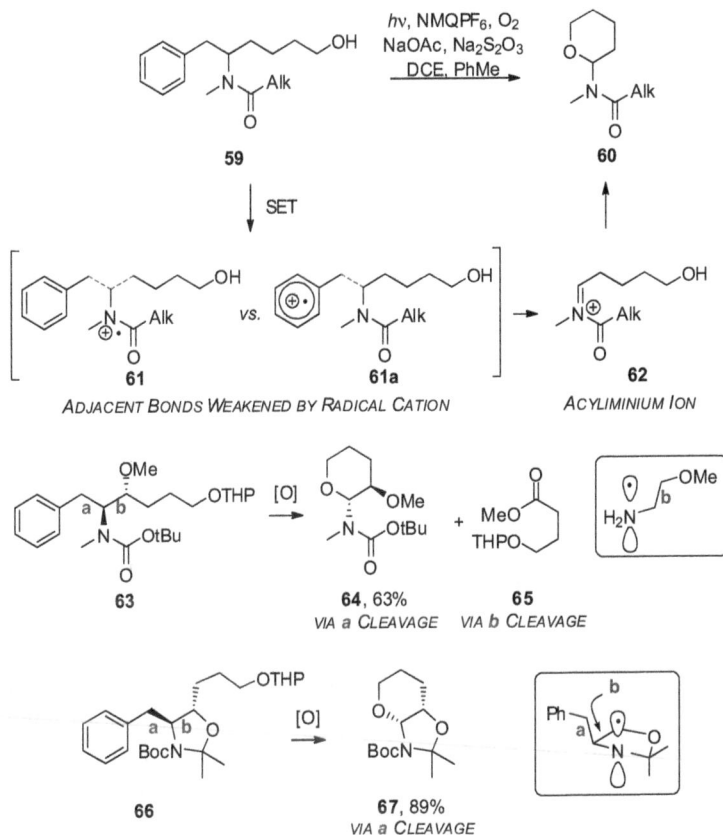

Scheme 11.23 Synthetic outcomes guided by the weakening of bonds adjacent to radical cations (DCE = dichloroethane, THP = tetrahydropyranyl, PhMe = toluene).

the prototypical reactions described previously) – sure enough **67** is formed with higher yield and greater selectivity. This result also addresses an interesting mechanistic question of this reaction (seldom asked in the field more generally): which group in the molecule is being oxidized first, and does this question matter? In the case of these substrates, though the benzylic bond is cleaved, a consideration of the oxidation potentials of the groups present would suggest the amide is oxidized first. The stereoelectronic effect, leading to the formation of product **65** is consistent with the radical cation remaining on the carbamate, rather than being transferred to the aromatic ring. The synthetic outcome arises from an interplay of oxidation potentials and the stereo-electronics of the partially filled orbitals with adjacent electron density.

Related cyclizations could be induced with ceric ammonium nitrate (CAN, Scheme 11.24).[157] Installation of a methoxy group on the aromatic ring of **68** led to decreased oxidation potential but increased the bond strength of the

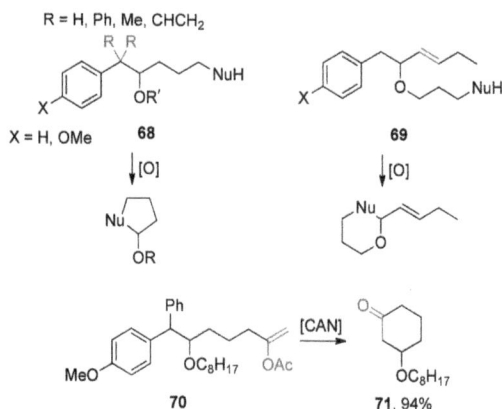

Scheme 11.24 Influence of substituents on the feasibility and selectivity of oxidative cyclizations.

benzylic C–C bond in comparison to the unsubstituted case (R = H), which translated into no observed cyclization products (or conversion of starting material beyond the SET) for **68** but smooth cyclization for **69**. The reactivity of the substrate containing the more electron-rich ring was regained through addition of groups (phenyl, vinyl, alkyl) to the benzylic position that encouraged fragmentation through stabilization of the benzyl radical. This allowed the reactions to proceed with the milder and more convenient oxidant CAN, and the pendant nucleophiles could become less reactive carbon-based groups such as enol acetates, in the case of **70** giving the cyclized product **71** in excellent yield, with such reactions giving high levels of stereocontrol when the pendant chain containing the internal nucleophile itself carries a stereocenter. It is interesting to note that consideration of the various oxidation potentials of the substrate and oxidant suggest that the SET is endergonic, but that if it is assumed that the process is an equilibrium then the thermodynamically favorable trapping of an intermediate oxocarbenium ion is what is driving the overall reaction to completion.

As is so often the case with CDC reactions, changing the oxidant gave a different outcome. DDQ gave rise to benzylic C–H bond cleavage rather than C–C bond cleavage, possibly due to the proximity of quinone's basic oxygen centers and the benzylic position when the reagent is associated with the substrate's aromatic groups. Use of alternative metal-based oxidants $(phen)_3Fe(PF_6)_3$ (phen = 1,10-phenanthroline) and CTAN $[(Bu_4N)_2Ce(NO_3)_6]$ gave no conversion despite their having similar oxidation potentials; the suggestion was that these ground-state oxidants require inner sphere electron transfer for bond cleavage to occur, which is prevented by the bulk of the surrounding ligands/ions. This is another important lesson in avoiding conclusions about a reaction's feasibility from the screening of a small number of oxidants, even when they have similar oxidizing power.

DDQ-Mediated oxidations combined with intramolecular trapping of benzylic cations were subsequently examined in more detail where the

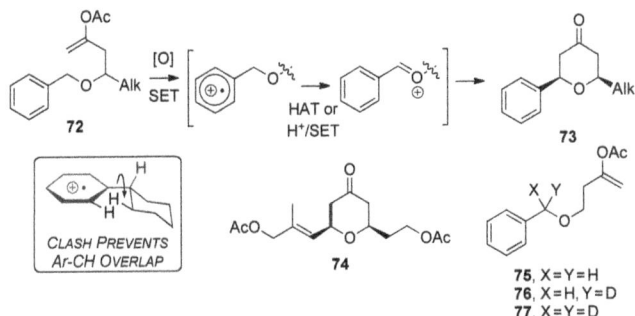

Scheme 11.25 Stereoselective intramolecular trapping of oxidatively generated oxonium ions.

substrates were benzylic, rather than homobenzylic, ethers (72, Scheme 11.25).[158] It is interesting that in this work, cleavage of the PMB (*para*-methoxybenzyl-) ether was a problematic side-reaction, requiring the use of powdered molecular sieves (plus 2,6-dichloropyridine and 1,2-dichloro-ethane as the solvent), whereas as described above (Klussmann's work and others) it is seen as advantageous for intermediates to be trapped by oxygen-based nucleophiles. The reaction was suggested to occur through a radical cation created by SET which lost a hydrogen atom to form an oxocarbenium ion which arranged itself in a chair conformation for cyclization, accounting for the observed product stereoselectivity. Of note is a stereo-electronic explanation for why the products of the reactions are themselves unreactive towards further conversion: product **73** lacks the necessary overlap between the aromatic ring (inset), which is presumably oxidized under the conditions of the reaction, and the benzylic C–H bond, preventing further atom transfer from the intermediate radical cation (we will address this again below). An illustration of the ability of CDC reactions to distinguish between several possible reaction pathways with the correct tuning of oxidant, conditions and substrate is shown by the successful formation of compound **74** from its allylic ether precursor in a remarkable 80% yield.

The mechanism was subsequently examined in more detail.[159] Evidence was found for a radical cation intermediate prior to H-atom abstraction in the formation of the oxocarbenium ion that is the precursor to cyclization. Inter- and intramolecular kinetic isotope effects were measured using the three substrates shown, where the mono-deuterated substrate **76** gives information on whether there is any preference for cleavage of C–H *vs.* C–D in the same molecule while a reaction containing a combination of the two substrates **75** and **77** allows measurement of an intermolecular kinetic isotope effect. The results were taken from an NMR-based assay where the reaction was run to approximately 10% conversion. The KIEs were largest with the most reactive substrates, but they were clearly significant in both the intra- and intermolecular cases. The clear meaning of this is that C–H bond cleavage is rate-determining, while formation of any reactive intermediate before that stage is not.

A question that arose was why the KIE should be higher for the more reactive substrates. This was taken as evidence against a hydride-transfer mechanism, but for an electron-transfer mechanism. As was seen in the earlier work,[157] installation of electron-donating groups (EDGs) on an aromatic ring lowers the oxidation potential, but in the radical cation resulting from oxidation, leads to stronger bonds in the benzylic position. This leads to an apparent contradiction: when an EDG is present, bond cleavage is harder, so why, then, are the reactions faster overall? Floreancig suggested that when an EDG is present there is a much higher concentration of radical ions, which dominates the slower rate of atom transfer.

The development of successful models for stereoselectivity in such cyclizations is indirect evidence for the intermediacy of an oxocarbenium ion[160] (and iminium ions in the related DDQ-mediated cyclizations of *e.g.*, *N*-vinyl amides[161,162] as well as thiocarbenium ions in the analogous sulfur-containing substrates[163]). Can we go further and resolve whether the mechanism is, in the earlier stages, SET then HAT, or some other combination? Essentially we cannot from the published results. It is challenging to interrogate such electron transfer without computational modeling, partly because the electron-transfer events are likely to be much faster than the relatively (glacially) slow movement of molecules and atoms.

Batista and co-workers investigated the mechanism of the DDQ-mediated oxidation of benzylaniline (**78**) to the corresponding imine (**79**, Scheme 11.26).[164] While the transformation is not a CDC reaction, the substrate bears resemblance to many others used in CDC reactions. The authors concluded from computational experiments that the oxidation occurred *via* the formation of a tight 1:1 complex **80** (formed through charge-transfer interactions, where approximately 0.25 *e* is transferred), with the transition state **81** leading to direct hydride transfer giving the ionic intermediate **82**. A second proton-transfer step completes the oxidation giving the imine. No reference was made to a possible radical-based mechanism for the

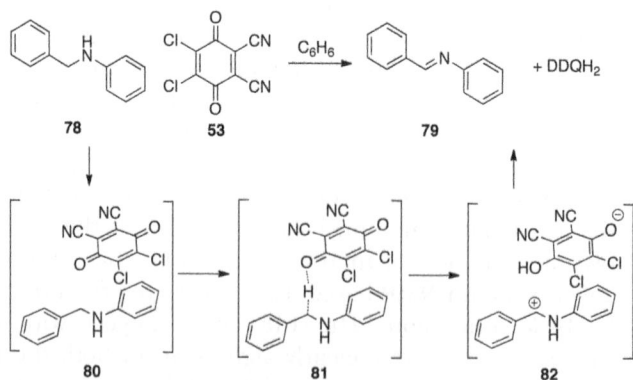

Scheme 11.26 DDQ-Mediated oxidation of benzylaniline, with computational predictions for the nature of the atom-transfer step.

transformation. A subsequent publication investigated the oxidation of toluene by DDQ, in an attempt to determine the mechanism of benzylic C–H activation by the oxidant.[165] With similar electron/proton- and hydride-transfer events proposed as possible mechanisms, experimental results and subsequent density functional theory (DFT) calculations to predict the energy of the transition states of the different pathways were undertaken. From kinetic isotope investigations, it was concluded that the C–H bond-breaking step was rate-determining, in line with investigations in related systems described by others above, but with the authors stating that this result was consistent with a one-step, direct hydride-transfer mechanism. The authors claimed that the free energy of the transition state for hydride transfer was lower than that for electron transfer from toluene and DDQ, according to the DFT calculations.

In recent work from our laboratory (under review) we have also examined the transition state of a related process (the CDC reaction shown in Scheme 11.22) and found evidence for a transition state (similar in general to **81**) with biradical character, implying a non-hydride mechanism; the experimental side of this study also generated spectroscopic evidence for initial SET.

11.4 Photochemical CDC Reactions

Iminium ions may be generated from amines using metal-mediated photocatalytic oxidation, and may then react with nucleophiles in the usual way in processes that may themselves be mediated by another metal salt. The mechanisms have been outlined and are broadly similar to those described earlier in this chapter (Scheme 11.27), with early reports using Ir-[166] or Ru-catalysis for the photo-oxidation step and copper for the nucleophilic addition step if needed[167,168] or the electron transfer may be mediated by an

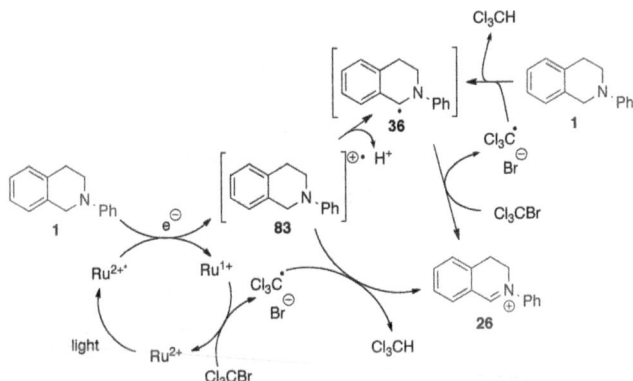

Scheme 11.27 Representative photocatalytic CDC reaction. Alternative pathways are direct formation of iminium ions *via* electron and atom transfer *vs.* a radical chain mechanism.

organic dye.[169,170] Ultimately the SET that starts the process is facilitated by the excited state of the metal complex. The radical cation that results (83) is able to participate in HAT (or PT/SET) to give the iminium ion 26 (which has been observed by NMR spectroscopy[167]) in the usual way. The electron gained by the metal complex may be transferred to a sacrificial oxidant, such as bromotrichloroethane, generating a new radical that carries out the HAT step to form the iminium ion. A further possibility[167] (top right of Scheme 11.27) is that the radical cation gives rise to an α-amino radical via proton loss which allows for a radical chain mechanism by converting the sacrificial reagent to a radical which then removes a H atom from the next molecule of substrate, sustaining the cycle.

Many metal complexes are known to be able to carry out the initial SET of typical CDC reactions, and the catalysts may also be heterogeneous.[171] Clearly also oxygen itself can be responsible for re-generation of the higher oxidation state of the metal complex.[172] Processes with similar mechanisms may operate that are not overall oxidative, such as alkylations with alkyl halides alpha to the carbonyl functionality, but these processes are often also generically referred to as *photoredox catalysis*.[173–176]

It has been noted that the excited states generated during the reaction could lead to side-reactions with other reagents, but this can be solved by adopting a stepwise reaction protocol in which iminium ions are generated photochemically to which are added nucleophiles in the absence of light, in a manner akin to some of the non-photochemical stepwise CDC-type reactions that have been reported.[167]

There have been several modifications of this general scheme published along with mechanistic investigations of intermediates, but these will not be exhaustively covered in this chapter since excellent, thorough reviews have recently been published describing this chemistry.[177,178] Notable examples in the field that have involved mechanistic analysis include CDC reactions based on a metal-free organic dye (eosin Y salt)/O_2 combination,[179] a Pt-complex/air system in which it was shown that the superoxide radical anion is responsible for removing the hydrogen atom to form the α-amino radical,[180] a Pd-porphyrin/O_2 combination in which the excited state of the catalyst generates singlet oxygen that is responsible for the oxidation of substrate,[181] the photo-excitation of an organic compound where electrons are donated to an added electron acceptor (1,4-dicyanonaphthalene) giving an intermediate carbocation that is trapped by oxygen-based nucleophiles intra- and intermolecularly,[182] the Ru-catalyzed photogeneration of acyliminium ions via intermolecular H-atom transfer by a sulfate radical anion,[183] and the generation of an azomethine ylid from an intermediate iminium ion that was able to participate in a cycloaddition reaction, where a large KIE was observed for H-atom abstraction from the radical cation to give the iminium ion.[184] There have also been recent reports of the intermediate α-amino radical itself being trapped via reaction with suitable electrophiles, both producing CDC reaction products and also providing strong evidence for their intermediacy in such reactions.[172,185–187]

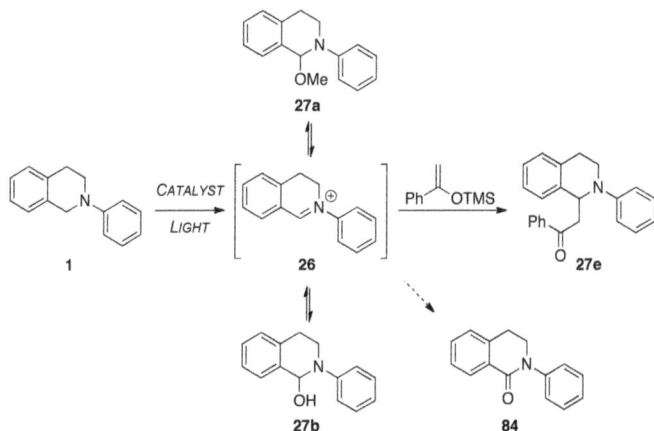

Scheme 11.28 Methoxy- (**27a**) and hydroxy-substituted (**27b**) intermediates implicated along the reaction pathway of the light-mediated CDC reaction of **1**.

In Xia's light-mediated coupling of silyl enol ethers to **1** (Scheme 11.28), the absence of the nucleophile led to the formation of the methoxy- (**27a**) and hydroxy-substituted intermediates (**27b**) in yields of 43% and 57% respectively in a similar fashion to that proposed by Klussmann for the copper-mediated CDC reaction, *i.e.*, off-cycle equilibria that trap the reactive intermediate.[188] The use of wet methanol gave complete conversion to the hemiaminal. The authors proposed that formation of these products prevented further irreversible oxidation to the corresponding amide **84**. Instead, **27a** and **27b** are in equilibrium with the iminium ion and this competes with the irreversible carbon–carbon bond-forming reaction to give **27e** upon addition of the nucleophile.

11.5 SOMO Activation

A further set of oxidative reactions are those dependent on so-called SOMO activation,[189] in which an *in situ* formed enamine (**87**, Scheme 11.29) is oxidized (preferentially) to a radical cation (**88**) that may couple with an added somophile such as an allylsilane or pyrrole, to give, after further electron loss from the initially formed product (**89**) the (typically enantio-enriched) umpolung product (**90**) where on the face of things a carbonyl-containing compound has behaved as an electrophile at the C2-position. This kind of chemistry was also reported by another group at the same time,[190] but the exact mechanism appears to be reagent-specific, with further investigation of this report suggesting the mechanism was more likely based on ionic (*via* enamine activation) intermediates.[191] A later report using Friedel–Crafts type reactivity of aromatic rings as the putative somophiles was suggested to operate by an ionic mechanism,[192] but DFT modeling for this reaction has been performed that shows a correlation between theory and experiment for a radical-based cyclization.[193]

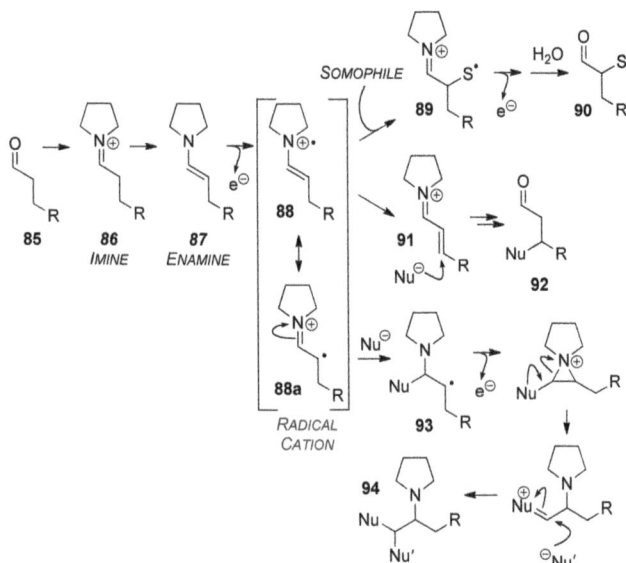

Scheme 11.29 The various pathways available to SOMO activation-type oxidative couplings in which it is the enamine that is oxidized.

The electron loss may be performed photochemically,[173] but in the original paper the oxidant ceric ammonium nitrate was employed. The radical cation has been observed by mass spectrometry[194] with DFT calculations supporting the electronic distribution shown (**88** and **88a**) arising from a mixture of two resonance structures but which may be best conceptualized as an alkyl radical conjugated to an iminium ion (**88a**).[193] It may also be possible for the enamine to react as is with an incoming radical cation (*i.e.*, the same reaction pathway but where the enamine is the somophile).[195]

These SOMO reaction types have been explored widely and have been reviewed in the context of other kinds of aminocatalysis.[196] In essence this reaction type is formally a kind of CDC process, though the mechanism is distinct. Radical clock experiments in the original studies strongly suggested a radical, rather than a polar, mechanism. Studies by Flowers *et al.*[197] showed that oxidation of the enamine is much faster than that of the amine catalyst, illustrating one of the most desirable features of CDC reactions: that the oxidant may be mixed with all the components in one pot, yet ideally should act on only one species. Kinetics experiments revealed the crucial and subtle role played by water (with respect to which the reaction is inverse first-order), its presence being crucial in preventing de-activation of the catalyst but at the same time slowing the forward reaction. The water is also important in solubilizing the oxidant, in which the reaction is apparently zero-order because of the oxidant's heterogeneous nature. Water coordinates to the oxidant, making it unavailable for its role in the cycle, explaining why the equivalent of endogenous water (produced during the reaction) is insufficient to keep the cycle going.

Looking at **88** in Scheme 11.29, and comparing the radical cation to those seen in the rest of this chapter makes us think whether in fact a proton and an electron could be removed by a base to give compound **91** that might then couple with nucleophiles to provide **92**. While this was not the mechanism proposed, this form of reaction, to give β-functionalized aldehydes, has been demonstrated through the oxidation, by *ortho*-iodoxybenzoic acid (IBX) or DDQ, of the enamine itself.[198–200] The mechanism may involve a SET process, *via* the radical cation, but may also be *via* polar oxidation processes mediated directly by the IBX or DDQ without the involvement of a radical cation intermediate. A catalytic version of this chemistry, using Pd(OAc)$_2$/O$_2$, was recently reported.[201] Rueping developed a similar reaction in which the oxidation (of an alcohol to an aldehyde) occurs first, followed by the involvement of the amine catalyst to effect the bond formation.[202] The direct loss of a proton from the radical cation to yield a neutral, conjugated radical was, however, recently proposed by MacMillan as part of a non-oxidative, photoredox/organocatalytic route to the β-arylation of carbonyl compounds.[203] An interesting question here is why this type of reactivity has not been seen more generally in such reactions, given that so many of the typical SOMO-type substrates have available Hs in alkyl chains.

Another mode of reactivity may be envisaged in which a resonance form of the radical cation (**88a**) could react with a polar nucleophile to give **93** and, following a further oxidation, be able to deliver functionalized product **94**. This mode has been observed by Loh (the nucleophile was methanol), who carried out mechanistic studies with isotopically labeled substrates that supported the mechanism shown.[204] The outcome will naturally depend on the conditions of the reaction and the nature of the trapping nucleophiles/somophiles, with the trapping of these interesting reactive intermediates gaining increasing attention.[205]

11.6 A Note on Stereoelectronic Effects

Reactions that do not proceed can give valuable clues to reaction mechanisms. In the case of organic oxidations there can be striking stereoelectronic barriers. For example it has been noted that, when a radical cation forms along a reaction pathway, subsequent loss of a hydrogen atom to generate an iminium ion intermediate is possible if there is orbital overlap between the partially filled orbital on nitrogen and the C–H bond being broken (Scheme 11.30). This was used to explain an 8 : 1 preference for formation of new bonds on the ring of *N*-methylpyrrolidine *vs.* formation on the exocyclic methyl in the previously described ClO$_2$-mediated oxidations of that ring system.[42] (This is also seen when the oxidant is DDQ, but the preference is the other way round in the DDQ-mediated cyanation of six-membered ring piperidines.[206]) We have seen arguments throughout this chapter along such lines and the relevant issues have been reviewed.[207] Such

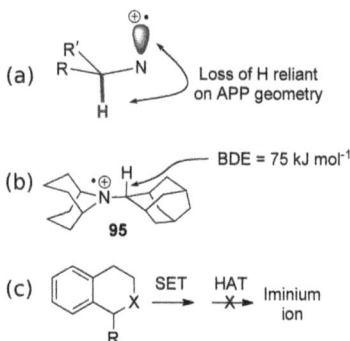

Scheme 11.30 Stereo-electronic effects on BDEs originating from orbital overlap with the radical cation (APP = antiperiplanar).

effects have been measured in cases where orbital overlap can be controlled through steric effects [e.g., **95**, Scheme 11.30(b)].[208] The extent of the influence is quite striking, with greatly weakened C–H bonds being observed adjacent to radical cations, with knock-on effects seen in the stereochemistry of reactions following the SET.[46,167,209–215]

Such effects probably explain another important but seldom-noted feature of many CDC reactions – their tendency not to produce over-oxidized products, i.e., that the product itself appears to be inert to subsequent reaction. This is extremely useful, synthetically. In our own work, for example, we have observed the total lack of reactivity of CDC products even when isolated and subjected to forcing reaction conditions [Scheme 11.30(c)] and this has been noted by others, e.g., ref. 96. In many CDC reactions, the products have substituents in the benzylic position of cyclic compounds. The steric demands of such molecules are such that frequently the substituent at the 1-position (i.e., in the benzylic position of the common substrate **1**) will occupy a pseudo-axial position, minimizing the clash with the neighboring C–H bond. This all but prevents the C–H in the 1-position, which is the most likely place for subsequent CDC chemistry, from overlapping with adjacent electron-deficient radical cations or other partially filled orbitals. While one might expect SET one would not expect HAT, with the requirement of C–H cleavage and in unpublished work we have seen electron but not atom loss in such cases. The corresponding disadvantage then is the difficulty of further functionalizing CDC products if that is so desired, but the advantage is a remarkably clean CDC reaction, and this probably explains the substrate scope of the field to date, and the tendency of studies to employ, for example, N-arylated THIQs.

11.7 Conclusions and Outlook

CDC reactions are synthetically elegant and permit access to molecules of clear biological/pharmaceutical value. With the exception of some of the peptide work described above, the substrate scope is still small. As we have seen, this is partly because the substrates investigated have few options following their oxidation, and can couple with nucleophiles directly in one pot. (There have, for example, been few examples of CDC reactions employing sulfur-containing compounds, though the precedents do imply that oxidation adjacent to a sulfur (rather than on the atom) and controlled reaction with a nucleophile in one pot is possible.[163,216]) The high yields and clean reactions are a feature of these typically limited options, for example the clear weakening of one bond near to the site of oxidation we saw in the previous section. The studies described in this chapter have provided clear insight into the reaction mechanisms and involve very basic considerations of electron transfer; these coupled with insights in the previous literature on the behavior of reactive intermediates provides our understanding of why the reactions perform so well, but also reveal the challenge of the way forward. If the substrate scope is to be broadened, and the substrates to permit the removal of an electron followed by the well-controlled loss of an atom and coupling of a nucleophile there will need to be additional means found to control the diversity of reactions that is possible. We saw an example of this above in the trapping of reactive iminium ions either by a physical process (crystallization in a solid) or as a chemical process in the reaction (equilibrium with methoxy/peroxy adducts). Even in those reactions that do behave well it is not always clear why some CDC reactions are successful while others are not, for example in the failure of *N*-acylated compounds to react where their *N*-arylated counterparts react smoothly. It is likely in the next few years we will see a combination of experimental and computational resources in teasing out the details of the reactive intermediates and their multifarious pathways in the understanding of these beautiful reactions.

References

1. C.-J. Li, *Acc. Chem. Res.*, 2009, **42**, 335.
2. C. J. Scheuermann and C. J. *Chem. Asian J.*, 2010, **5**, 436.
3. C. S. Yeung and V. M. Dong, *Chem. Rev.*, 2011, **111**, 1215.
4. O. Daugulis, H.-Q. Do and D. Shabashov, *Acc. Chem. Res.*, 2009, **42**, 1074.
5. G. Deng, L. Zhao and C. J. Li, *Angew. Chem., Int. Ed.*, 2008, **47**, 6278.
6. N. Dastbaravardeh, M. Schnürch and M. D. Mihovilovic, *Org. Lett.*, 2012, **14**, 1930.
7. A. E. Wendlandt, A. M. Suess and S. S. Stahl, *Angew. Chem., Int. Ed.*, 2011, **50**, 11062.
8. C. Zhang, C. Tang and N. Jiao, *Chem. Soc. Rev.*, 2012, **41**, 3464.
9. E. A. Mitchell, A. Peschiulli, N. Lefevre, L. Meerpoel and B. U. W. Maes, *Chem.–Eur. J*, 2012, **18**, 10092.
10. K. R. Campos, *Chem. Soc. Rev.*, 2007, **36**, 1069.
11. C. Liu, H. Zhang, W. Shi and A. Lei, *Chem. Rev.*, 2011, **111**, 1780.
12. T. Shono, *Tetrahedron*, 1984, **40**, 811.

13. J.-I. Yoshida, S. Suga, S. Suzuki, N. Kinomura, A. Yamamoto and K. Fujiwara, *J. Am. Chem. Soc.*, 1999, **121**, 9546.
14. T. Chiba and Y. Takata, *J. Org. Chem.*, 1977, **42**, 2973.
15. S. Andreades and E. W. Zahnow, *J. Am. Chem. Soc.*, 1969, **91**, 4181.
16. E. Le Gall, J.-P. Hurvois, T. Renaud, C. Moinet, A. Tallec, P. Uriac, S. Sinbandhit and L. Toupet, *Liebigs Ann. Chem.*, 1997, 2089.
17. J.-I. Yoshida, K. Kataoka, R. Horcajada and A. Nagaki, *Chem. Rev.*, 2008, **108**, 2265.
18. O. Baslé, N. Borduas, P. Dubois, J. M. Chapuzet, T.-H. Chan, J. Lessard and C.-J. Li, *Chem.-Eur. J*, 2010, **16**, 8162.
19. J.-I. Yoshida and S. Suga, *Chem.-Eur. J*, 2002, **8**, 2650.
20. M. E. Wolff, *Chem. Rev.*, 1963, **63**, 55.
21. S. Murata, K. Suzuki, M. Miura and M. Nomura, *J. Chem. Soc., Perkin Trans. 1*, 1990, 361.
22. G. Galliani, B. Rindone and P. L. Beltrame, *J. Chem. Soc., Perkin Trans. 2*, 1976, 1803.
23. E. H. White and D. J. Woodcock, in *The Amino Group*, ed. S. Patai, Wiley Interscience, London, 1968, ch. 8.
24. G. Kok, T. D. Ashton and P. J. Scammells, *Adv. Synth. Catal.*, 2009, **351**, 283.
25. Y. Su, L. Zhang and N. Jiao, *Org. Lett.*, 2011, **13**, 2168.
26. R. N. Butler, *Chem. Rev.*, 1984, **84**, 249.
27. S.-I. Murahashi, *Angew. Chem., Int. Ed.*, 1995, **34**, 2443.
28. S.-I. Murahashi and D. Zhang, *Chem. Soc. Rev.*, 2008, **37**, 1490.
29. M. A. Ischay and T. P. Yoon, *Eur. J. Org. Chem.*, 2012, 3359.
30. Z. Li and C.-J. Li, *J. Am. Chem. Soc.*, 2004, **126**, 11810.
31. C. M. R. Volla and P. Vogel, *Org. Lett.*, 2009, **11**, 1701.
32. P. Li, Y. Zhang and L. Wang, *Chem.-Eur. J*, 2009, **15**, 2045.
33. C. J. Li, *Acc. Chem. Res.*, 2010, **43**, 581.
34. H. Mitsudera and C.-J. Li, *Tetrahedron Lett.*, 2011, **52**, 1898.
35. V. A. Peshkov, O. P. Pereshivko and E. V. van de Eycken, *Chem. Soc. Rev.*, 2012, **41**, 3790.
36. C. Zhang, C. K. De, R. Mal and D. Seidel, *J. Am. Chem. Soc.*, 2008, **130**, 416.
37. S. Murarka, I. Deb, C. Zhang and D. Seidel, *J. Am. Chem. Soc.*, 2009, **131**, 13226.
38. A. Dieckmann, M. T. Richers, A. Y. Platonova, C. Zhang, D. Seidel and K. N. Houk, *J. Org. Chem.*, 2013, **78**, 4132.
39. B. Zhang, Y. Cui and N. Jiao, *Chem. Commun.*, 2012, **48**, 4498.
40. O. G. Mancheno and T. Stopka, *Synthesis*, 2013, **45**, 1602.
41. M. C. Paderes, J. B. Keister and S. R. Chemler, *J. Org. Chem.*, 2013, **78**, 506.
42. C.-K. Chen, A. G. Hortmann and M. R. Marzabadi, *J. Am. Chem. Soc.*, 1988, **110**, 4829.
43. D. H. Rosenblatt, L. A. Hull, D. C. De Luca, G. T. Davis, R. C. Weglein and H. K. R. Williams, *J. Am. Chem. Soc.*, 1967, **89**, 1158.

44. L. A. Hull, G. T. Davis, D. H. Rosenblatt, H. K. R. Williams and R. C. Weglein, *J. Am. Chem. Soc.*, 1967, **89**, 1163.
45. L. A. Hull, G. T. Davis and D. H. Rosenblatt, *J. Am. Chem. Soc.*, 1969, **91**, 6247.
46. N. J. Leonard and W. K. Musker, *J. Am. Chem. Soc.*, 1960, **82**, 5148.
47. S.-I. Murahashi, T. Naota and K. Yonemura, *J. Am. Chem. Soc.*, 1998, **110**, 8256.
48. H. Volz and H. Gartner, *Eur. J. Org. Chem.*, 2007, 2791.
49. S.-I. Murahashi, T. Naota, N. Miyaguchi and T. Nakato, *Tetrahedron Lett.*, 1992, **33**, 6991.
50. S.-I. Murahashi, N. Komiya, H. Terai and T. Nakae, *J. Am. Chem. Soc.*, 2003, **125**, 15312.
51. S.-I. Murahashi, N. Komiya and H. Terai, *Angew. Chem., Int. Ed.*, 2005, **44**, 6931.
52. S.-I. Murahashi, T. Nakae, H. Terai and N. Komiya, *J. Am. Chem. Soc.*, 2008, **130**, 11005.
53. L. T. Burka, F. P. Guengerich, R. J. Willard and T. L. Macdonald, *J. Am. Chem. Soc.*, 1985, **107**, 2549.
54. M.-Z. Wang, C.-Y. Zhou, M.-K. Wong and C.-M. Che, *Chem.–Eur. J*, 2010, **16**, 5723.
55. M. North, *Angew. Chem., Int. Ed.*, 2004, **43**, 4126.
56. S. Murata, M. Miura and M. Nomura, *J. Chem. Soc., Chem. Commun.*, 1989, 116.
57. S. Murata, K. Teramoto, M. Miura and M. Nomura, *Bull. Chem. Soc. Jpn.*, 1993, **66**, 1297.
58. S. Murata, K. Suzuki, A. Tamatani, M. Miura and M. Nomura, *J. Chem. Soc., Perkin Trans. 1*, 1992, 1387.
59. Z. Li, L. Cao and C.-J. Li, *Angew. Chem., Int. Ed.*, 2007, **46**, 6505.
60. X. Guo, R. Yu, H. Li and Z. Li, *J. Am. Chem. Soc.*, 2009, **131**, 17387.
61. E. Shirakawa, N. Uchiyama and T. Hayashi, *J. Org. Chem.*, 2011, **76**, 25.
62. P. Liu, Z. Wang, J. Lin and X. Hu, *Eur. J. Org. Chem.*, 2012, 1583.
63. M. Ghobrial, K. Harhammer, M. D. Mihovilovic and M. Schnürch, *Chem. Commun.*, 2010, **46**, 8836.
64. M. O. Ratnikov, X. Xu and M. P. Doyle, *J. Am. Chem. Soc.*, 2013, **135**, 9475.
65. M. O. Ratnikov and M. P. Doyle, *J. Am. Chem. Soc.*, 2013, **135**, 1549.
66. A. J. Catino, J. M. Nichols, B. J. Nettles and M. P. Doyle, *J. Am. Chem. Soc.*, 2006, **128**, 5648.
67. E. C. McLaughlin, H. Choi, K. Wang, G. Chiou and M. P. Doyle, *J. Org. Chem.*, 2009, **74**, 730.
68. O. Baslé and C.-J. Li, *Org. Lett.*, 2008, **10**, 3661.
69. G. Zhang, Y. Zhang and R. Wang, *Angew. Chem., Int. Ed.*, 2011, **50**, 10429.
70. Y. Li, L. Ma, F. Jia and Z. Li, *J. Org. Chem.*, 2013, **78**, 5638.
71. S. Hashizume, K. Oisaki and M. Kanai, *Chem. Asian J.*, 2012, 7, 2600.
72. T. Sonobe, K. Oisaki and M. Kanai, *Chem. Sci.*, 2012, **3**, 3249.

73. S. Singhal, S. L. Jain and B. Sain, *Chem. Commun.*, 2009, 2371.
74. J. Xie, H. Li, J. Zhou, Y. Cheng and C. Zhu, *Angew. Chem., Int. Ed.*, 2012, **51**, 1252.
75. K. Alagiri and K. R. Prabhu, *Org. Biomol. Chem.*, 2012, **10**, 835.
76. K. M. Jones, P. Karier and M. Klussmann, *ChemCatChem*, 2012, **4**, 51.
77. K. Alagiri, G. S. R. Kumara and K. R. Prabhu, *Chem. Commun.*, 2011, **47**, 11787.
78. W. Han and A. R. Ofial, *Chem. Commun.*, 2009, 5024.
79. A. Sud, D. Sureshkumar and M. Klussmann, *Chem. Commun.*, 2009, 3169.
80. S. Murata, A. Tamatani, K. Suzuki, M. Miura and M. Nomura, *Chem. Lett.*, 1990, 757.
81. T. Tsuchimoto, Y. Ozawa, R. Negoro, E. Shirakawa and Y. Kawakami, *Angew. Chem., Int. Ed.*, 2004, **43**, 4231.
82. Z. Li and C.-J. Li, *J. Am. Chem. Soc.*, 2005, **127**, 6968.
83. Z. Li and C.-J. Li, *J. Am. Chem. Soc.*, 2005, **127**, 3672.
84. Z. Li and C.-J. Li, *Eur. J. Org. Chem.*, 2005, 3173.
85. O. Baslé and C.-J. Li, *Chem. Commun.*, 2009, 4124.
86. Y. Shen, M. Li, S. Wang, T. Zhan, Z. Tan and C.-C. Guo, *Chem. Commun.*, 2009, 953.
87. O. Baslé and C.-J. Li, *Green Chem.*, 2007, **9**, 1047.
88. Z. Li and C.-J. Li, *J. Am. Chem. Soc.*, 2006, **128**, 56.
89. Z. Li, D. S. Bohle and C.-J. Li, *Proc. Natl. Acad. Sci., U. S. A.*, 2006, **103**, 8928.
90. L. Zhao, O. Basle and C.-J. Li, *Proc. Natl. Acad. Sci., U. S. A.*, 2009, **106**, 4106.
91. L. Zhao and C.-J. Li, *Angew. Chem., Int. Ed.*, 2008, **47**, 7075.
92. E. Boess, D. Sureshkumar, A. Sud, C. Wirtz, C. Farès and M. Klussmann, *J. Am. Chem. Soc.*, 2011, **133**, 8106.
93. K. M. Jones and M. Klussmann, *Synlett*, 2012, **23**, 159.
94. Y. Shen, Z. Tan, D. Chen, X. Feng, M. Li, C.-C. Guo and C. Zhu, *Tetrahedron*, 2009, **65**, 158.
95. E. Boess, C. Schmitz and M. Klussmann, *J. Am. Chem. Soc.*, 2012, **134**, 5317.
96. H. Richter and O. García Mancheño, *Eur. J. Org. Chem.*, 2010, 4460.
97. Database of Reactivity Parameters, http://www.cup.lmu.de/oc/mayr/reaktionsdatenbank/.
98. H. Mayr, B. Kempf and A. R. Ofial, *Acc. Chem. Res.*, 2003, **36**, 66.
99. Z. Li, P. D. MacLeod and C.-J. Li, *Tetrahedron: Asymm.*, 2006, **17**, 590.
100. G.-J. Cheng, L.-J. Song, Y.-F. Yang, X. Zhang, O. Wiest and Y.-D. Wu, *ChemPlusChem*, 2013, **78**, 943.
101. M. Ghobrial, M. Schnürch and M. D. Mihovilovic, *J. Org. Chem.*, 2011, **76**, 8781.
102. M. L. Deb, S. S. Dey, I. Bento, M. T. Barros and C. D. Maycock, *Angew. Chem., Int. Ed.*, 2013, **52**, 9791.
103. S. J. Park, J. Price and M. H. Todd, *J. Org. Chem.*, 2012, **77**, 949.

104. S. Schulz, *Chem.-Eur. J,* 2010, **16**, 6416.
105. F. Minisci, A. Citterio, E. Vismara and C. Giordano, *Tetrahedron,* 1985, **41**, 4157.
106. R. J. Phipps, N. P. Grimster and M. J. Gaunt, *J. Am. Chem. Soc.,* 2008, **130**, 8172.
107. X. Ribas, R. Xifra, T. Parella, A. Poater, M. Sola and A. Llobet, *Angew. Chem., Int. Ed.,* 2006, **45**, 2941.
108. C. A. Correia, L. Yang and C.-J. Li, *Org. Lett.,* 2011, **13**, 4581.
109. R. A. Kumar, G. Saidulu, K. R. Prasad, G. S. Kumar, B. Sridhar and K. R. Reddy, *Adv. Synth. Catal.,* 2012, **354**, 2985.
110. Z.-Q. Lao, W.-H. Zhong, Q.-H. Lou, Z.-J. Li and X.-B. Meng, *Org. Biomol. Chem.,* 2012, **10**, 7869.
111. T. Nobuta, N. Tada, A. Fujiya, A. Kariya, T. Miura and A. Itoh, *Org. Lett.,* 2013, **15**, 574.
112. J. M. Allen and T. H. Lambert, *J. Am. Chem. Soc.,* 2011, **133**, 1260.
113. Y.-P. Zhu, M.-C. Liu, F.-C. Jia, J.-J. Yuan, Q.-H. Gao, M. Lian and A.-X. Wu, *Org. Lett.,* 2012, **14**, 3392.
114. W. Wei, Y. Shao, H. Hu, F. Zhang, C. Zhang, Y. Xu and X. Wan, *J. Org. Chem.,* 2012, **77**, 7157.
115. Á. Pintér, A. Sud, D. Sureshkumar and M. Klussmann, *Angew. Chem., Int. Ed.,* 2010, **49**, 5004.
116. Á. Pintér and M. Klussmann, *Adv. Synth. Catal.,* 2012, **354**, 701.
117. B. Zhang, S.-K. Xiang, L.-H. Zhang, Y. Cui and N. Jiao, *Org. Lett.,* 2011, **13**, 5212.
118. W. A. Waters, *Trans. Faraday Soc.,* 1946, **42**, 184.
119. D. H. Reid, M. Fraser, B. B. Molloy, H. A. S. Payne and R. G. Sutherland, *Tetrahedron Lett.,* 1961, **2**, 530.
120. E. A. Braude, L. M. Jackman and R. P. Linstead, *J. Chem. Soc.,* 1954, 3548.
121. J. R. Barnard and L. M. Jackman, *J. Chem. Soc.,* 1960, 3110.
122. B. M. Trost, *J. Am. Chem. Soc.,* 1967, **89**, 1847.
123. E. S. Lewis, J. M. Perry and R. H. Grinstein, *J. Am. Chem. Soc.,* 1970, **92**, 899.
124. C. Höfler and C. Rüchardt, *Liebigs Ann. Chem.,* 1996, 183.
125. J. Weiss, *J. Chem. Soc.,* 1942, 245.
126. R. S. Mulliken, *J. Am. Chem. Soc.,* 1952, **74**, 811.
127. R. Foster, *Organic Charge-Transfer Complexes,* Academic Press, New York, 1969.
128. H. D. Becker, *J. Org. Chem.,* 1969, **34**, 1203.
129. H.-D. Becker, *J. Org. Chem.,* 1965, **30**, 982.
130. S. Fukuzumi, N. Nishizawa and T. Tanaka, *J. Org. Chem.,* 1984, **49**, 3571.
131. S. Fukuzumi, S. Koumitsu, K. Hironaka and T. Tanaka, *J. Am. Chem. Soc.,* 1987, **109**, 305.
132. R. Rathore, S. M. Hubig and J. K. Kochi, *J. Am. Chem. Soc.,* 1997, **119**, 11468.
133. T. M. Bockman, S. M. Hubig and J. K. Kochi, *J. Am. Chem. Soc.,* 1998, **120**, 2826.

134. S. Fukuzumi, K. Ohkubo, Y. Tokuda and T. Suenobu, *J. Am. Chem. Soc.*, 2000, **122**, 4286.
135. K. M. Zaman, S. Yamamoto, N. Nishimura, J. Maruta and S. Fukuzumi, *J. Am. Chem. Soc.*, 1994, **116**, 12099.
136. S. Yamamoto, T. Sakurai, L. Yingjin and Y. Sueishi, *Phys. Chem. Chem. Phys.*, 1999, **1**, 833.
137. A. Ohno, H. Yamamoto and S. Oka, *Bull. Chem. Soc. Jpn.*, 1981, **54**, 3489.
138. B. W. Carlson and L. L. Miller, *J. Am. Chem. Soc.*, 1985, **107**, 479.
139. L. L. Miller and J. R. Valentine, *J. Am. Chem. Soc.*, 1988, **110**, 3982.
140. R. Rathore, S. V. Lindeman and J. K. Kochi, *J. Am. Chem. Soc.*, 1997, **119**, 9393.
141. S. M. Hubig, R. Rathore and J. K. Kochi, *J. Am. Chem. Soc.*, 1999, **121**, 617.
142. E. Baciocchi, T. Del Giacco, F. Elisei and O. Lanzalunga, *J. Am. Chem. Soc.*, 1998, **120**, 11800.
143. F. D'Souza and G. R. Deviprasad, *J. Org. Chem.*, 2001, **66**, 4601.
144. D. Bijl, H. Kainer and A. C. Rose-Innes, *Nature*, 1954, **174**, 830.
145. Y. Hayashi and T. Mukaiyama, *Chem. Lett.*, 1987, 1811.
146. Y.-C. Xu, C. Roy and E. Lebeau, *Tetrahedron Lett.*, 1993, **34**, 8189.
147. Y.-C. Xu, D. T. Kohlman, S. X. Liang and C. Erikkson, *Org. Lett.*, 1999, **1**, 1599.
148. D. J. Clausen and P. E. Floreancig, *J. Org. Chem.*, 2012, **77**, 6574.
149. W. Muramatsu, K. Nakano and C.-J. Li, *Org. Lett.*, 2013, **15**, 3650.
150. A. S.-K. Tsang and M. H. Todd, *Tetrahedron Lett.*, 2009, **50**, 1199.
151. A. S.-K. Tsang, K. Ingram, J. Keiser, B. Hibbert and M. H. Todd, *Org. Biomol. Chem.*, 2013, **11**, 4921.
152. A. S.-K. Tsang, P. Jensen, J. M. Hook, A. S. K. Hashmi and M. H. Todd, *Pure Appl. Chem.*, 2011, **83**, 655.
153. G. Zhang, Y. Ma, S. Wang, W. Kong and R. Wang, *Chem. Sci.*, 2013, **4**, 2645.
154. Y. Cui, L. A. Villafane, D. J. Clausen and P. E. Floreancig, *Tetrahedron*, 2013, **69**, 7618.
155. D. L. Aubele, J. C. Rech and P. E. Floreancig, *Adv. Synth. Catal.*, 2004, **346**, 359.
156. P. E. Floreancig, *Synlett*, 2007, 191.
157. L. Wang, J. R. Seiders and P. E. Floreancig, *J. Am. Chem. Soc.*, 2004, **126**, 12596.
158. W. Tu, L. Liu and P. E. Floreancig, *Angew. Chem., Int. Ed.*, 2008, **47**, 4184.
159. H. H. Jung and P. E. Floreancig, *Tetrahedron*, 2009, **65**, 10830.
160. L. Liu and P. E. Floreancig, *Angew. Chem., Int. Ed.*, 2010, **49**, 5894.
161. G. J. Brizgys, H. H. Jung and P. E. Floreancig, *Chem. Sci.*, 2012, **3**, 438.
162. L. Liu and P. E. Floreancig, *Org. Lett.*, 2009, **11**, 3152.
163. Y. Cui and P. E. Floreancig, *Org. Lett.*, 2012, **14**, 1720.
164. O. R. Luca, T. Wang, S. J. Konezny, V. S. Batista and R. H. Crabtree, *New J. Chem.*, 2011, **35**, 998.

165. V. S. Batista, R. H. Crabtree, S. J. Konezny, O. R. Luca and J. M. Praetorius, *New J. Chem.*, 2012, **36**, 1141.
166. A. G. Condie, J. C. González-Gómez and C. R. J. Stephenson, *J. Am. Chem. Soc.*, 2010, **132**, 1464.
167. D. B. Freeman, L. Furst, A. G. Condie and C. R. J. Stephenson, *Org. Lett.*, 2012, **14**, 94.
168. M. Rueping, R. M. Koenigs, K. Poscharny, D. C. Fabry, D. Leonori and C. Vila, *Chem.–Eur. J*, 2012, **18**, 5170.
169. D. P. Hari and B. König, *Org. Lett.*, 2011, **13**, 3852.
170. Y. Pan, C. W. Kee, L. Chen and C.-H. Tan, *Green Chem.*, 2011, **13**, 2682.
171. C. Wang, Z. Xie, K. E. deKrafft and W. Lin, *J. Am. Chem. Soc.*, 2011, **133**, 13445.
172. S. Zhu, A. Das, L. Bui, H. Zhou, D. P. Curran and M. Rueping, *J. Am. Chem. Soc.*, 2013, **135**, 1823.
173. D. A. Nicewicz and D. W. C. MacMillan, *Science*, 2008, **322**, 77.
174. K. Zeitler, *Angew. Chem., Int. Ed.*, 2009, **48**, 9785.
175. T. P. Yoon, M. A. Ischay and J. Du, *Nat. Chem.*, 2010, **2**, 527.
176. C. Dai, J. M. R. Narayanam and C. R. J. Stephenson, *Nat. Chem.*, 2011, **3**, 140.
177. J. Hu, J. Wang, T. H. Nguyen and N. Zheng, *Beilstein J. Org. Chem.*, 2013, **9**, 1977.
178. J. M. R. Narayanam and C. R. J. Stephenson, *Chem. Soc. Rev.*, 2011, **40**, 102.
179. Q. Liu, Y.-N. Li, H.-H. Zhang, B. Chen, C.-H. Tung and L.-Z. Wu, *Chem.–Eur. J*, 2012, **18**, 620.
180. J.-J. Zhong, Q.-Y. Meng, G.-X. Wang, Q. Liu, B. Chen, K. Feng, C.-H. Tung and L.-Z. Wu, *Chem.–Eur. J.*, 2013, **19**, 6443.
181. W.-P. To, Y. Liu, T.-C. Lau and C.-M. Che, *Chem.–Eur. J.*, 2013, **19**, 5654.
182. G. Pandey, S. Pal and R. Laha, *Angew. Chem., Int. Ed.*, 2013, **52**, 5146.
183. C. Dai, F. Meschini, J. M. R. Narayanam and C. R. J. Stephenson, *J. Org. Chem.*, 2012, **77**, 4425.
184. Y.-Q. Zou, L.-Q. Lu, L. Fu, N.-J. Chang, J. Rong, J.-R. Chen and W.-J. Xiao, *Angew. Chem., Int. Ed.*, 2011, **50**, 7171.
185. J. D. Nguyen, J. W. Tucker, M. D. Konieczynska and C. R. J. Stephenson, *J. Am. Chem. Soc.*, 2011, **133**, 4160.
186. L. R. Espelt, E. M. Wiensch and T. P. Yoon, *J. Org. Chem.*, 2013, **78**, 4107.
187. Y. Miyake, K. Nakajima and Y. Nishibayashi, *J. Am. Chem. Soc.*, 2012, **134**, 3338.
188. G. Zhao, C. Yang, L. Guo, H. Sun, C. Chen and W. Xia, *Chem. Commun.*, 2012, **48**, 2337.
189. (a) T. D. Beeson, A. Mastracchio, J. B. Hong, K. Ashton and D. W. C. MacMillan, *Science*, 2007, **316**, 582; (b) K. Narasaka, T. Okauchi, K. Tanaka and M. Murakami, *Chem. Lett.*, 1992, **21**, 2099.
190. M. Sibi and M. Hasegawa, *J. Am. Chem. Soc.*, 2007, **129**, 4124.
191. J. F. Van Humbeck, S. P. Simonovich, R. R. Knowles and D. W. C. MacMillan, *J. Am. Chem. Soc.*, 2010, **132**, 10012.

192. K. C. Nicolaou, R. Reingruber, D. Sarlah and S. Bräse, *J. Am. Chem. Soc.*, 2009, **131**, 2086.

193. J. M. Um, O. Gutierrez, F. Schoenebeck, K. N. Houk and D. W. C. MacMillan, *J. Am. Chem. Soc.*, 2010, **132**, 6001.

194. R. Beel, S. Kobialka, M. L. Schmidt and M. Engeser, *Chem. Commun.*, 2011, **47**, 3293.

195. J. E. Wilson, A. D. Casarez and D. W. C. MacMillan, *J. Am. Chem. Soc.*, 2009, **131**, 11332.

196. M. Nielsen, D. Worgull, T. Zweifel, B Gschwend, S. Bertelsen and K. A. Jørgensen, *Chem. Commun.*, 2011, **47**, 632.

197. J. J. Devery, J. C. Conrad, D. W. C. MacMillan and R. A. Flowers, *Angew. Chem., Int. Ed.*, 2010, **49**, 6106.

198. S.-L. Zhang, H.-X. Xie, J. Zhu, H. Li, X.-S. Zhang, J. Li and W. Wang, *Nat. Commun.*, 2011, **2**, 211.

199. Y. Hayashi, T. Itoh and H. Ishikawa, *Angew. Chem., Int. Ed.*, 2011, **50**, 3920.

200. J. Xiao, *ChemCatChem*, 2012, **4**, 612.

201. Y.-L. Zhao, Y. Wang, X.-Q. Hu and P.-F. Xu, *Chem. Commun.*, 2013, **49**, 7555.

202. M. Rueping, H. Sundén, L. Hubener and E. Sugiono, *Chem. Commun.*, 2012, **48**, 2201.

203. M. T. Pirnot, D. A. Rankic, D. B. C. Martin and D. W. C. MacMillan, *Science*, 2013, **339**, 1593.

204. J.-S. Tian and T.-P. Loh, *Angew. Chem., Int. Ed.*, 2010, **49**, 8417.

205. J. M. Campbell, H.-C. Xu and K. D. Moeller, *J. Am. Chem. Soc.*, 2012, **134**, 18338.

206. (a) R. J. Sundberg, M.-H. Théret and L. Wright, *Org. Prep. Proc. Int.*, 1994, **26**, 386; (b) W. C. Groutas, M. Essawi and P. S. Portoghese, *Synth. Commun*, 1980, **10**, 495; (c) R. T. Dean, H. C. Padgett and H. Rapoport, *J. Am. Chem. Soc.*, 1976, **98**, 7448.

207. E. Baciocchi, M. Bietti and O. Lanzalunga, *Acc. Chem. Res.*, 2000, **33**, 243.

208. (a) S. F. Nelsen and J. T. Ippoliti, *J. Am. Chem. Soc.*, 1986, **108**, 4879; (b) Y. L. Chow, W. C. Danen, S. F. Nelsen and D. H. Rosenblatt, *Chem. Rev*, 1978, **78**, 243.

209. C. J. Schlesener, C. Amatore and J. K. Kochi, *J. Am. Chem. Soc.*, 1984, **106**, 7472.

210. F. D. Lewis, *Acc. Chem. Res.*, 1986, **19**, 401.

211. D. D. M. Wayner, J. J. Dannenberg and D. Griller, *Chem. Phys. Lett.*, 1986, **131**, 189.

212. L. M. Tolbert, R. K. Khanna, A. E. Popp, L. Gelbaum and L. A. Bottomley, *J. Am. Chem. Soc.*, 1990, **112**, 2373.

213. A. L. Perrott and D. R. Arnold, *Can. J. Chem.*, 1992, **70**, 272.

214. A. L. Perrott, H. J. P. de Lijser and D. R. Arnold, *Can. J. Chem.*, 1997, **75**, 384.

215. M. Freccero, A. Pratt, A. Albini and C. Long, *J. Am. Chem. Soc.*, 1998, **120**, 284.

216. Z. Li, H. Li, X. Guo, L. Cao, R. Yu, H. Li and S. Pan, *Org. Lett.*, 2008, **10**, 803.

Subject Index

References to figures and schemes are in *italic* type.

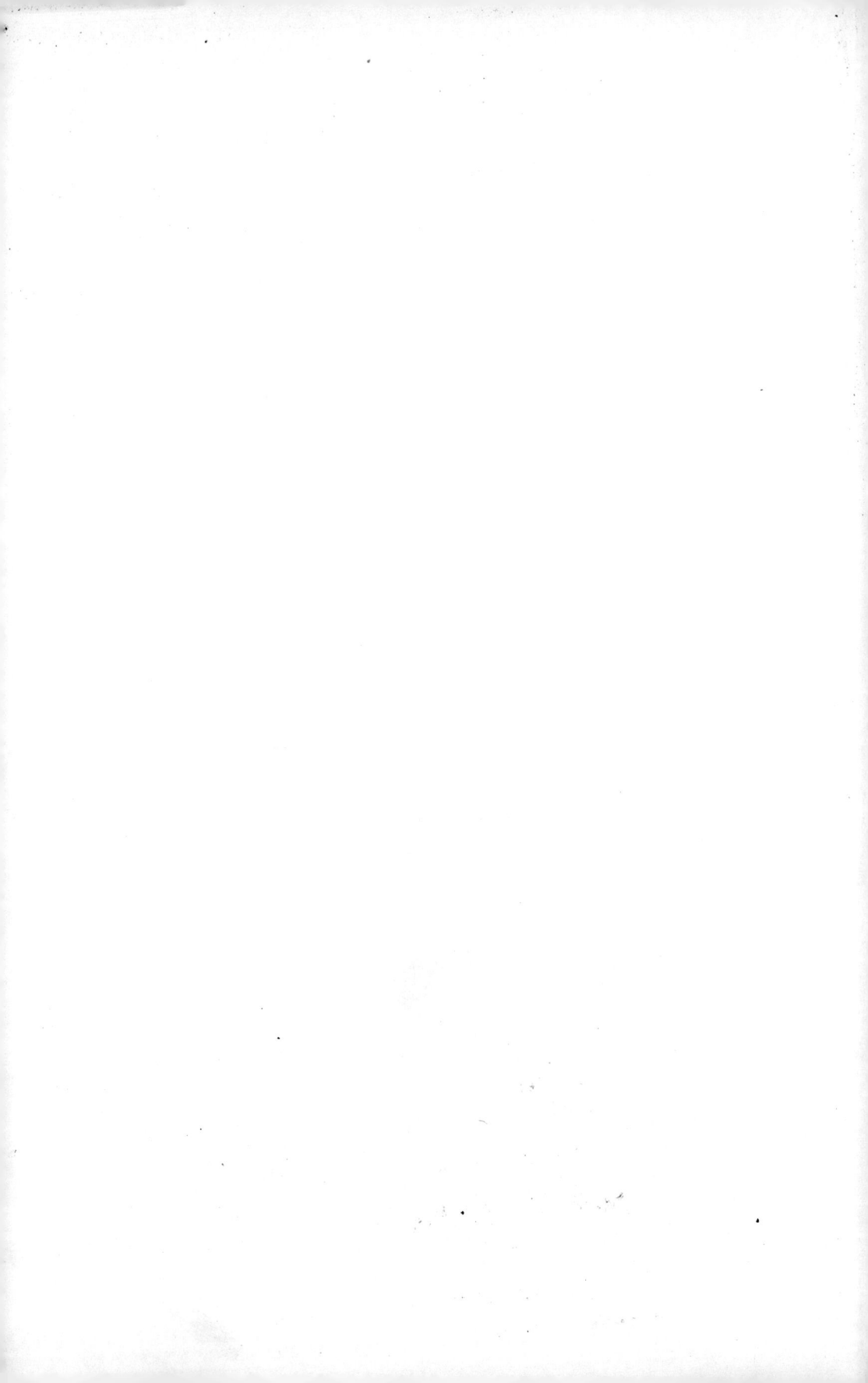